# Dinosaurs

Spencer G. Lucas

# DINO

The Textbook, Sixth Edition

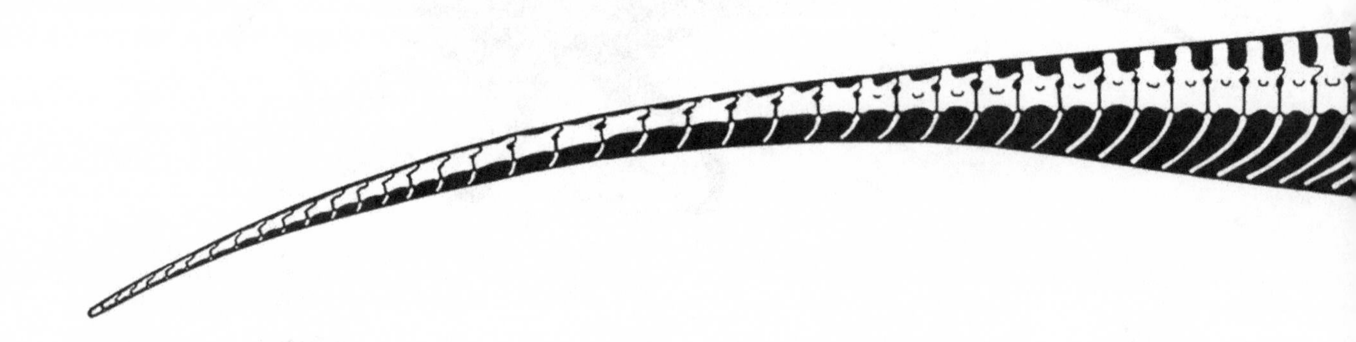

# SAURS

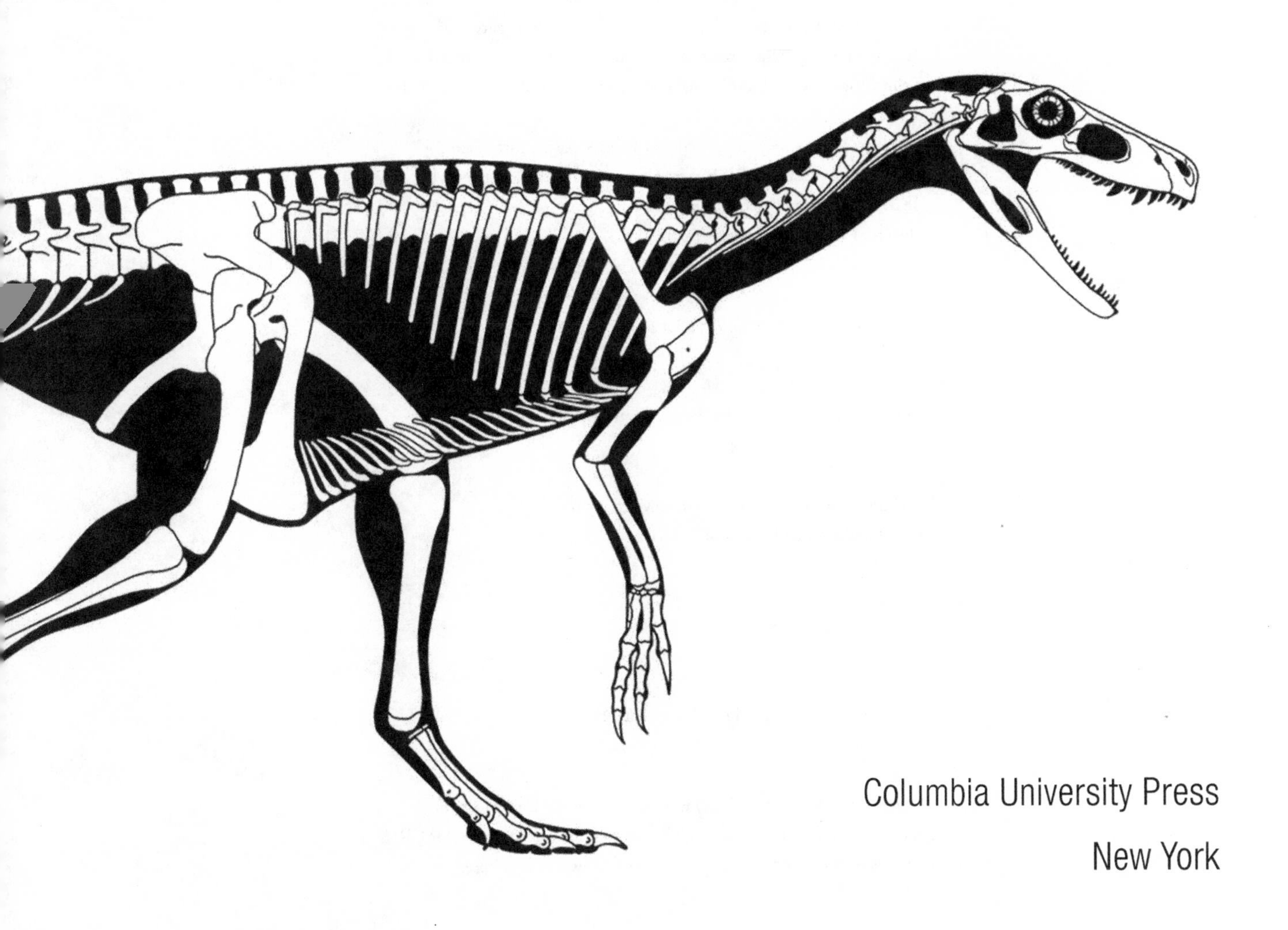

Columbia University Press

New York

Columbia University Press
*Publishers Since 1893*
New York Chichester, West Sussex
cup.columbia.edu



Library of Congress Cataloging-in-Publication Data
Names: Lucas, Spencer G., author.
Title: Dinosaurs : the textbook / Spencer G. Lucas.
Description: Sixth edition. | New York : Columbia University Press, [2016] | Includes bibliographical references and index. | Description based on print version record and CIP data provided by publisher; resource not viewed.
Identifiers: LCCN 2015048842 (print) | LCCN 2015044367 (ebook) | ISBN 9780231541848 () | ISBN 9780231173100 (cloth : alk. paper) | ISBN 9780231173117 (pbk. : alk. paper)
Subjects: LCSH: Dinosaurs—Textbooks.
Classification: LCC QE861.4 (print) | LCC QE861.4 .L94 2016 (ebook) | DDC 567.9—dc23
LC record available at http://lccn.loc.gov/2015048842

Columbia University Press books are printed on permanent and durable acid-free paper.

This book is printed on paper with recycled content.
Printed in the United States of America

COVER DESIGN: Lisa Hamm
COVER IMAGE: © Mark Hallet/Mark Hallett Paleo Art
HALF-TITLE AND TITLE PAGE ART: © Scott Hartman

For Yami

ཿ

# CONTENTS IN BRIEF

# CONTENTS IN DETAIL

# BOXED READINGS

# PREFACE

IN the 1980s, the geology faculty at the University of New Mexico, at my suggestion, initiated an introductory-level course on dinosaurs. As the lone vertebrate paleontologist on campus, I, of course, was to teach this course. I had several years of teaching introductory geology—both physical and historical geology—under my belt. But now a problem faced me: no textbook existed for a dinosaur course. Furthermore, in a decade-long stint as a university student—from freshman to doctorate—I had never taken a course on dinosaurs. Few colleagues were teaching dinosaur courses at that time, and all they could offer was a syllabus with a list of suggested readings. Not fully satisfied with their offerings, I set out to design a course and provide reading material from available sources to suit my own ideas about how to teach college freshmen and sophomores about dinosaurs.

The book I have written is for the semester-long course I have taught as it has been honed by years of experimentation and student feedback to a lean but comprehensive introduction to the dinosaurs. This book thus fulfills the needs of the many faculty teaching introductory-level dinosaur courses across the United States. My reviewers share this belief, and I hope we are right.

There is, however, a second reason why I wrote this book. It represents my attempt to slog through the available morass of information and ideas about dinosaurs, some controversial, others ridiculous, to establish a firm ground of established facts and reasonable inference. Much of what Americans think they know about dinosaurs is wrong, and some of what they are being told today in some popular books is hype. This book tries to right the wrongs and avoid the hype by going out of its way not to promote unreasonable speculation about dinosaurs. Not everything in it is above debate, but nothing here is science fiction. As such, I want this book to teach many people about dinosaurs and the science of studying dinosaurs as few other books do.

These are heady times for dinosaur science. Almost daily, new discoveries, novel methods, and innovative ideas are pushing forward the frontiers of our knowledge of dinosaurs. Americans seem to have an insatiable appetite for information on dinosaurs.

This book provides a "first course," and I hope it fosters an accurate understanding of the dinosaurs and a deep appreciation of dinosaur science in all who read it.

## ORGANIZATION

The book is essentially divided into three parts. The first part, chapters 1 to 3, is designed to provide the beginning student with the minimum background in geological and biological concepts necessary to understand the remainder of the text. In the second part, chapters 4 to 9, I have you "meet the dinosaurs." These chapters review each group of dinosaurs. Each chapter focuses on two or three well-known taxa that are exemplary of the group. The remaining discussion covers aspects of phylogeny, diversity, distribution, and functional morphology. The third part, chapters 10 to 17, covers a variety of thought-provoking topics. These chapters discuss everything from the history of the great dinosaur hunters to the extinction of the dinosaurs. The emphasis in many of the chapters is on concepts of broad applicability; in other words, concepts also relevant to subjects other than dinosaurs. Thus, for example, I believe that the history of dinosaur collecting and study can be used to tell the student much about how scientific perceptions change over time. Finally, I have included an appendix of dinosaur anatomy, a dinosaur dictionary, and a glossary for ease in understanding and using anatomical terms, locating definitions, and identifying key information.

I have strived to present a balanced review of competing ideas in controversial areas. For example, I believe the weight of evidence suggests that some dinosaurs had a higher metabolic rate than that of living ectotherms, whereas there is no evidence of such a heightened metabolic rate in other dinosaur groups. I intend to present the range of evidence on this subject and not to promote a particular point of view not justified by the evidence.

The sixth edition of this textbook updates many areas, large and small, to keep current in one of the most rapidly evolving fields of scientific discovery and research that I know of. All reference lists at the end of each chapter have been updated. This new edition also corrects as many sins of commission and omission as I could beat out of the fifth edition; there were a few!

## ACKNOWLEDGMENTS

First, I want to thank Patrick Fitzgerald, who acquired this book for Columbia University Press. I also want to thank the staff at and contractors for Columbia for diverse help. I wish to extend my thanks and appreciation to the reviewers whose thoughtful comments, criticisms, and encouragement have helped tremendously in revising and improving the final draft. Several colleagues, museums, and other institutions provided photographs that add to the quality of instruction in this textbook. Their contributions are acknowledged, where appropriate, throughout the text.

Over the years, I have learned much about dinosaurs from my colleagues and students. To the rest of you who collect dinosaurs, do the research, give the talks, and write the papers, thanks for teaching me so much.

Finally, I thank my wife, Yami Lucas, for her support, encouragement, and editorial help.

## INSTRUCTOR'S MANUAL

Many who teach dinosaur courses are not vertebrate paleontologists, and few, if any, of the instructors have ever had the opportunity to enroll in a dinosaur course during their college-student careers. The first edition of this book was the first textbook written specifically for a dinosaur course. For these and other reasons, I have written an Instructor's Manual to accompany the text.

An Instructor's Manual is available for professors and teachers who adopt the text for use with students. Instructors can request a copy by sending an e-mail to coursematerials@columbiauniversitypress.com. Please provide instructor's name and title, name of the institution, name of the course, and number of students in the course. More information is available on the book's web page at cup.columbia.edu.

The Instructor's Manual includes a suggested syllabus along with a description of the text's organization and chapter interdependence to assist instructors in planning how best to use the text to meet the needs of their courses. The manual provides a description of the material covered in each chapter as well as suggestions for how to present the material. In the manual, I discuss what material should be emphasized and provide methods for overcoming potential difficulties. Answers to all review questions are also provided for each chapter.

Finally, the Instructor's Manual includes a test item file with 25 to 30 multiple-choice and true/false questions for each chapter.

# Dinosaurs

# 1

# INTRODUCTION

In 1842, British comparative anatomist **Richard Owen** (1804–1892) coined the word **dinosaur**. Owen constructed this word from the Greek words *deinos*, meaning "terrible" (though Owen considered it to mean "fearfully great"), and *sauros*, meaning "lizard" or "reptile." To Owen, the "fearfully great lizards" were large, extinct reptiles known from only a handful of fossils discovered in western Europe since the 1820s. Today, dinosaur fossils are known from all continents and represent hundreds of distinct types of dinosaurs.

In this chapter, I briefly answer some basic questions about dinosaurs and introduce some topics discussed at greater length in this book.

## WHAT ARE DINOSAURS?

Many people apply the term "dinosaur" to any large, extinct animal. To most people, any large extinct reptile qualifies as a dinosaur. Many even identify large, extinct mammals, such as wooly mammoths, as dinosaurs (figure 1.1). Some authors and toy manufacturers perpetuate incorrect ideas about what a dinosaur is by presenting a variety of nondinosaurs as dinosaurs. Examples include the flying reptile *Pteranodon* (a pterosaur) and the mammal-like reptile *Dimetrodon* (a pelycosaur).

Dinosaurs are most easily thought of as a group of extinct reptiles that had an **upright posture.** They first appeared about 225 to 230 million years ago and became extinct 66 million years ago. Birds, the descendants of dinosaurs, are still with us. Dinosaurs can be identified as reptiles because of their reptilian skeletal features and because dinosaurs, like many other reptiles, reproduced by laying hard-shelled eggs. The upright posture of dinosaurs, in which the legs extend directly underneath the body, distinguishes them from reptiles that hold their limbs out to the side of the body in a **sprawling posture** (figure 1.2). Large size is not a prerequisite for being a dinosaur,

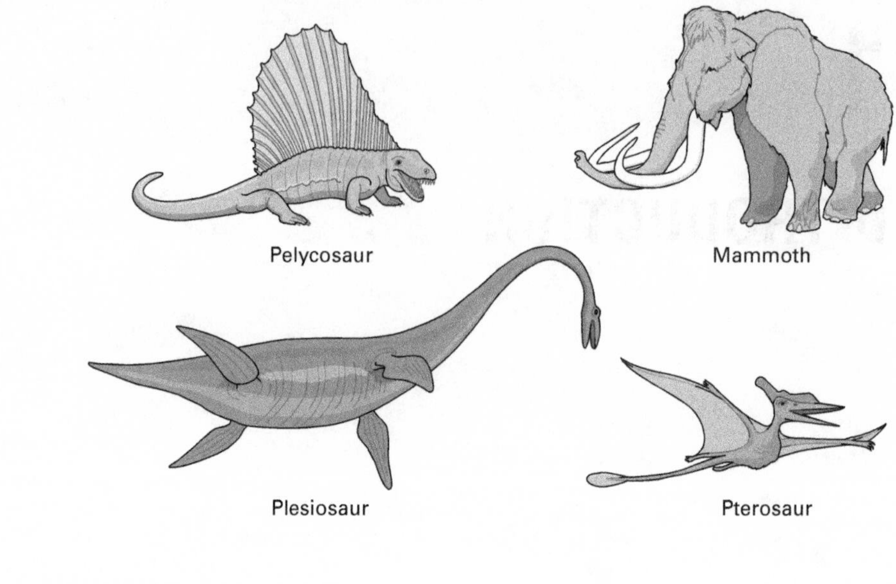


**FIGURE 1.1**
Although these kinds of animals are thought to be dinosaurs by many people, they are not. (Drawing by Network Graphics)

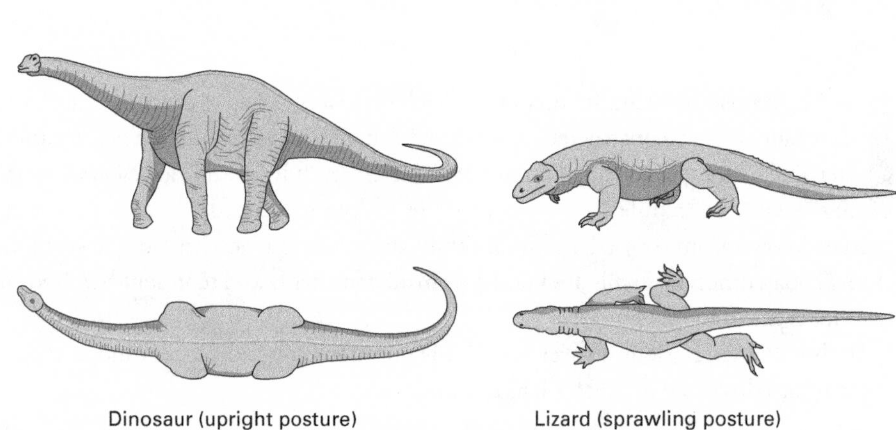


**FIGURE 1.2**
Dinosaurs are a group of extinct reptiles having an upright limb posture. In contrast, the limbs of other reptiles, such as lizards, are positioned at the sides of the body, in a sprawling posture. (Drawing by Network Graphics)

as some dinosaurs were no larger than a chicken. Skeletal features of one group of dinosaurs are remarkably like those of birds. These features indicate that dinosaurs were the ancestors of birds. Other skeletal features unique to dinosaurs will be discussed in chapter 4. But, for now, we can define dinosaurs as reptiles with an upright posture and thereby identify them as a natural group with an evolutionary history distinct from that of other reptiles.

## WHEN AND WHERE DID DINOSAURS LIVE?

As mentioned, dinosaurs first appeared about 225 to 230 million years ago and became extinct 66 million years ago. So, dinosaurs lived on Earth for approximately 160 million years (figure 1.3). Most paleontologists date the origin of humans at 2 or 3 million years ago. This means that dinosaurs persisted from 50 to 80 times as long as we have currently been on Earth. Movies and cartoons that portray humans and dinosaurs living side by side are very wrong.

Dinosaur fossils have long been collected in North America, Europe, Asia, South America, Africa, and Australia, and fossils were also discovered in Antarctica in 1989. So, we now have dinosaur fossils from all continents (figure 1.4). At least 500 different kinds of dinosaurs have received scientific names. Their broad geographic distribution, their survival for about 160 million years, their variety in shape and size, and, in many instances, their extremely large size, identify dinosaurs as one of the most successful groups of land animals in the history of life.

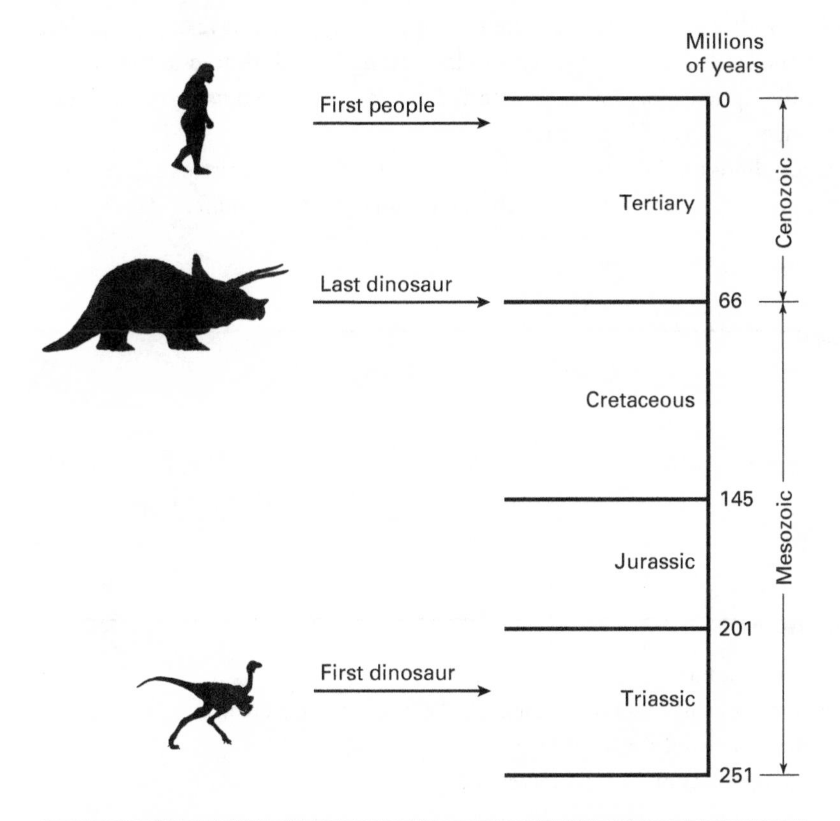


**FIGURE 1.3**
Dinosaurs became extinct approximately 66 million years ago, more than 60 million years before humans appeared on Earth. (Drawing by Network Graphics)

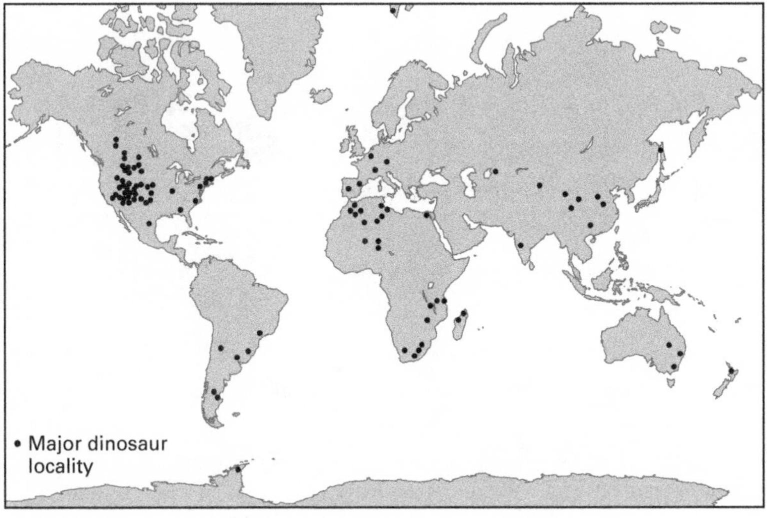


**FIGURE 1.4**
Dinosaur fossils have been collected on all continents, including Antarctica. (Drawing by Network Graphics)

## WHY STUDY DINOSAURS?

Dinosaurs fascinate most people, including young children. This fascination stems from their large size, strange shapes, and long-ago extinction. Some dinosaurs, such as *Tyrannosaurus rex*, one of the largest meat-eating land animals to have walked the earth, and certainly the most famous, terrify us. Other dinosaurs, such as *Stegosaurus*, puzzle us with their unusual body shape or armor. Clearly, one reason to study dinosaurs is because they are interesting.

Dinosaurs also are worth studying because they represent a unique episode in the history of life on this planet. They appeared some 225 to 230 million years ago, evolved into some of the largest and most successful land animals of all time, and then disappeared 66 million years ago. Dinosaurs clearly have much to teach us about evolution and extinction, especially of large animals.

So, we study dinosaurs for two reasons: first, because they interest us and, second, because they were an important part of the evolutionary history of life.

## Key Terms

dinosaur
Richard Owen
sprawling posture
upright posture

## Review Questions

1. What is a dinosaur?
2. Name some animals commonly thought to be dinosaurs that are not.
3. When and where did dinosaurs live?
4. What can the study of dinosaurs teach us?

## Find a Dinosaur!

To find a dinosaur, you need only visit a natural history museum or a state or national park. East of the Mississippi, your chances of finding a dinosaur are best served by a visit to the American Museum of Natural History (New York), the Carnegie Museum of Natural History (Pittsburgh, Pennsylvania), the Field Museum of Natural History (Chicago, Illinois), or the Smithsonian Institution's National Museum of Natural History (Washington, D.C.). West of the Mississippi, try the Museum of the Rockies (Bozeman, Montana), the Wyoming Dinosaur Center (Thermopolis), the Denver Museum of Nature & Science, the New Mexico Museum of Natural History and Science (Albuquerque, New Mexico), and the Natural History Museum of Los Angeles County. In Canada, best bets are the Royal Ontario Museum (Toronto, Ontario), the Canadian Museum of Nature (Ottawa, Ontario), and the Royal Tyrrell Museum of Palaeontology (Drumheller, Alberta).You can also find dinosaurs in some state and national parks. These include Dinosaur State Park and Arboretum (Rocky Hill, Connecticut) and Clayton Lake State Park (Clayton, New Mexico), both of which have numerous dinosaur tracks (preserved dinosaur footprints). Probably the very best dinosaur fossil park in the United States is Dinosaur National Monument near Vernal, Utah. Here, Late Jurassic skeletons accumulated along a riverbank and are now displayed in a huge building erected over the bonebed. This is one of the world's great dinosaur localities. There are also many dinosaur parks in cities across the country. For example, take a trip back in time at Dinosaur Park in Rapid City, South Dakota, with its vintage sculptures of dinosaurs created in the 1930s.

# 2

# EVOLUTION, PHYLOGENY, AND CLASSIFICATION

PALEONTOLOGISTS, scientists who study fossils, base their understanding of the history of life on the **fossil record**, which comprises all fossils discovered as well as those awaiting discovery. The fossil record of dinosaurs indicates that they existed for about 160 million years. During that time, many kinds of dinosaurs evolved. Key to understanding this **evolution** is determining the family tree, or genealogy, of dinosaurs. Which dinosaurs were closely related to each other, and which were only distant relatives? Answering these questions requires some understanding of the principles of evolution and phylogeny. Knowing how scientific names are given to dinosaurs or to groups of dinosaurs requires an understanding of the basic ideas and methods of biological classification.

## EVOLUTION

Evolution, simply defined, is the origin and change of organisms over time. Today, biologists and paleontologists concluded that evolution occurs in the specific way first formulated by **Charles Darwin** (1809–1882).

In his book *On the Origin of Species by Means of Natural Selection* (1859), Darwin argued that organisms adapt to the environments in which they live. In other words, each organism has a specific way of living and interacting with its environment. When the environment changes, those organisms that are better able to cope with the changed environment are more successful—which, in Darwinian terms, means they reproduce more—than those organisms less able to cope with the change. The most successful organisms are thus "selected for," and those least successful are "selected against," in the jargon of what is called **natural selection**. Another phrase that has been used to describe this is "survival of the fittest," where "fittest" refers to the most successful organisms: those producing the most offspring.

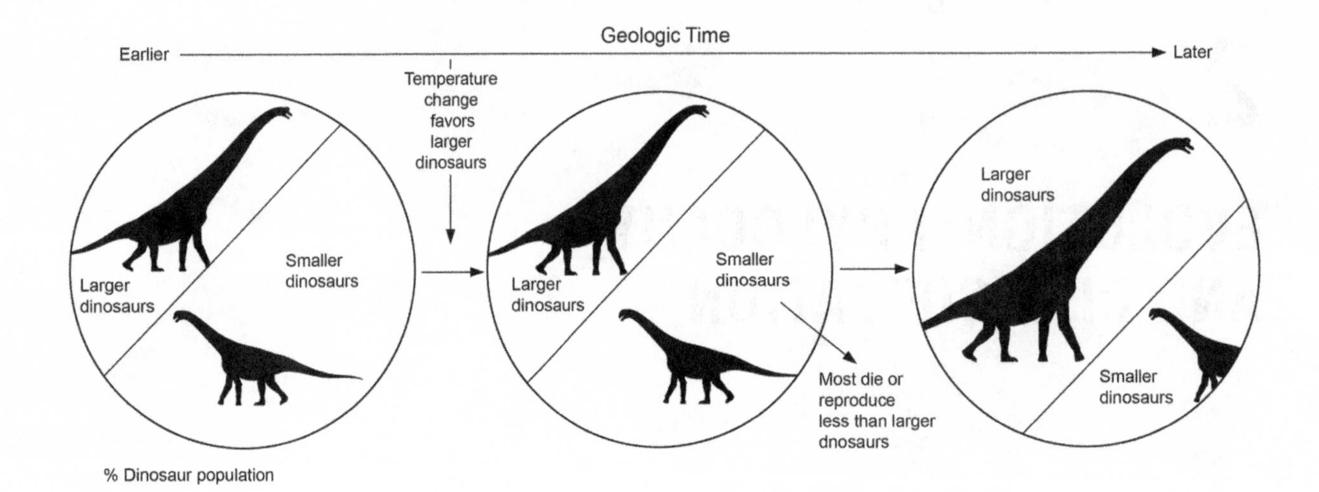


**FIGURE 2.1**
Natural selection can lead to the evolution of larger dinosaurs. (© Scott Hartman)

In Darwin's view, certain organisms are more successful than others when faced with environmental change because of the variation that occurs in any population of organisms. We see this variation most easily by recognizing that no two people are alike, and the same is true in any population of animals or plants. So, when the environment changes, some individuals in a population are better able to cope with the change than others. Which organisms are better able to cope, however, is not something that the organisms themselves control, because they cannot anticipate what kind of environmental change might occur.

For example, in a population of one species of dinosaurs, we would expect to find a range of body sizes (figure 2.1). If climate were to change to favor larger dinosaurs in the population, the larger variants would be selected for. Because they would reproduce more successfully than the smaller dinosaurs, we would expect the dinosaurs to become larger in the next generation, provided that the trait of larger size could be passed on from one generation of dinosaurs to the next. In this way, larger dinosaurs might evolve by the process of natural selection.

Evolution by natural selection is also called "Darwinian" evolution. In Darwinian evolution, the evolutionary history of a group of organisms is diagrammed by a family tree (or genealogy) of populations undergoing natural selection. The family tree emphasizes the fact that there are ancestors, descendants, and other relationships among a group of organisms. Because organisms change as a result of natural selection, Darwinian evolution can also be described in Darwin's own phrase as "descent with modification."

The family tree of a group of evolving organisms is also termed its **phylogeny**, from the Greek words for "tribe" (*phylum*) and "birth" (*genos*). On a phylogeny, each branch, or each bundle of branches with a common stem, is called a **clade** (*clados* is Greek for

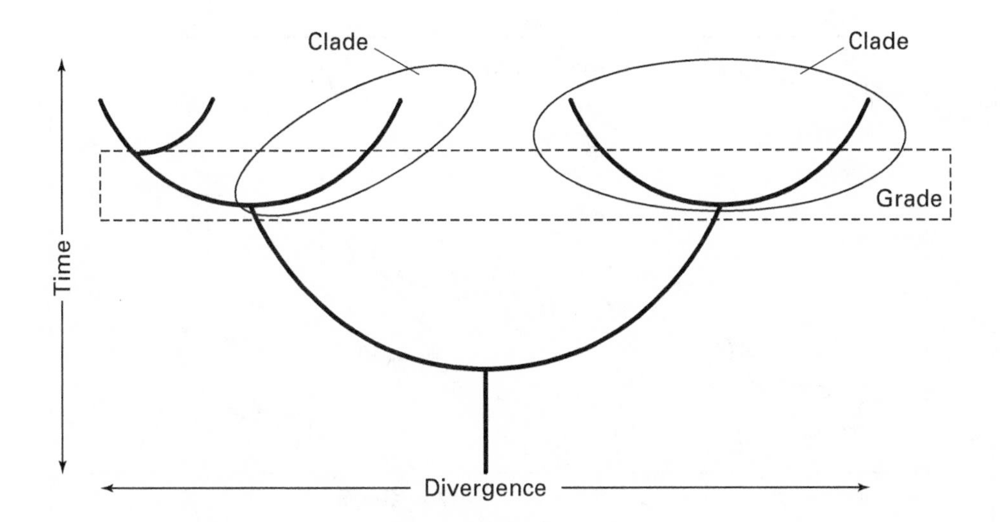


**FIGURE 2.2**
A phylogeny is an evolutionary family tree consisting of distinct clades (branches). A horizontal slice through a phylogeny is a grade. (Drawing by Network Graphics)

"branch"), whereas a horizontal slice through a phylogeny is called a **grade** (figure 2.2). When a new kind of organism appears, we speak of **origination**, and a new clade or segment of a clade is added to the phylogeny. When a kind of organism disappears, we speak of **extinction**. When one clade splits into two we speak of **divergence**; the evolution of similar features in two unrelated clades is called **convergence**. The populations of organisms shown on a phylogeny are usually assigned to groups called **taxa** (singular: **taxon**). For example, each different kind of dinosaur in the phylogeny of dinosaurs is a dinosaur taxon, and these taxa can be grouped into larger dinosaur taxa based on their relationships to each other. When we speak of how taxa are related to each other, we are referring to their **phylogenetic relationships**.

## PHYLOGENY

A phylogeny is a family tree, or genealogy, of taxa. How do paleontologists construct a phylogeny of a group of extinct taxa like dinosaurs?

For more than a century, the most common method was to consider the time span and overall similarity of the fossils of taxa as keys to their phylogenetic relationships. This method, called stratophenetic, from the words for "layer" (of rock) and "population" (of organisms), identifies ancestral taxa as those older than descendant taxa that resemble their ancestors closely in one or more features (figure 2.3). Critical to a **stratophenetic phylogeny** is the notion that the differences between the ancestor and its descendant are not so great as to seem implausible given the time interval between when the ancestor and descendant lived.

The problem many paleontologists have with stratophenetic phylogeny, especially those who study dinosaurs, is that it makes a great assumption about the completeness of the fossil record: the ancestor will be represented by fossils older than the

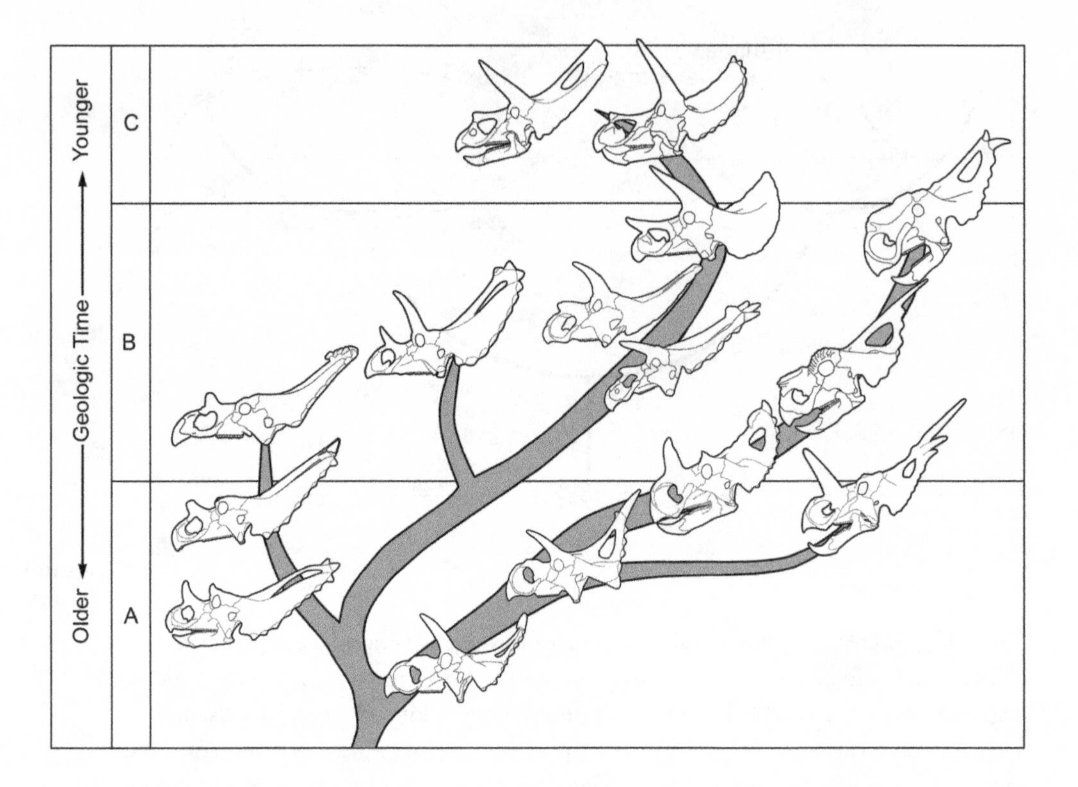


**FIGURE 2.3**
A stratophenetic approach to phylogeny identifies ancestors and descendants as time-successive, similar taxa. This diagram of skulls of horned dinosaurs is a stratophenetic phylogeny. (© Scott Hartman)

fossils of the descendant. In theory, this would be the case, but in practice, the fossil record of many taxa, especially dinosaurs, is very incomplete. Because of this, we can almost never be certain when older fossils represent the ancestors of younger fossils, and we cannot assume that we have discovered all the ancestors and descendants in a phylogeny.

Because the known fossil record of dinosaurs lacks many ancestors and descendants, most paleontologists who study dinosaurs no longer use the stratophenetic method to construct phylogenies. Instead, they use a method that makes fewer assumptions about the completeness of the fossil record. This method is called **cladistics**, and the resultant **cladistic phylogeny** is called a **cladogram** (figure 2.4). The phylogenies of dinosaurs presented in this book are cladograms. Cladistics is the dominant method of phylogeny reconstruction used by vertebrate paleontologists.

A cladogram does not incorporate information about the time ranges of taxa. Instead, it is based only on the similarities of taxa. But neither overall similarity nor just any similarities are used to construct a cladogram. Those similarities of value to cladistics are those features that are **evolutionary novelties**: inherited changes from previously existing structures. Cladistics argues that two taxa are closely related when they share evolutionary novelties. Taxa sharing the most such novelties are most closely related and are shown on the cladogram as diverging from a common ancestor (see figure 2.4). In cladistic terms, these taxa form a **monophyletic group** and share two or more clades with a single ancestor. In contrast, groups lacking a common ancestor on a cladogram are termed **polyphyletic** (see figure 2.4).

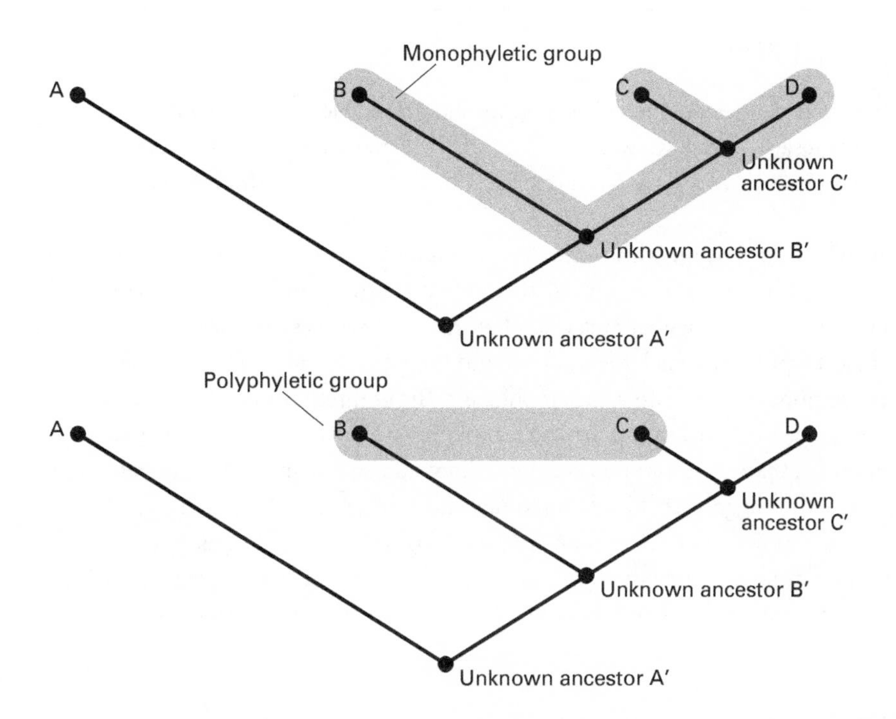


**FIGURE 2.4**
A cladistic phylogeny is called a cladogram. A **monophyletic group** consists of all the branches that share a stem, whereas a **polyphyletic** group is formed by uniting branches with different stems. (Drawing by Network Graphics)

A good example of a cladistic phylogeny is provided by trying to construct a cladogram of the three taxa trout, horse, and whale (box 2.1; figure 2.5). Based on many similarities, we might conclude that a trout and a whale are more closely related to each other than either is to a horse. But, if we focus on evolutionary novelties that distinguish mammals, such as the horse and whale, from bony fishes, such as the trout, then the cladogram that identifies the horse and whale as members of a monophyletic group is well founded. These evolutionary novelties include warm-bloodedness, hair, and giving live birth, which are features shared by horses and whales but not by trout.

Thus, evolutionary novelties allow us to distinguish a monophyletic group on the cladogram for whales and horses distinct from the clade for trout. But, how do we identify the evolutionary novelties to be used in constructing cladograms? In other words, which features shared by two taxa are evolutionary novelties and which features are not? No simple answers to these questions exist, but we can get a feel for how evolutionary novelties are identified by examining the theoretical basis for their identification.

When we view the origination and evolution of a taxon, we can see that it must inherit some features from its ancestor. The inherited features can be identified as **primitive**. But, the features that arise for the first time in a new taxon—its evolutionary novelties—are thought of as **derived**. These derived features, not the primitive features, unite the organisms into a group of closely related organisms. But, they do so only if the evolutionary novelties in question arose only once. Evolutionary convergence occurs when evolutionary novelties arise more than once in separate taxa not

## Box 2.1

### A Cladogram of Dinosaurs and Birds

Once we understand the principles and methods behind cladistics, we should be able to construct a cladogram for any taxon, including the dinosaurs. Here, we will construct a basic cladogram of dinosaurs, birds, and nondinosaurian "reptiles." To do so, we are asking the question, which of these taxa are most closely related to each other?

Because we have to start somewhere, let's begin with some long-standing ideas, treating them as reasonable assumptions with which to direct our cladistic efforts. These ideas are that reptiles are descended from amphibians and that dinosaurs are descended from some other group of reptiles. This allows us to identify the sprawling posture of most reptiles, in which the limbs are positioned at the sides of the body, as a primitive feature of reptiles inherited from their ancestors, the amphibians. The upright posture of dinosaurs, in which the limbs are positioned under the body, thus must be an evolutionary novelty of dinosaurs with respect to their reptilian ancestors. Birds share an upright posture with dinosaurs, so this shared evolutionary novelty (if it did not evolve convergently) suggests that birds and dinosaurs are more closely related to each other than either is to nondinosaurian reptiles. We thus can construct a cladogram in which nondinosaurian reptiles form one clade and birds plus dinosaurs the other (box figure 2.1). For the sake of convenience, we can indicate the evolutionary novelty shared by birds and dinosaurs (upright posture) on the cladogram.

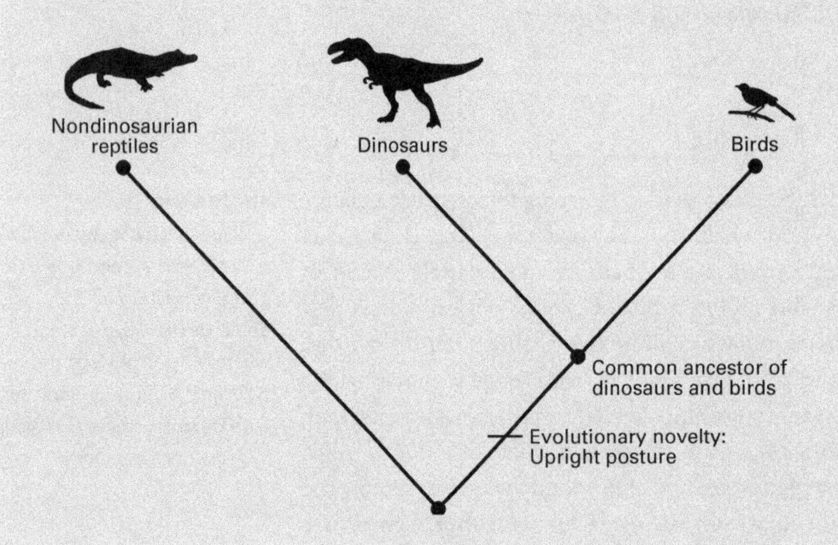


**BOX FIGURE 2.1** This cladogram indicates that birds and dinosaurs are more closely related to each other than either is to nondinosaurian reptiles. (Drawing by Network Graphics)

descended from a single close ancestor. For example, the wings of bats and of birds, though similar evolutionary novelties, arose in two taxa with very different ancestors and thus demonstrate evolutionary convergence. Convergence presents the greatest threat to arriving at the correct cladistic phylogeny of a group of taxa.

Once we have constructed a cladogram, we can add information to this phylogeny to turn it into a **phylogenetic tree**. The information we typically add is a geological time scale on the vertical axis and an indication of which taxa may have been ancestors and which may have been descendants (figure 2.6).

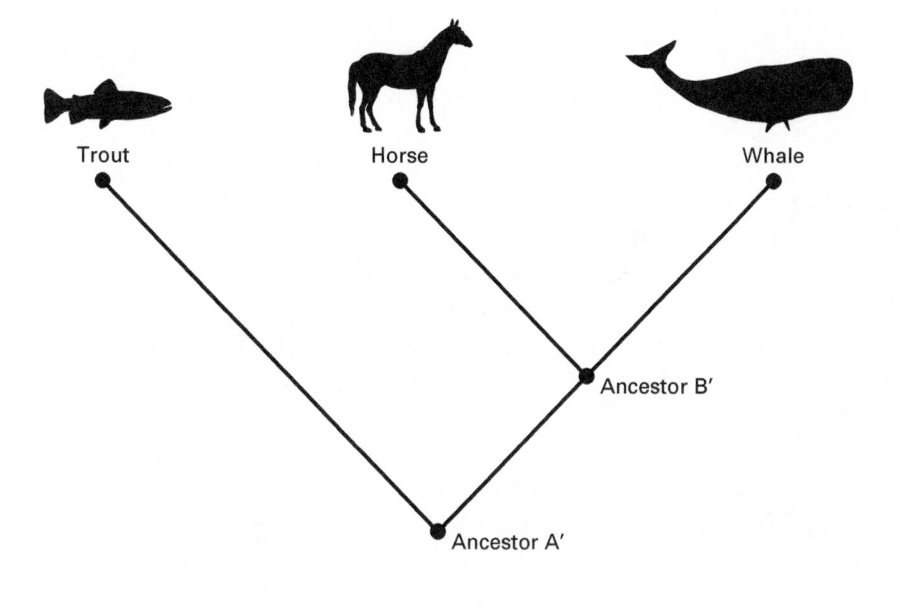


**FIGURE 2.5**
This cladogram indicates that horses and whales are more closely related to each other than either is to trouts. (Drawing by Network Graphics)

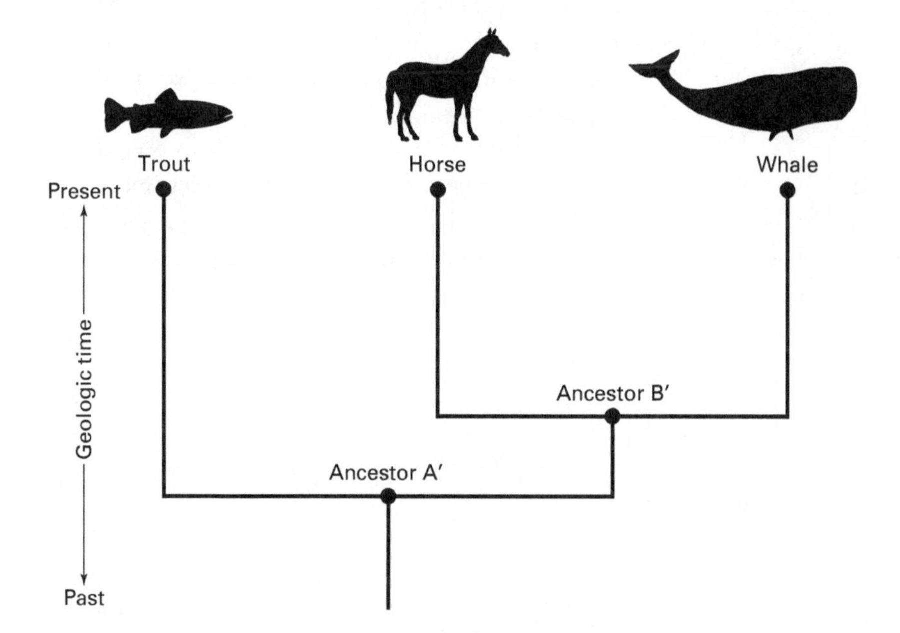


**FIGURE 2.6**
A cladogram can be turned into a phylogenetic tree by making its vertical axis into a geologic time scale and indicating possible ancestors and descendants. (Drawing by Network Graphics)

Cladistics makes no assumptions about the completeness of the fossil record; it simply tries to identify closely related taxa by their shared evolutionary novelties. However, cladistics does not lack problems. Some question the method itself, arguing that it is a method of identifying clades by clustering character states to identify clades that runs contrary to our understanding of the evolutionary process (box 2.2). Other problems arise from convergence, when a shared evolutionary novelty thought to unite two taxa turns out not to be that at all (figure 2.7). Also, not all paleontologists may agree that a particular feature is an evolutionary novelty and not a primitive feature

## Box 2.2

### Cladistics Refuted?

Cladograms of dinosaurs are generated by computer programs. The computer algorithm used is a kind of cluster analysis that analyzes a matrix of character states created for a group of taxa. The cluster analysis groups the taxa (identifies clades) by shared derived similarities of the character states. In other words, those taxa with the most shared derived similarities are clustered to form clades (box table 2.2).

However, in 1999, University of California at Davis paleontologist Geerat Vermeij pointed out a serious problem with such computer-assisted cladistic analyses. Vermeij noted that a well-accepted conclusion about the evolutionary process is that if the evolution of animals takes place by a branching phylogeny, then evolution produces clades that are genetically distinct. This means that evolution within one clade cannot be affected genetically by other clades. However, cluster-analysis cladograms are based on all the character states included in the data matrix. This means that the character states are treated as interdependent, not independent, and therefore that the identification of a given clade by the cluster analysis is affected by the character states of the other clades.

This is most easily seen when taxa are added or removed from a character matrix and the shape of the cladogram changes, which often happens. In such a case, the cluster analysis is not identifying clades that are genetically independent of each other. Thus, it does not produce clades that mirror the well-accepted evolutionary process.

What to do? Criticisms of Vermeij's observations have been few and have not challenged his central point. What we need is a method of phylogenetic analysis that identifies clades that are genetically independent of each other. At present, dinosaur phylogeny is wholly cladistic, so the phylogenies of dinosaurs presented in this book are cladograms. Clearly, a new or modified cladistic approach to produce phylogenies consistent with evolutionary theory is needed.

**BOX TABLE 2.2**

| Taxon/characters | 1 | 2 | 3 | 4 | 5 | 6 | 7 | 8 | 9 | 10 |
|---|---|---|---|---|---|---|---|---|---|---|
| *Gryposaurus* | 0 | 0 | 0 | 0 | 0 | 0 | 1 | 0 | 0 | 0 |
| *Tsintaosaurus* | 0 | 0 | 0 | 1 | 0 | 0 | 1 | 0 | 0 | 0 |
| *Parasaurolophus* | 0 | 0 | 1 | 1 | 0 | 0 | 1 | 0 | 2 | 0 |
| *Lambeosaurus* | 1 | 1 | 1 | 1 | 1 | 0 | 1 | 1 | 1 | 0 |
| *Corythosaurus* | 1 | 1 | 2 | 1 | 1 | 1 | 1 | 1 | 0 | 2 |
| *Hypacrosaurus* | 1 | 1 | 1 | 2 | 1 | 1 | 2 | 1 | 0 | 2 |

This character matrix lists taxa of hadrosaurs and the scored character-states of some features of their skulls. For example, character 6 is "skull with helmet-shaped crest" and scored as no crest (0) or crested (1). Such character matrices are the basis of computer-generated cladograms.

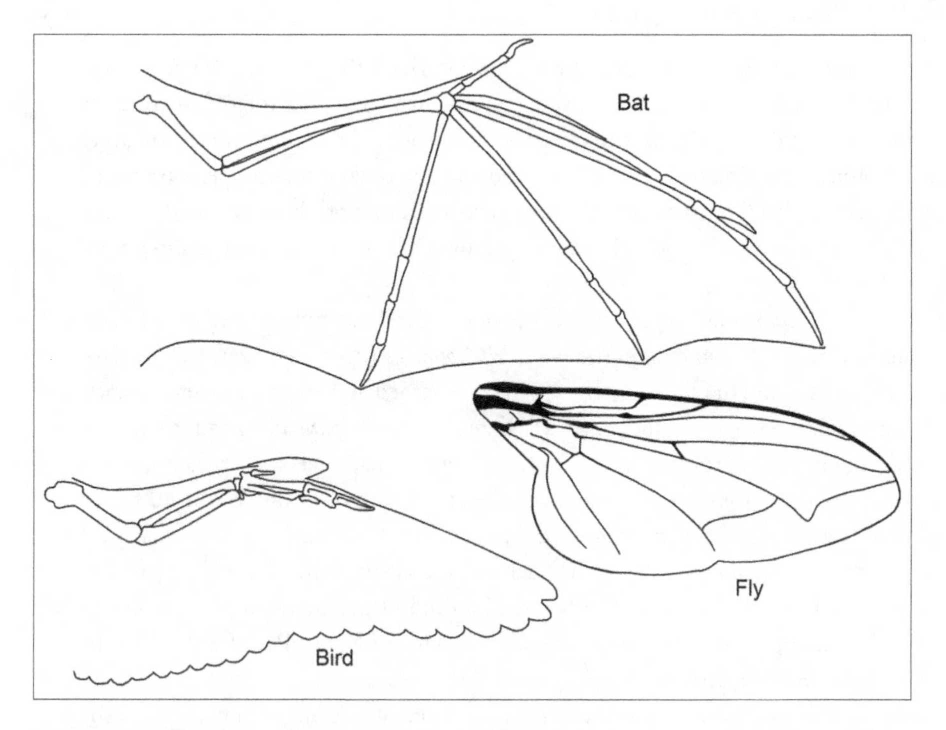


**FIGURE 2.7**
Convergence can produce evolutionary novelties in distinct clades that do not indicate close relationship. The wings of birds, bats, and flies are superficially similar but evolved in distinct clades. They are a classic example of evolutionary convergence. (© Scott Hartman)

inherited from an ancestor. This leads to debate and disagreement among paleontologists attempting to construct cladistic phylogenies of extinct organisms. As we shall see in this book, such debates and disagreements are taking place over certain aspects of dinosaur phylogeny.

## CLASSIFICATION

Constructing a phylogeny based on shared evolutionary novelties is the fundamental basis for classifying taxa. This is because the **biological classification** of organisms is the grouping of them into taxa that share an ancestry. In other words, closely related taxa are placed in the same group so that the classification reflects their phylogenetic relationships.

Thus, the principal information paleontologists use to classify extinct organisms is their phylogeny. Based on their phylogeny, organisms are grouped into categories called taxa, and those taxa receive names. For the classification of animals, living and extinct, the categorical system used is a hierarchy first formulated in the eighteenth century by **Carolus Linnaeus** (1707–1778), a Swedish naturalist. A set of international rules governs the method of naming the categories (taxa) (box 2.3).

## Box 2.3

### The International Rules: Priority and Synonyms

The *International Code of Zoological Nomenclature* spells out the rules by which the scientific names of animals are constructed. A central rule is the **principle of priority**. This rule mandates that the oldest correctly proposed scientific name for an animal has priority over later names proposed for that animal. In other words, if an animal received a proper scientific name in, say, 1900, and a second name was properly proposed for the same animal in 1980 (mistakes like this do occur!), the name proposed in 1900 should be used because it has priority. The name proposed in 1980 is then identified as a **synonym** of the 1900 name and should not be used.

The dinosaur generic names *Apatosaurus* and *Brontosaurus* provide a good example of the principle of priority. Paleontologist Othniel Charles Marsh coined the name *Apatosaurus* (deceptive lizard) in 1877, for gigantic dinosaur hip and backbones found in Colorado. In 1879, Marsh named an almost complete gigantic dinosaur skeleton from Wyoming *Brontosaurus* (thunder lizard). Marsh believed *Apatosaurus* and *Brontosaurus* were different kinds of dinosaur, so his proposal of two different names made sense. But, comparisons many years later revealed that the bones from Colorado and the skeleton from Wyoming belonged to the same kind of dinosaur. So, *Brontosaurus* is a synonym of *Apatosaurus*, which is the correct scientific name for this type of dinosaur. Despite this, *Brontosaurus*, which is a much more colorful name for this gargantuan dinosaur than is *Apatosaurus*, gained wide use, especially in popular books. Indeed, many people continue to use the name *Brontosaurus*—it even appeared on a stamp issued by the United States Postal Service in the 1980s—although the name is technically incorrect.

The rules in the *International Code of Zoological Nomenclature* also cover the details of how to spell the scientific names of animals. These names must take Latin endings, and many of them have Greek or Latin roots. The genus and species names are always italicized. Such rules may seem a bit arbitrary, but they ensure uniformity in the scientific naming of dinosaurs and other animals, which facilitates communication among scientists worldwide. So, even though a dinosaur may be discovered in China and described by a Chinese paleontologist in an article written in Chinese, the dinosaur's scientific name will consist of two Latinized words (box figure 2.3). This makes it possible for those paleontologists who do not read or write Chinese to understand and use the dinosaur's name.

**重庆龙属(新属) *Chungkingosaurus* gen. nov.**

(图版 40—43)

**属的特征** 中小个体的剑龙。头骨较高，下颌骨厚实。牙齿细小，排列密集，互不重叠。齿冠不对称。背椎和尾椎为双平型的椎体。荐椎 4—5 个完全愈合，附有一加强椎体，荐部背面封闭不全。骨板呈棘板状，大而厚实。股骨骨干圆直，第四转节不明显，股骨与胫骨之比: 1.61—1.68。

该属目前包括一个种，两个未定种:

1. 江北重庆龙(新属新种) *Chungkingosaurus jiangbeiensis* gen. et sp. nov.
2. 峨岭的重庆龙(未定种一) *Chungkingosaurus* sp. 1
3. 重庆的重庆龙(未定种二) *Chungkingosaurus* sp. 2

**江北重庆龙(新属新种) *Chungkingosaurus jiangbeiensis* gen. et sp. nov.**

(图版 40—42)

**种属名称解释** 化石因发现于重庆市江北区而命名:江北重庆龙(新属新种)(*Chungkingosaurus jiangbeiensis* gen. et sp. nov.)。

**特征** 小个体的剑龙。头骨吻端较高，下颌骨厚实，牙齿细小，排列紧密，齿冠不对称，齿环不发育。背椎和尾椎均为双平型的椎体，荐椎 4 个和一个加强腰椎完全愈合一起，荐部背面封闭不全。骨板棘板状，大而厚实。肱骨有不明显的骨干。肠骨髋臼窝浅平，股骨圆直，无第四转节。胫骨近端显著扩粗，关节面圆形，胫、腓骨远端同距骨和跟骨完全愈合。

**BOX FIGURE 2.3**
This portion of an article describing a new dinosaur from China follows the *International Code of Zoological Nomenclature* by naming the new genus and species in Latinized words. (From Z. Dong et al. 1983. *The Dinosaurian Remains from Sichuan Basin, China.* Beijing: Science Press)

The **Linnaean hierarchy** is a system of classification in which smaller (lower) taxa are nested into larger (higher) taxa (figure 2.8). The categories of this classification extend from large and very inclusive to small and very exclusive, and each category has a category name. Thus, a phylum is a very large category in the Linnaean hierarchy. It contains one or more classes, and each class contains one or more orders and so on. The name assigned to a taxon follows the international rules that mandate the endings and forms of Latinization of most names (see box 2.3). For example, the rules specify that **family** names must end with the suffix "-idae." So, the name of the family containing *Tyrannosaurus* and closely related dinosaurs is Tyrannosauridae.

The name of a species of organisms, the lowest rank in the Linnaean hierarchy, always has two parts and is called a **binomen.** These are the **genus** (plural: **genera**) name followed by the **species** (plural: **species**) name. Thus, the name of the species to which all readers of this book belong is *Homo* (the generic name) *sapiens* (the trivial name), a Latin name that literally means "man the wise." Another way to explain this is to say that living humans belong to the genus *Homo* and to the species *Homo sapiens.* The specific name *sapiens* cannot stand alone as the name of a species, which must be a binomen—therefore *Homo sapiens.*

The international rules of zoological nomenclature set standards for naming taxa that allow all scientists worldwide to name taxa in a consistent way so as to facilitate communication (see box 2.3). But, because different ideas exist about the phylogenetic relationships of many organisms, the process of constructing classifications from phylogenies is full of disagreements that are useful in an active science. Once it comes time to name taxa, however, the rules that govern their naming ensure uniformity in the construction of the names and their endings. Nevertheless, the naming of taxa is a creative enterprise in which biologists and paleontologists are free to base the name on some attribute of the organism or organisms to be named or on anything else (box 2.4).

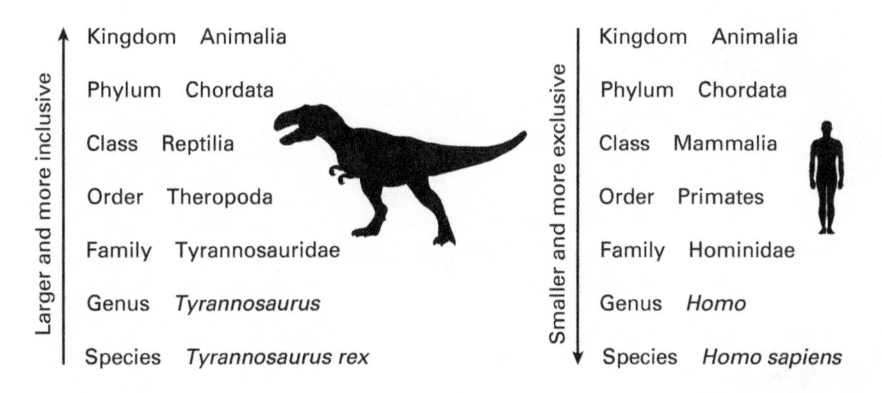


**FIGURE 2.8**
The Linnaean hierarchy groups organisms into progressively more inclusive groups. The major categories in the hierarchy—kingdom, phylum, class, order, family, genus, species—can be easily memorized with this mnemonic: "King Phillip Can Order Fresh Green Salad." (Drawing by Network Graphics)

Box 2.4

## Unusual Dinosaur Names

The *International Code of Zoological Nomenclature* describes the rules by which the names of dinosaurs are constructed. But, just what goes into those names is left to the creativity of the paleontologist(s) coining the name. Some dinosaurs are named after a distinctive anatomical feature. A good example is *Diplodocus*, which means "double beam," in reference to the double-beamed chevron bones beneath the dinosaur's tail. Other dinosaurs are named for the places where they were found. *Tuojiangosaurus* takes its name from the Tuo Jiang, a river in Sichuan Province, China near the discovery site. And, a few dinosaurs are named after people, such as *Piatnitzkysaurus*, named for Alejandro Piatnitzky (1879–1959), a Russian-born Argentinian geologist. However, it is most often the species name that is named after people, such as *Diplodocus carnegii*, named after multimillionaire and museum patron Andrew Carnegie.

Many dinosaur names are based on Greek or Latin roots, especially those named after an anatomical feature. But, this need not be the case, and there are a variety of unusual dinosaur names based on other linguistic root words. A good example is *Khaan*, a Cretaceous theropod from Mongolia. The genus name is from the Mongol word for "lord." An unusual dinosaur name based on a person's name is *Drinker*, a tiny Jurassic ornithopod. This dinosaur was named for paleontologist Edward Drinker Cope. (Drinker was the maiden name of Cope's mother.) Another unusual name is *Irritator*, a Cretaceous theropod from South America, named after the state of mind of the paleontologists who named it. Most unusual and difficult to pronounce is *Nqwebasaurus*, a theropod from Jurassic–Cretaceous boundary strata in South Africa. This name is taken from the Bushman language of southern Africa and includes a click sound characteristic of the language. Thus, it is pronounced N-(click with tongue)-kwe-bah-SORE-us.

More traditionally constructed names can also be quite creative. Thus, there is *Bambiraptor*, from the Upper Cretaceous of North America. This tiny theropod was named after the Disney cartoon character Bambi. Also, there is *Colepiocephale*, a Late Cretaceous pachycephalosaur from North America (box figure 2.4). This name was constructed from the Greek roots *colepio* (knuckle) and *cephale* (head), a perfect name for a dome-headed dinosaur!

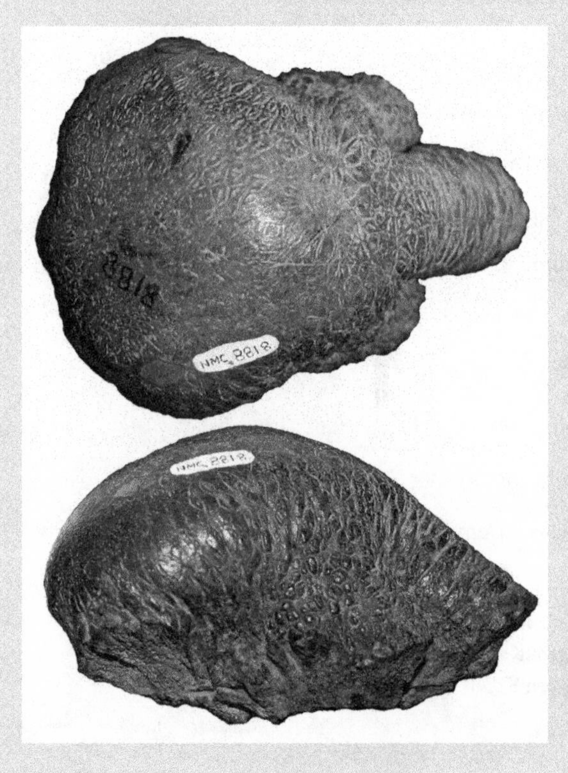


**BOX FIGURE 2.4**
This is the small (about 13 centimeters long) skullcap of the Cretaceous pachycephalosaur *Colepiocephale*. (Courtesy Robert M. Sullivan)

As an example, place yourself in the shoes of paleontologist Henry Fairfield Osborn, who, in 1905, was presented with the huge skeleton of a meat-eating dinosaur found in Montana. In the Linnaean hierarchy, dinosaurs belong to the phylum Chordata, animals with a flexible cord or rod running down the middles of their backs. Within Chordata, dinosaurs belong to the subphylum Vertebrata, in which the rod is segmented into vertebrae. The class Reptilia, vertebrates that lay hard-shelled eggs and have scaly skin and limb structures for fully terrestrial locomotion, includes the dinosaurs. Reptilia is one of several classes in the subphylum Vertebrata. Dinosaurs can be more precisely classified within Reptilia as members of the infraclass Archosauria (the "ruling reptiles," discussed in chapter 4). Most meat-eating dinosaurs belong to an order called Theropoda (see chapter 5).

But, Osborn's new meat-eating dinosaur differed from all other known theropods, so he had to name a new genus and species for this huge predator. The name he coined, *Tyrannosaurus rex*, was constructed in accordance with the rules of zoological nomenclature. It is Latin for "king" (*rex*) of the "tyrant lizards" (*Tyrannosaurus*), an appropriate and colorful name for what was then considered to be the largest land-living predator of all time. In the following year, 1906, Osborn created the family Tyrannosauridae for this dinosaur, because he realized it was quite different from most other dinosaurs in the order Theropoda.

## DINOSAURS AND EVOLUTION

Paleontologists view evolution as taking place at two levels. Evolution at the species level, or **microevolution**, is the evolution of populations of organisms that results in the origination of new species. Evolution above the species level, or **macroevolution**, is the origination and evolution of taxa larger than species, including genera, families, and orders, and may take millions of years. Macroevolution also refers to the origin of major adaptive features.

To decipher microevolution from fossils requires a large number of fossils that are close to each other in geologic time. Only with such a dense fossil record can paleontologists document the variation and small changes of extinct organisms during relatively short intervals of geologic time (100,000 years or less)—the types of changes that lead to the origination of new species.

In contrast, to study macroevolution does not require such an extensive fossil record. A good example of a macroevolutionary problem relevant to dinosaurs is the origin of birds. Fossils of dinosaurs and those of the earliest birds provide extensive insight into the origin of a higher taxon, the birds (class Aves), from the dinosaurs, as will be discussed in chapter 15. This is the case even though a dense fossil record of the evolutionary changes from a species of dinosaur to the first bird species remains to be documented.

The fossil records of many shelled invertebrates (animals without backbones, such as clams) and of some mammals are extensive enough to allow paleontologists to study microevolution. But, the fossil record of most dinosaurs is not. Dinosaur fossils are rarely abundant enough or close enough to each other in time to allow paleontologists to study the microevolution of dinosaurs. Instead, dinosaur fossils do document the macroevolution of some of Earth history's largest and most bizarre animals.

Dinosaur fossils thus are evidence of the evolutionary diversification of one of the most successful groups of land animals to have existed on Earth. This diversity can best be understood as a variety of adaptations to different environments and ways of life. Dinosaur extinction ended a complex and prolonged episode of evolution.

## Summary

1. Evolution is the origin and change of groups of organisms over time. It can also be described by Darwin's phrase as "descent with modification."
2. Darwinian evolution occurs by natural selection. This means that variants in a population of organisms better adapted to their environment will reproduce more successfully than less adapted variants.
3. The stratophenetic method of phylogeny reconstruction requires a much more complete fossil record than exists for dinosaurs. Therefore, most paleontologists who study dinosaurs do not use this method.
4. The cladistic method of phylogeny, which consists of constructing cladograms, is favored over the stratophenetic method. Cladistics identifies closely related taxa as those that share evolutionary novelties.
5. Problems with using the cladistic method arise from evolutionary convergence and difficulties in identifying features as evolutionary novelties.
6. Biological classification groups organisms into taxa according to their phylogenetic relationships.
7. The names of taxa are constructed according to international rules.
8. The fossil record of dinosaurs is best applied to problems of macroevolution.

## Key Terms

binomen
biological classification
clade
cladistic phylogeny
cladistics
cladogram
convergence
Charles Darwin
derived
divergence
evolution
evolutionary novelty
extinction
family
fossil record
genus (plural: genera)
grade
Linnaean hierarchy
Carolus Linnaeus
macroevolution
microevolution
monophyletic group
natural selection
origination
paleontologist
phylogenetic relationship
phylogenetic tree
phylogeny
polyphyletic
polyphyletic group
primitive
principle of priority
species (plural: species)
stratophenetic phylogeny
synonym
taxon (plural: taxa)

## Review Questions

1. Explain the central ideas behind Darwinian evolution.
2. Why do paleontologists who study dinosaurs favor cladistics instead of the stratophenetic method of phylogeny reconstruction?
3. Why does convergence pose a problem to cladists? What are other potential problems in constructing cladistic phylogenies?
4. Construct a cladogram for the three taxa: frog, dog, and cat. What shared evolutionary novelties led you to construct the cladogram?
5. What is the basis of biological classification?
6. What is the principle of priority and how is it applied to the scientific names of dinosaurs? What problems are alleviated by applying this principle?
7. What are the strengths and weaknesses of the dinosaur fossil record for studying evolution?

## Further Reading

Benton, M. J. 2014. *Vertebrate Palaeontology*. 4th ed. Hoboken, N.J.: Wiley-Blackwell. (College textbook on the evolution of all vertebrates)

Futuyma, D. J. 2013. *Evolution*. 3rd ed. Sunderland, Mass.: Sinauer. (College textbook on evolution)

International Commission on Zoological Nomenclature. 2000. *International Code of Zoological Nomenclature*. 4th ed. London: International Trust for Zoological Nomenclature. (The international rule book for the construction of the scientific names of animals)

Prothero, D. R. 2007. *Evolution: What the Fossils Say and Why It Matters*. New York: Columbia University Press. (Explains the basics of evolution and its relationship to the fossil record)

Prothero, D. R. 2013. *Bringing Fossils to Life: An Introduction to Paleobiology*. 3rd ed. New York: Columbia University Press. (College textbook on all aspects of paleontology)

Wiley, E. O., and B. S. Lieberman. 2011. *Phylogenetics: Theory and Practice of Phylogenetic Systematics*. Hoboken, N.J: Wiley-Blackwell. (A complete introduction to cladistics)

# 3

# FOSSILS, SEDIMENTARY ENVIRONMENTS, AND GEOLOGIC TIME

DINOSAURS have been extinct for 66 million years, so we must study their fossils to understand the history of this fascinating group of animals. Dinosaur fossils were formed in the environments in which the dinosaurs lived and died. The rocks that contain dinosaur fossils provide clues to the climate and habitats in which they lived. The reign of the dinosaurs began 225 to 230 million years ago and ended 66 million years ago. In this chapter, I examine the nature of dinosaur fossils, the relationship between rocks and ancient sedimentary environments, and the geologic time scale of dinosaur evolution.

## FOSSILS

The word **fossil** literally means "something dug up." Today, the word generally refers to any physical trace of past life. Most dinosaur fossils are mineralized bones. But, dinosaur fossils also include tracks, eggs, skin impressions, tooth marks, stomach stones (**gastroliths**), and fossilized feces (**coprolites**) (figure 3.1). These fossils are called **trace fossils** (or ichnofossils, from the Greek word *ichnos*, which means "trace") in contrast to mineralized dinosaur bones, which are termed **body fossils**. Chapter 12 of this book discusses dinosaur trace fossils.

How are fossils formed? How long does it take for a bone to become a fossil? These are common questions asked of paleontologists, the scientists who study fossils. Key to the process of fossilization is for the dinosaur bone (or footprint, egg, etc.) to be buried in sediment (rock particles) (figure 3.2). When a dinosaur died, it was either buried in sediment immediately or its remains, perhaps after scavenging, weathering, or being transported by running water, were buried somewhat later (box 3.1). The soft tissues of the dinosaur—skin, muscles, and internal organs—rotted away quickly, whether the dinosaur was buried immediately upon or long after death. Only rarely are soft tissues preserved, as in the case of the duck-billed dinosaur tail fossil with skin impressions shown in figure 3.3.

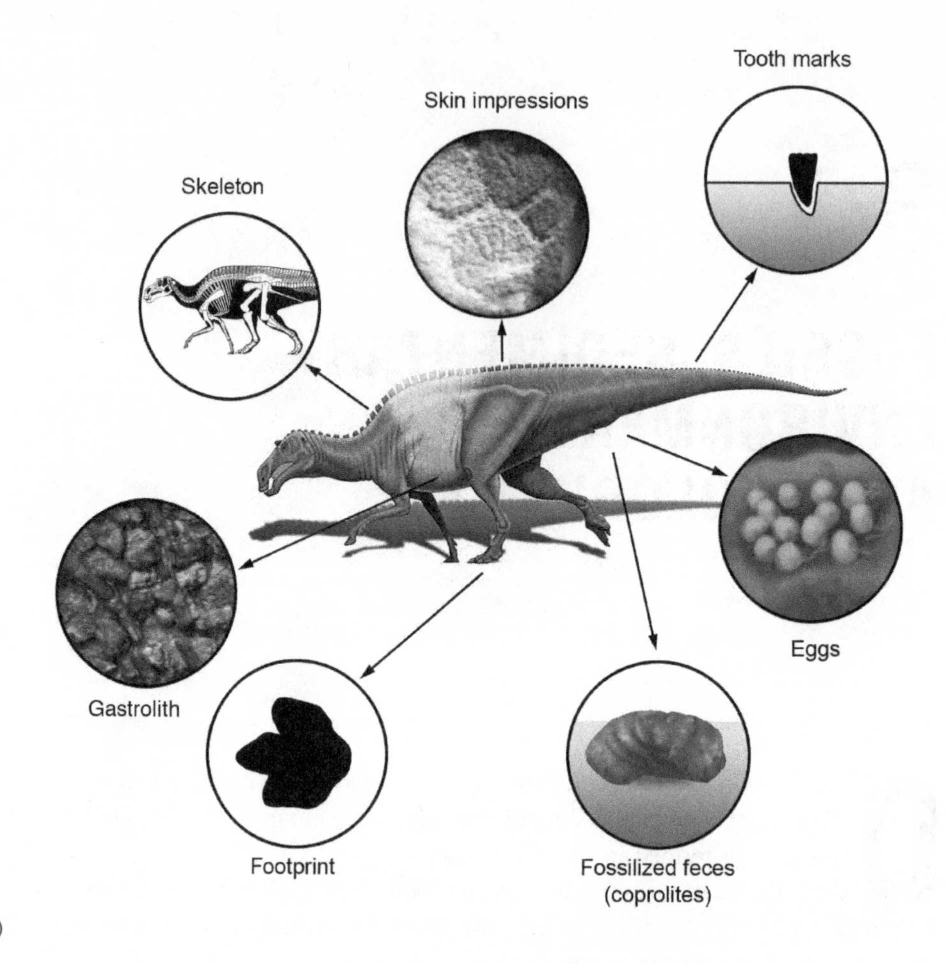


**FIGURE 3.1**
Dinosaur fossils comprise bones and skeletons, tracks, eggs, skin impressions, tooth marks, gastroliths, and coprolites. (© Scott Hartman)

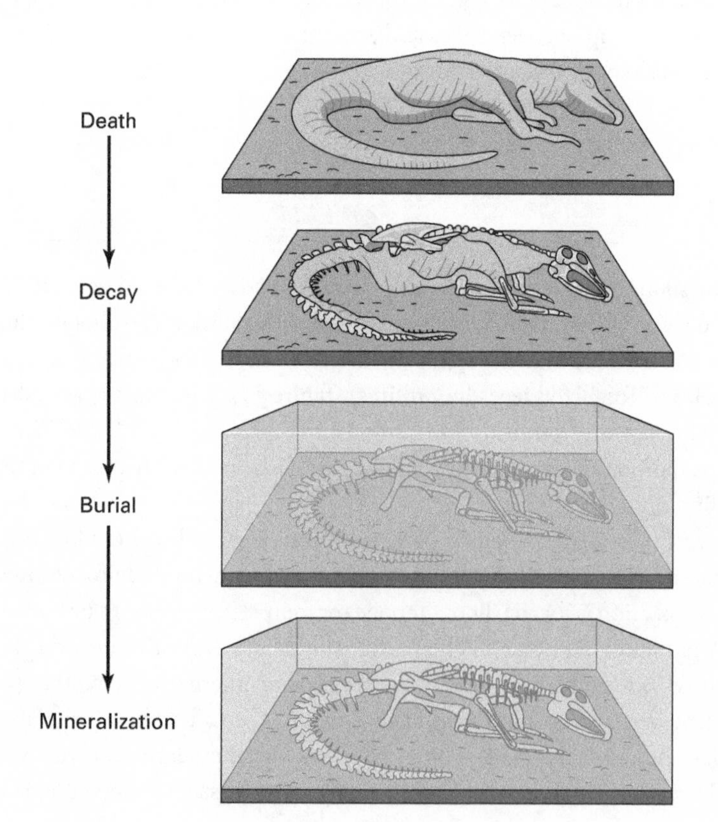


**FIGURE 3.2**
The process of fossilization involves death, decay, burial, and mineralization. (Drawing by Network Graphics)

**FIGURE 3.3**
These skin impressions surround part of the tail of a duck-billed dinosaur found in the Upper Cretaceous of New Mexico and are a rare instance of the fossilization of dinosaur soft tissues.

Box 3.1

## Taphonomy

In 1943, Soviet paleontologist Ivan Yefremov (1908–1972) combined the Greek words *taphos*, meaning "burial," and *nomos*, meaning "laws," to describe a branch of paleontology he called **taphonomy**. Taphonomy is the study of how fossils are formed and what biases are inherent in the fossil record. By undertaking taphonomic studies, paleontologists gain insight into the behavior and ecology of extinct organisms. They also come to grips with the kinds of information contained in a collection of fossils. They do these things by trying to understand the processes that have taken place between the time an organism, or group of organisms, was alive and the present, when we collect fossils of that organism or group of organisms.

If we imagine a population of dinosaurs alive, say, 100 million years ago, the first step toward fossilization of these animals is, of course, their death (box figure 3.1). If they die in an accident and their bodies are buried rapidly, such as in a flash flood, then the dinosaurs immediately become a "burial assemblage." Most dinosaurs, however, probably died in other ways that did not result in the immediate burial of their bodies. They may have been preyed upon by other dinosaurs or died from disease or old age, or in other types of accidents. Also, many dinosaur carcasses must have been scavenged by other animals long before burial. Because most forms of death for dinosaurs did not include burial, taphonomists speak of a "death assemblage" made up of all the dead dinosaurs—whether they were whole carcasses or partial and preyed upon and/or scavenged carcasses—that were not buried at the time of death. Only later, usually through geological processes, such as the accumulation of river sediments around and over bones, did burial occur. The death assemblage is eventually converted into a burial assemblage.

(*continued*)

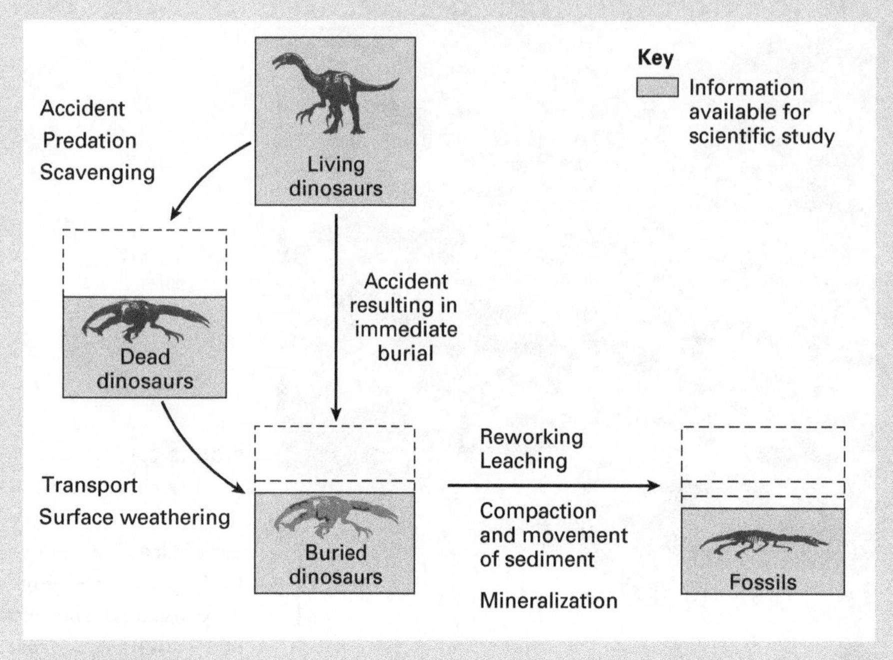


**BOX FIGURE 3.1**
Taphonomic processes destroy information between the life of a dinosaur and the dinosaur fossils we collect and study. (Drawing by Network Graphics)

This is an important distinction. We would expect the fossils of dinosaurs that enter the burial assemblage immediately upon death to consist of complete or virtually complete skeletons. And, by contrast, we would expect the fossils of dinosaurs that did not immediately enter the burial assemblage to consist of much less complete skeletons. Indeed, the fossils of such dinosaurs would consist mainly of parts of skeletons, isolated bones, and bone fragments. This is because any number of taphonomic processes, including surface weathering of bone, destruction of bone by predators and scavengers, and damage to bone by running water and moving sediment, would have reduced the integrity of a dinosaur skeleton prior to burial. These simple observations show that complete dinosaur skeletons are mostly those of animals buried immediately upon death or very soon after. Many dinosaurs are represented by less complete specimens, whereas the overwhelming majority of dinosaurs left no fossils at all, simply because their skeletons were totally destroyed before they were buried.

Just because a dinosaur skeleton or bone was buried does not mean it necessarily became a fossil. Other taphonomic processes may have intervened between the burial assemblage and the fossil assemblage (see box figure 3.1). For example, buried dinosaur bones could be exhumed by erosion and weather away before being fossilized. Or, even though buried, a skeleton might not become a fossil because of the fact that the chemistry of the soil and ground water were of the wrong type to lead to fossilization.

We can see that many factors may prevent a dinosaur from becoming a fossil that we can collect and study. Taphonomists stress this aspect of the fossil record by pointing out how much information taphonomic processes have destroyed between the time when the dinosaur was alive and when it became a fossil in a museum. When a dinosaur was alive—a flesh-and-blood, moving animal—it was in a state of 100 percent information. But, as death, burial, and fossilization took place, much information was lost, leaving us with only a small percentage of the original information available from the living dinosaur. Taphonomy helps paleontologists to understand how much information is removed between the life of a dinosaur and the dinosaur fossils we collect and study.

After burial, the remaining bones and teeth typically must undergo some form of **mineralization** to become fossils. Bones and teeth are not just inorganic minerals, but contain a significant amount of organic matter. This matter either decays and is replaced by minerals, such as silica, calcite, or iron, or forms complex compounds by combining with these minerals. The mineral replacement or combination takes place on a microscopic scale, so that tiny spaces around the inorganic matrix of the bone are filled with the new minerals. The resulting mineralized bone is thus a combination of the original inorganic bone matrix (a mineral called **calcium phosphate**) and the new minerals. However, in some cases, the original inorganic bone matrix is also partly or completely replaced by the new mineral. In most mineralized bones, the quality of microscopic structures is very high, allowing them to be studied under the microscope (figure 3.4).

The length of time it takes to mineralize bone is not well established. Most paleontologists believe it takes 10,000 years or more for a bone to fossilize. This is because most bones that are 10,000 years old or younger—mainly those excavated at archaeological sites—show little or no mineralization. However, some bones much older than 10,000 years also show little mineralization. Clearly, the rate at which a bone is mineralized depends on the type and chemistry of the sediment within which it is buried, the amount of water in the sediment, and other aspects of the sedimentary environment in which the bone is buried. This makes it difficult to generalize about the rate at which fossilization takes place. But, 10,000 years still remains a good minimum estimate of the time it takes for most bones to fossilize.

The preservation of delicate organic structures, such as skin or eggs, requires a special type of sedimentary environment in which these structures will not be damaged or destroyed. Dinosaurs must have made literally trillions of footprints during the

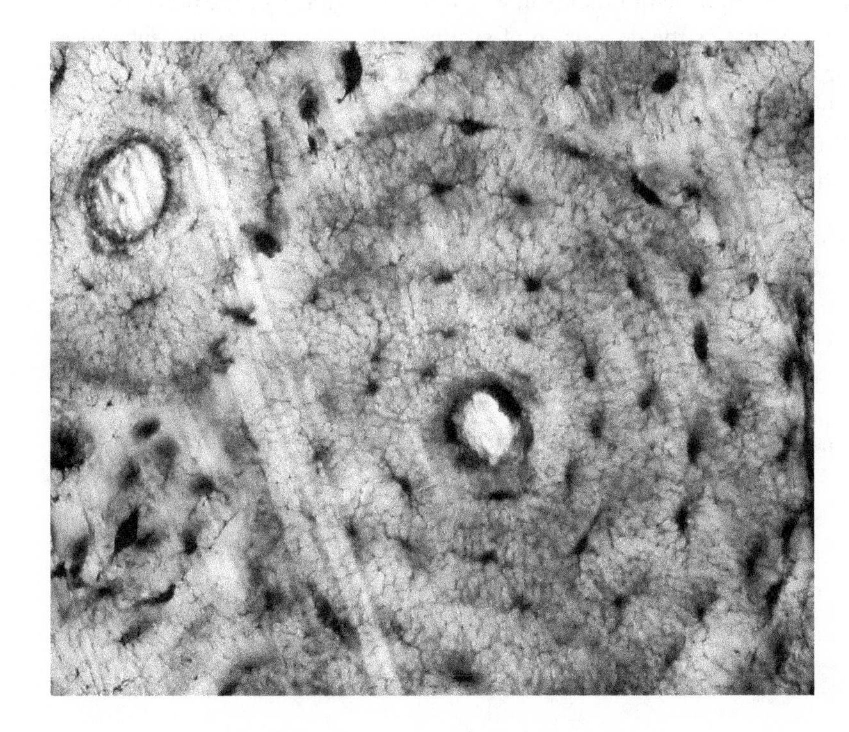

**FIGURE 3.4**
This cross-section of a fossilized dinosaur bone seen under high magnification shows the microscopic detail preserved. Here you can see small vascular canals (dark oval or round holes) that were filled with blood and soft tissue. The lighter-colored areas are mostly the inorganic mineral matrix of the bone. (Courtesy J. M. Rensberger)

160 or so million years of their existence. But, with few exceptions, only those footprints made in moist sediment and buried quickly, before they could erode, were preserved as fossil footprints.

Dinosaur fossils of all types are discovered and collected by fossil hunters worldwide. People who collect and study fossils professionally are called paleontologists, from the Greek words for "ancient-life studies." Dinosaur fossils collected by paleontologists are cleaned and studied in laboratories in museums and universities.

## SEDIMENTARY ENVIRONMENTS

Geologists recognize three types of rock on the surface of the earth. **Igneous rocks** are those that cool from a molten state. They include all volcanic rocks. **Metamorphic rocks** are those that have been altered (metamorphosed) by temperature and/or pressure. Common metamorphic rocks are marble, which is limestone before metamorphism, and quartzite, which is sandstone before metamorphism. **Sedimentary rocks** are those formed by the accumulation and cementation of mineral grains or by chemical (including organic) precipitation at the surface of the Earth. Sedimentary rocks are often the by-products of the erosion of igneous or metamorphic rocks. Sandstone, limestone, shale, and siltstone are common sedimentary rocks. Fossils are very rarely found in igneous or metamorphic rocks. Almost all fossils, and those of dinosaurs are no exception, are found in sedimentary rocks.

The type of sedimentary rock within which a dinosaur fossil is found contains much information about the environment in which the dinosaur lived and died. This is because different types of sedimentary rocks are characteristic of the distinct **sedimentary (depositional) environments** in which the rocks formed. The process of rock formation encompasses the deposition of sediment (sedimentation) and subsequent compaction and cementation.

The study of sedimentary rocks and their formation is called **sedimentology. Sedimentologists** make a broad distinction between **marine** (in the sea) and nonmarine (on the land) sedimentary environments (figure 3.5). Dinosaurs did not live in the sea,

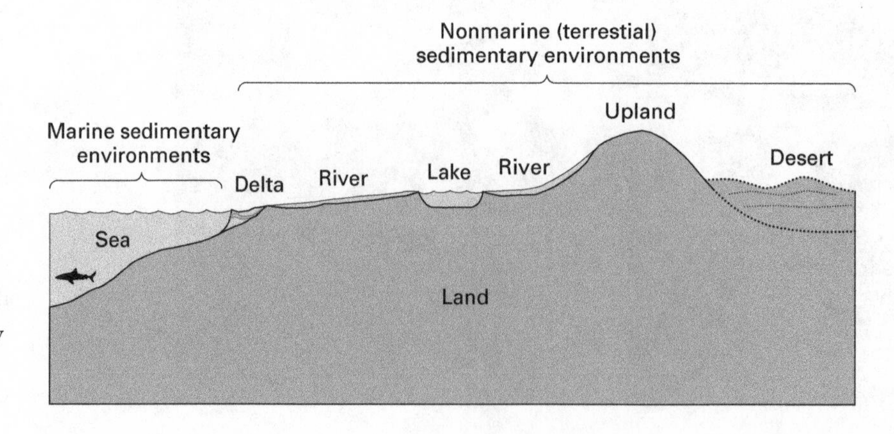


**FIGURE 3.5**
Sedimentologists distinguish marine (in the sea) from nonmarine or terrestrial (on the land surface) sedimentary environments. (Drawing by Network Graphics)

so their fossils are seldom found in marine sedimentary rocks. Therefore, we need not concern ourselves very much with marine sedimentary environments. Instead, we can focus mainly on the nonmarine (also called terrestrial) sedimentary environments—fluvial, lacustrine, eolian, and deltaic—where almost all dinosaurs were fossilized.

## Fluvial Environments

**Fluvial** environments are those in which rivers and streams are the dominant agents of sedimentation (deposition) (figure 3.6). Because the water is moving in one direction, often quite rapidly, sediments deposited in fluvial environments are typically made up of relatively large particles of sand and gravel. These particles, when cemented, become sandstone and conglomerate. Fluvial sandstones and conglomerates often display structures characteristic of running water, such as ripple marks formed by currents or cross-beds formed on underwater dunes (figure 3.7). Coarse particle size, ripple marks, and cross-beds are the primary features by which geologists identify fluvial sedimentary rocks.

Most dinosaur fossils are found in fluvial sedimentary rocks. They are especially common in river **floodplain** deposits. This is because rapid deposition of sediments is typical of fluvial sedimentary environments. These environments allowed for the large bones of dinosaurs to be buried most rapidly, thus making these bones the most likely to fossilize.

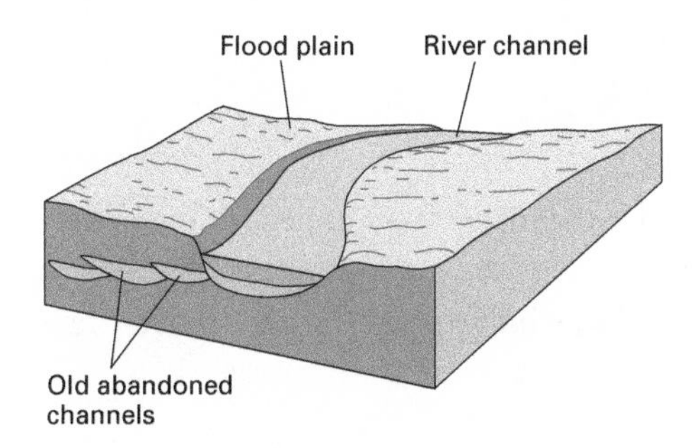


**FIGURE 3.6**
Fluvial environments are those in which rivers dominate sedimentation. (Drawing by Network Graphics)

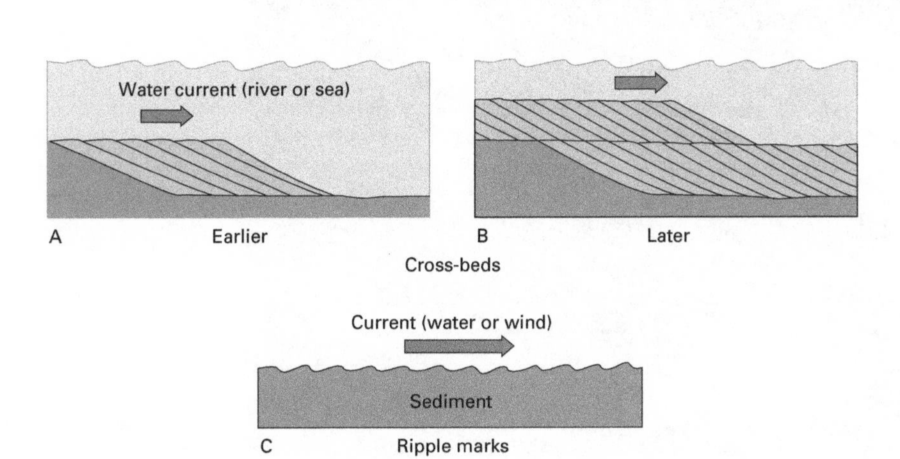


**FIGURE 3.7**
Running water produces cross-beds (*A* and *B*) and ripple marks (*C*). (Drawing by Network Graphics)

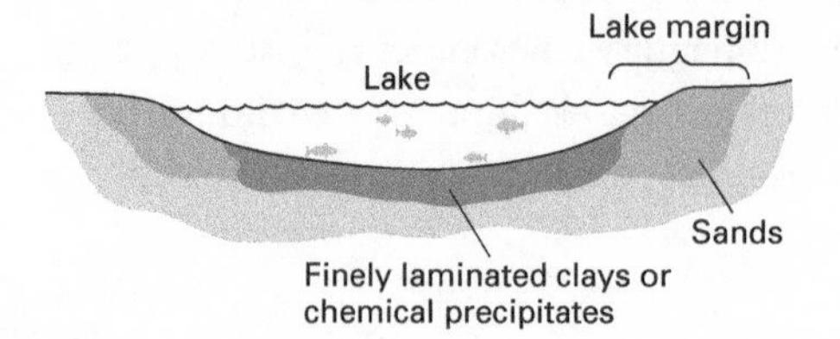


**FIGURE 3.8**
Lacustrine environments are those in which sedimentation takes place in a lake. (Drawing by Network Graphics)

## Lacustrine Environments

**Lacustrine** sedimentary environments are those in which deposition takes place in or on the margins of lakes (figure 3.8). There, the action of water is much gentler than in rivers, so lacustrine sedimentary rocks (consisting of mud and silt) have much smaller particle sizes than fluvial sedimentary rocks. Lacustrine rocks also have different sedimentary structures than fluvial rocks. Typically, they are laminated in thin layers of varying thicknesses that reflect periodic (sometimes seasonal) fluctuations in deposition. And, in lakes where not much mud or silt is involved in deposition, chemical precipitation of calcium carbonate, silica, or gypsum can become the dominant mode of deposition.

A fair number of dinosaur fossils have been found in lacustrine sedimentary rocks. Most of these dinosaur fossils, however, were found in the rocks that represent shoreline or river-delta deposits on lake margins. This suggests that few, if any, dinosaurs were actually aquatic and lived in lakes.

## Eolian Environments

**Eolian** environments are those in which wind is the major agent of deposition. Windblown sediments typically display the same types of sedimentary structures—ripple marks and cross-beds—as fluvial sediments. But, wind generally cannot move larger

**FIGURE 3.9**
This Middle Jurassic cross-bedded sandstone in New Mexico was deposited in an eolian environment.

particles like running water can, so eolian sedimentary rocks are fine-grained sandstones and siltstones. Also, wind tends to sort and abrade the particles it moves, giving the particles a more even size distribution and rounder shapes than particles moved by water. Cross-bedded, ripple-marked, fine-grained, rounded, well-sorted (consisting of grains of similar size) sandstones are typical rocks formed in ancient deserts (figure 3.9).

Few body fossils of dinosaurs were formed in eolian sediments because very little water was available for mineralization, but fossil footprints of dinosaurs abound in some eolian sedimentary rocks.

## Deltaic Environments

A delta is a triangular body of sediment formed where a river enters a large, quiet body of water, either a sea or lake. So, **deltaic** sediments are similar to fluvial sediments in many features. But the overall shape (geometry) of deltaic sediments and their proximity to lacustrine or **marine** sediments reveal their place of origin. Because deposition is often rapid in deltas, many dinosaur fossils, especially those of dinosaurs that lived near or along seashores, are preserved in deltaic deposits (figure 3.10).

**FIGURE 3.10**
These rocks in New Mexico, formed by river deltas 73 million years ago, are loaded with dinosaur fossils.

## GEOLOGIC TIME

When we talk about the "when" and "how fast" (or slow) of dinosaur evolution, we have to think in terms of geologic time—millions of years. Paleontologists and geologists, however, don't think about geologic time just in terms of numbers, but also in terms of a **relative time scale.** In a relative time scale, the goal is simply to determine whether one event occurred before (is older than) or after (is younger than) another. The statement of the age of an event is thus an age relative to the age of another event. For example, *Stegosaurus* lived before, or is geologically older than, *Tyrannosaurus*. This is a statement of the relative geologic age of these two dinosaurs.

This contrasts with a **numerical time scale** in which the goal is to assign a numerical age, usually expressed in millions of years, to an event. In a numerical time scale, we say *Stegosaurus* lived 150 million years ago, whereas *Tyrannosaurus* lived 67 million years ago. It might seem simplest and most precise to assign numerical ages to all the events in the age of dinosaurs. But the fact is that this is not possible, although most events in dinosaur evolution can be dated to within about 5 million years. Because of this general inability to assign more precise numerical ages, paleontologists use a relative geologic time scale to discuss the age of a dinosaur or an event involving dinosaurs.

The relative geologic time scale is a hierarchy of names applied to intervals of geologic time (figure 3.11). These names resemble the names of the months of the year or the days of the week. In a given week, we know Thursday comes after Tuesday, and,

| Era | Period | Epoch | Millions of years ago |
|---|---|---|---|
| Cenozoic | Neogene | Holocene | 0.01 |
| | | Pleistocene | 2.6 |
| | | Pliocene | 5 |
| | | Miocene | 23 |
| | Paleogene | Oligocene | 34 |
| | | Eocene | 56 |
| | | Paleocene | 66 |
| Mesozoic | Cretaceous | | 145 |
| | Jurassic | | 201 |
| | Triassic | | 251 |
| Paleozoic | Permian | | 299 |
| | Pennsylvanian | | 323 |
| | Mississippian | | 359 |
| | Devonian | | 419 |
| | Silurian | | 443 |
| | Ordovician | | 485 |
| | Cambrian | | 541 |
| Precambrian | | | |

**FIGURE 3.11**
This geologic time scale shows both the major divisions of relative geologic time and the approximate numerical ages of the division boundaries. (Drawing by Network Graphics)

in a given year, we know March follows February. The names and their succession on the relative geologic time scale similarly allow paleontologists to be certain that the Cretaceous followed the Jurassic and that the Triassic came before the Cenozoic.

The construction of a relative geologic time scale by geologists and paleontologists was heavily rooted in two principles. The first, the **principle of superposition**, states that in layered rocks (strata), the oldest rocks are at the bottom and younger layers are on top. The second, the **principle of biostratigraphic correlation**, states that rocks that contain the same types of fossil are of the same age (box 3.2). This principle is one of the basic ideas behind **biostratigraphy**, the identification and organization of strata based on their fossil content and the use of fossils in stratigraphic correlation. Stratigraphic correlation is the process of determining the equivalence of age or position of strata in different areas. It is one of the fundamental goals of **stratigraphy**, the study of layered rocks.

Geologists and paleontologists long ago realized that much of the vast thickness of **strata** exposed on Earth's surface is full of fossils, many of which are present only in specific layers and are thus considered to represent organisms distinctive of a particular interval of geologic time. The study of strata and fossils began in western Europe during the late eighteenth century, so many of the names used in the relative geologic time scale were coined by European geologists and based on European rocks and fossils (see figure 3.11).

One useful aspect of the time scale is that it is a hierarchy of time intervals from long to short. For the purposes of this book, the longest intervals used are the eras. During the past 542 million years, there were three eras: the **Paleozoic** (ancient life), **Mesozoic** (intermediate life), and **Cenozoic** (recent life). The boundaries of these eras coincide with major extinctions, such as the Permo-Triassic extinction, which eliminated many characteristic Paleozoic organisms, and the terminal Cretaceous, or Cretaceous–Tertiary extinction, which eliminated the dinosaurs and some other characteristic Mesozoic organisms. Because dinosaurs lived only during the Mesozoic Era, we need concern ourselves here with just the three geologic periods of the Mesozoic: the **Triassic, Jurassic,** and **Cretaceous.**

## The Triassic Period

German geologist Friedrich August von Alberti (1795–1878) coined the term "Triassic" in 1834. In studying the salt deposits of Germany, Alberti found three different rock sequences, an older one dominated by sandstone, an intermediate one mostly of limestone, and a younger one mostly of shale. All three sequences were younger than rocks identified as Permian but older than those identified as Jurassic. Thus, Alberti established a distinct time interval between the Permian and Jurassic Periods and called it the Triassic for the three rock sequences he had noted (*triad* being Latin for "three"). The Triassic is itself subdivided into three time intervals: Early, Middle, and Late Triassic. Geologists, however, refer to Triassic rocks as being Lower, Middle, or Upper Triassic. This convention—using "Lower" and "Upper" for rocks, but "Early" and "Late" for time ("Middle" is used for both)—is applied to all portions of the rock record and geologic time.

## Box 3.2

### A Dinosaur-Based Biostratigraphic Correlation

Today, Tanzania in eastern Africa, Portugal in western Europe, and the western United States are many thousands of kilometers apart. But, during the Late Jurassic, when dinosaurs lived in these three areas, they were somewhat closer together because of continental drift, though still thousands of kilometers apart. In the western United States—especially Wyoming, Colorado, Utah, and Oklahoma—Late Jurassic dinosaur fossils are now found in a sequence of rocks called the Morrison Formation. In Portugal, the Lourinhã and Alcobaça formations form a rock sequence that contains some of the same kinds of dinosaur fossils. In Tanzania, similar dinosaurs have been collected from a different sequence of rocks geologists call the Tendaguru Series. It is mostly because of the similarity of the dinosaur fossils found in the Morrison Formation, the Portuguese rock formations, and the Tendaguru Series that paleontologists believe all three rock sequences are of approximately the same age. The dinosaur fossils provide an important biostratigraphic correlation of three rock sequences widely separated geographically, then and now (box figure 3.2).

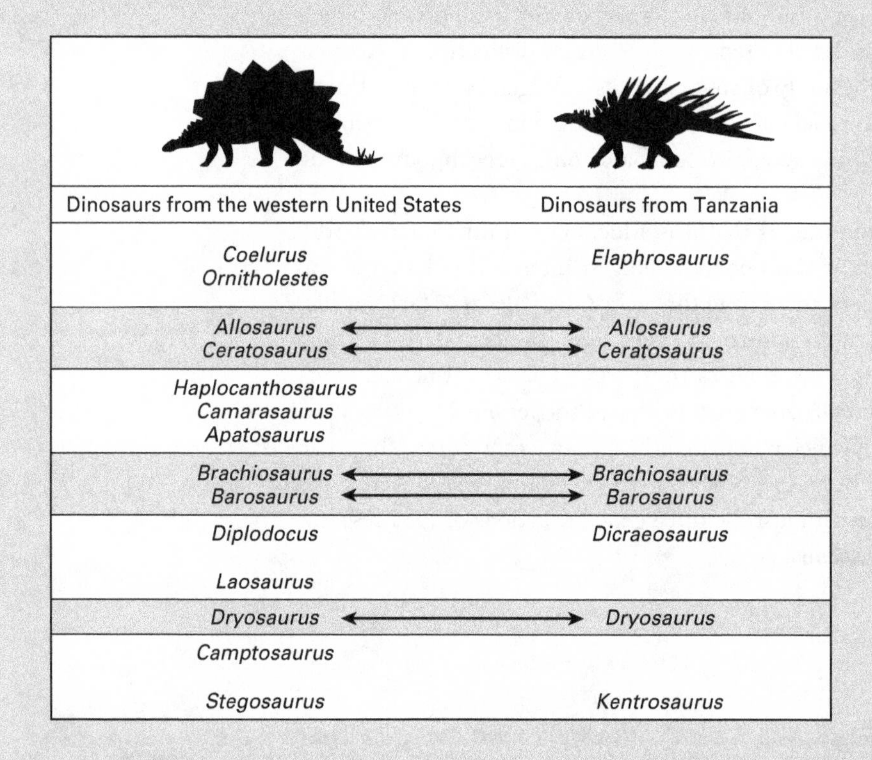


**BOX FIGURE 3.2**
Dinosaur fossils from rocks as far apart as the western United States and Tanzania are similar and support the correlation of these rocks. (Drawing by Network Graphics)

Dinosaur fossils from the Morrison Formation are among the most familiar to the American public: "*Brontosaurus*" (properly called *Apatosaurus*), *Diplodocus*, *Brachiosaurus*, *Allosaurus*, and *Stegosaurus*. These and other Morrison Formation fossils were first collected by American paleontologists during the late nineteenth century. Imagine the surprise of many paleontologists when some of the same types of dinosaurs, such as *Brachiosaurus* and *Allosaurus*, were discovered in the Tendaguru Series in eastern Africa by German paleontological expeditions just before the First World War. More recently, four of the dinosaur genera known from the Morrison Formation have been identified in Portugal: *Allosaurus*, *Apatosaurus*, *Ceratosaurus*, and *Torvosaurus*. According to the principle of biostratigraphic correlation, this means that the dinosaur-bearing sequences from the western United States, Portugal, and Tanzania are of the same relative age.

The Tendagaru Series, however, has additional importance to the correlation of these dinosaur faunas. This is because the dinosaur-bearing rocks in Tanzania were deposited along the shore of a Late Jurassic sea, and some of the dinosaur bones from the Tendaguru Series have the fossils of marine clams attached to them. Apparently, the clams grew on these bones in the shoreline lagoons and estuaries after the dinosaurs died. These clams are considered Late Jurassic in age because they are the same types as clams found in other marine rocks of well-accepted Late Jurassic age. So, we can now say that the same types of dinosaurs lived in the western United States, Portugal, and eastern Africa during the Late Jurassic when certain types of clams lived in the sea. These discoveries have given paleontologists a more complete picture of the Late Jurassic world, on the land and in the sea.

Today, rocks of Triassic age are recognized worldwide. The Triassic Period lasted 50 million years, from 201 to 251 million years ago (see figure 3.11). Dinosaurs did not appear until the Late Triassic, about 225 to 230 million years ago. They appeared at almost exactly the same time as the first turtles, crocodiles, pterosaurs (flying reptiles), plesiosaurs (long-necked marine reptiles), and mammals.

## The Jurassic Period

Many of us know Alexander von Humboldt (1769–1859) as a famous explorer and geographer of the late eighteenth century. But von Humboldt was also a trained geologist who, in 1799, first used the name "Jura" for a distinctive limestone found in the Jura Mountains of Switzerland. This became the basis for the word "Jurassic," used by geologists to refer to the time period between the Triassic and the Cretaceous, 145 to 201 million years ago (see figure 3.11).

Like the Triassic, the Jurassic is divided into Early, Middle, and Late intervals of time and Lower, Middle, and Upper rock intervals. Dinosaurs flourished everywhere on Earth during the Jurassic and, judging by their abundance and diversity, were extremely successful.

## The Cretaceous Period

In parts of western Europe, especially in Great Britain, France, and Belgium, rocks that are younger than Jurassic and older than Tertiary are mostly chalk. For this reason, a Belgian geologist, Jean-Baptiste-Julien D'Omalius d'Halloy (1783–1875) used the French term *Terrain Crétacé* (Cretaceous System) in 1822 to refer to these rocks (*creta* being Latin for "chalk").

The Cretaceous Period is the interval of geologic time between the Jurassic and Tertiary, 66 to 145 million years ago (see figure 3.11). It has traditionally been divided only into Early and Late time intervals, but, during the past 30 years, a Middle Cretaceous time interval has also been distinguished by some geologists. Dinosaurs are known from rocks of Cretaceous age on all the world's continents. They became extinct at the end of the Cretaceous.

## Numerical Ages

The geologic time scale used here (see figure 3.11) includes not just the names of the eras, periods, and epochs, but also assigns numerical ages, in millions of years, to their boundaries. How are these numerical ages calculated?

Before the discovery of radioactivity, calculating numerical ages for events in geologic history was little more than guesswork. But radioactivity, the spontaneous decay or falling apart of some types of atoms, provides a natural clock for estimating numerical geologic ages, because the rate of decay is constant for a given type of atom (figure 3.12). So, geologists can use this **radioactive clock** to determine the numerical age of rocks, provided that the rate of decay is known for the radioactive atoms and that the rock contains a known quantity of such atoms.

In laboratories, geochemists have calculated with great precision the rates at which radioactive atoms decay. Some of these rates are so slow that the decay products can be measured to estimate ages as old as billions of years. However, the rate of decay of some radioactive atoms—carbon-14 is a good example—is too fast to be used to estimate ages older than a few tens of thousands of years. In contrast, the rate of decay of uranium-238 is so slow that it takes 4.5 billion years for half the uranium atoms in a sample of uranium to decay. Calculating the amount of decay of uranium atoms by measuring the product of the decay, which is lead-206, in a rock from the age of dinosaurs involves measuring a very small amount. Avoiding errors in measurement is difficult; sometimes mistakes are made, and incorrect numerical ages are calculated.

These problems, however, do not prevent the best laboratories from calculating accurate numerical ages. Instead, what poses the largest problem for a numerical time scale is the fact that most rocks, indeed nearly all sedimentary rocks, do not contain enough decaying radioactive atoms to accurately determine age. Sufficient quantities of these

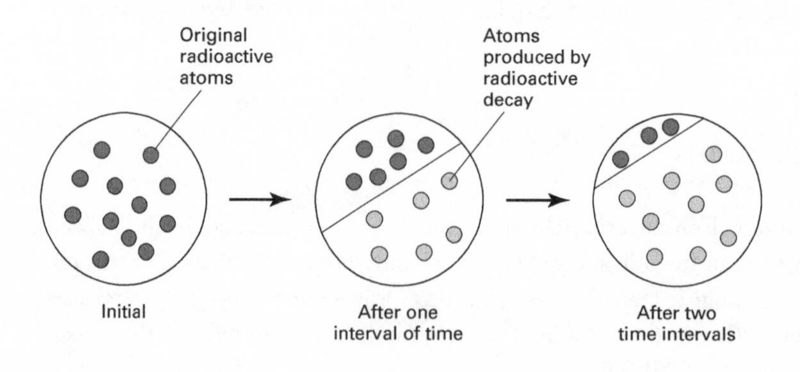


**FIGURE 3.12**
The slow decay of radioactive atoms provides a radioactive clock for measuring geologic time in millions of years. (Drawing by Network Graphics)

types of atoms are found almost exclusively in igneous rocks. However, because igneous rocks typically do not contain fossils, the usual way to calculate a numerical age for a fossil is to find an igneous rock layer close to the sedimentary rock that does contain fossils. In ideal circumstances, there are sheets of lava above and below a layer of sedimentary rock containing dinosaur fossils, which yield numerical ages bracketing the age of the fossils (figure 3.13). But, this does not happen often, and paleontologists can only estimate the numerical age of a dinosaur fossil by evaluating its proximity to the nearest rocks yielding a numerical age.

Such evaluation relies on several lines of evidence used in stratigraphic correlation. But, the fact remains that we cannot assign precise numerical ages to all dinosaur fossils, and we may never be able to do so. This is why we continue to use the divisions of the relative geologic time scale, supplemented by the best numerical-age estimates available, when discussing the "when" and "how fast" of the age of dinosaurs and their evolution.

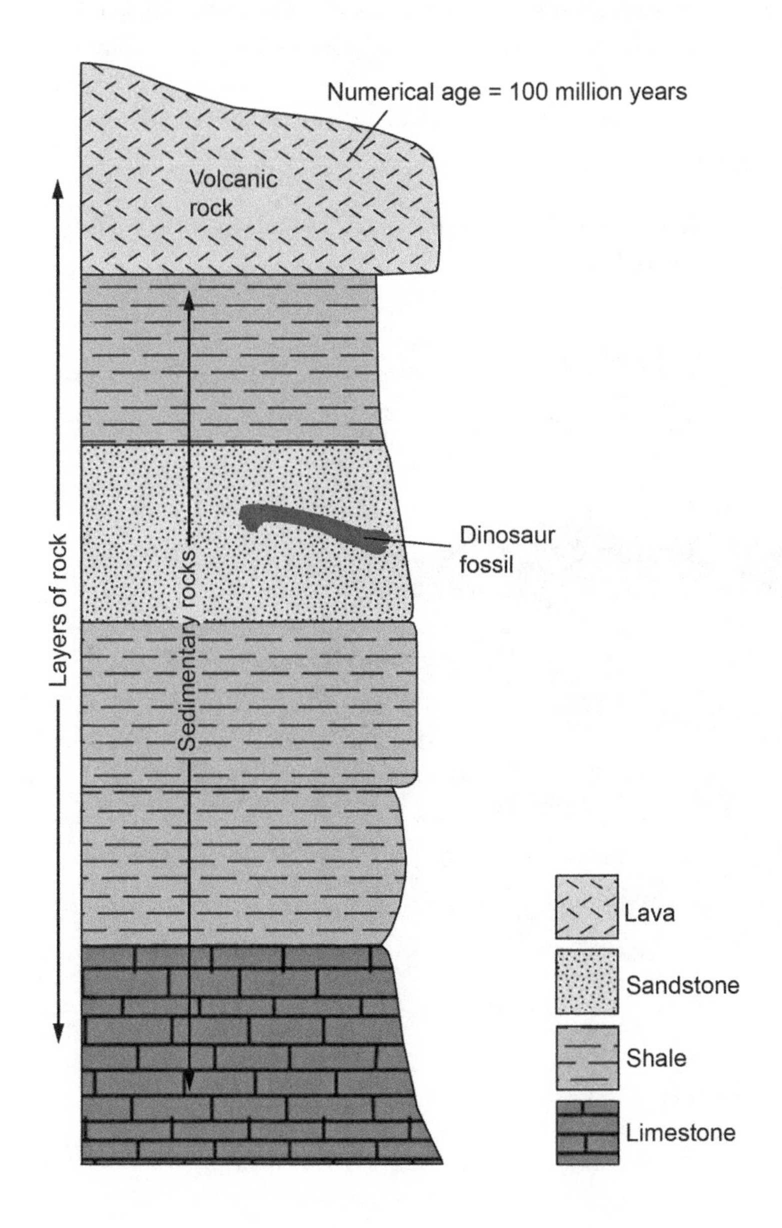


**FIGURE 3.13**
A numerical age from a volcanic rock only approximates the age of dinosaur fossils found in nearby sediments. In this example, the volcanic rock yields a numerical age of 100 million years. Because the volcanic rock is above the layer with the dinosaur bone, the principle of superposition indicates the bones must be older than 100 million years. (© Scott Hartman)

Box 3.3

## How to Find a Dinosaur Fossil

Every year, thousands of dinosaur fossils are discovered and collected worldwide. Here's how to find a dinosaur fossil:

1. Dinosaurs lived only during the Mesozoic Era, from the Late Triassic to the end of the Cretaceous. Limit your search to rocks that formed during that geologic time interval.
2. Dinosaurs lived on land, so almost all of their fossils are found in sedimentary rocks deposited in fluvial, lacustrine, or deltaic sedimentary environments. Focus on such rocks.
3. Dinosaur fossils can be discovered only once they begin to erode from rock on the land surface. So, find places where rock is actively eroding, not places where rock is covered by soil and vegetation. Badlands are ideal.
4. You need a search image, which means you need to know what a dinosaur fossil—bone, tooth, or footprint, for example—looks like. Look at photographs or go to a museum and familiarize yourself with dinosaur fossils before you begin your search.
5. Hike and search. At this point, luck may also be needed to help you find that rare place where a dinosaur fossil is eroding from the rock.

If you reach that place, then you will discover a dinosaur fossil (box figure 3.3). You then become the first human to see this dinosaur since it became entombed in rock, many millions of years ago!

**BOX FIGURE 3.3**
You may find a dinosaur fossil like this one—the ulna of a Late Cretaceous ceratopsian.

## COLLECTING DINOSAUR FOSSILS

Dinosaur fossils are present in Mesozoic sedimentary rocks formed in nonmarine environments. So, the search for dinosaur fossils focuses on these rocks (box 3.3). Paleontologists will examine such rocks, relying in part on the locations of previously discovered dinosaur fossils and in part on sheer luck, to find dinosaur fossils.

Once a dinosaur fossil is found, if it is small enough and sufficiently solid it may be collected by simply picking it up (if it is on the surface) or digging it right out of the rock. But most dinosaur fossils are so large and/or fragile that they must be encased in a plaster jacket before being collected. This is done by carefully digging around the fossil (figure 3.14), covering it with paper (so the plaster does not adhere directly to the fossil), and then wrapping the fossil with strips of burlap soaked in wet plaster. When the wet plaster hardens, it forms a solid case that protects the fossil from damage during transport to the laboratory.

**FIGURE 3.14**
Excavating a dinosaur takes much care, patience, and hard work.

In the laboratory, technicians remove the plaster jacket and surrounding rock from the dinosaur fossil. This process of preparation has to be undertaken with great care and precision so as not to damage the fossil. Often the cleaned dinosaur fossil needs further stabilization and is coated with shellac-like hardeners. Complete preparation of a dinosaur fossil may take months or years!

After preparation, the dinosaur fossil is ready for scientific study and display. Paleontologists and technicians at museums and universities worldwide undertake the collecting, preparation, and study of dinosaur fossils.

## Summary

1. Fossils are evidence of past life. Dinosaur fossils are not only bones, but also footprints, eggs, skin impressions, tooth marks, stomach stones, and feces.
2. Dinosaur bones were almost always mineralized as they were fossilized, a process believed to take at least 10,000 years.
3. Taphonomy is the study of the processes that intervened between the life of a dinosaur and the fossils of the dinosaur we collect and study. Taphonomy particularly concerns the information lost via these processes.
4. Dinosaur fossils are preserved almost exclusively in sedimentary rocks formed on the continents by rivers, lakes, and deltas.
5. Geologic time is measured by two time scales: a relative one and a numerical one.
6. On the relative time scale, dinosaurs lived during the Mesozoic Era, from the Late Triassic Period through the entire Jurassic and Cretaceous Periods, until their extinction at the end of the Cretaceous.
7. On the numerical time scale, dinosaurs lived about 66 to 225 or 230 million years ago.
8. The numerical time scale is based primarily on the decay of radioactive atoms found in sufficient quantity, with few exceptions, in igneous rocks.
9. Because dinosaur fossils are not found in igneous rocks, numerical ages for dinosaurs are estimates with varying degrees of accuracy. This is why most statements about the ages of dinosaurs employ the relative time scale.
10. Collecting and preparing dinosaur fossils is a complex process that can take months or years.

## Key Terms

biostratigraphy
body fossil
calcium phosphate
Cenozoic Era
coprolite
Cretaceous Period
deltaic
eolian
floodplain
fluvial
fossil
gastrolith
igneous rock
Jurassic Period
lacustrine
marine
Mesozoic Era
metamorphic rock
mineralization
numerical time scale
Paleozoic Era
principle of biostratigraphic correlation
principle of superposition

radioactive clock
relative time scale
sedimentary (depositional) environment
sedimentary rock
sedimentologist
sedimentology
strata
stratigraphy
taphonomy
trace fossil
Triassic Period

## Review Questions

1. What types of dinosaur fossils can paleontologists collect and study?
2. How does bone fossilize? How long might fossilization take?
3. What taphonomic processes intervene between the time a dinosaur is alive and when its remains have fossilized? What effect do these processes have on the information available to paleontologists?
4. Which types of rock and which types of sedimentary environment contain most dinosaur fossils? Why are few fossils found in igneous and metamorphic rocks?
5. Compare and contrast the relative and the numerical geologic time scales.
6. Why can't we assign precise numerical ages to all dinosaur fossils?
7. When, in terms of the relative and the numerical time scales, did dinosaurs live?

## Further Reading

Berry, W. B. N. 1987. *Growth of a Prehistoric Time Scale.* Palo Alto, Calif.: Blackwell Scientific. (Reviews the history of the relative geologic time scale and the principles behind it)

Gradstein, F. M., J. G. Ogg, M. D. Schmitz, and G. M. Ogg. 2012. *The Geologic Time Scale 2012.* 2 vols. Amsterdam: Elsevier. (Recent grand synthesis of the geologic time scale)

Martin, R. E. 1999. *Taphonomy: A Process Approach.* Cambridge: Cambridge University Press. (A comprehensive review of the theory and practice of taphonomy)

Prothero, D. R., and F. Schwab. 1990. *Sedimentary Geology.* New York: Freeman. (Textbook on stratigraphic principles and sedimentary environments)

Stanley, S. M., and J. A. Luczaj. 1989. *Earth System History.* New York: Freeman. (Textbook that provides an overview of fossils and the history of life)

# 4

# THE ORIGIN OF DINOSAURS

THE oldest dinosaur fossils are of Late Triassic age, about 225 to 230 million years old. These fossils, and those of other Triassic reptiles closely related to dinosaurs, provide paleontologists with a relatively good understanding of the origin of dinosaurs. This understanding, however, is far from complete, and many details of dinosaur origins remain to be discovered. In this chapter, I delineate the relationships of dinosaurs to other reptiles and review the current understanding of the origin of dinosaurs.

## DINOSAURS AS REPTILES

For many years, scientists have used the word **reptile** to refer to living turtles, snakes, lizards, and **crocodilians**, as well as their fossil relatives. "Reptile" traditionally referred to a tetrapod (four-legged vertebrate) with scaly skin that reproduces by laying an amniotic egg. An **amniotic egg** (figure 4.1) is usually a hard-shelled egg (some have leathery shells), such as the chicken egg we are familiar with. More precisely, it is an egg in which the developing embryo is almost totally surrounded by a liquid-filled cavity that is enclosed by a membrane called the amnion. This egg can be laid on dry land, unlike the eggs of fishes and amphibians, which must be laid in water or in a very moist place. The appearance of the amniotic egg freed reptiles from the dependence on water that characterizes amphibians. As added evidence of that freedom, we see in reptiles a much more solidly connected vertebral column and more powerful limb skeletons than we do in amphibians (figure 4.2). Thus, the amniotic egg and skeletal modifications for terrestrial locomotion are the key evolutionary novelties that distinguish reptiles from their ancestors among the amphibians.

Viewed cladistically, we can thus identify reptiles as a single clade or monophyletic group. But, most of the key features of reptiles are also seen in their descendants, which

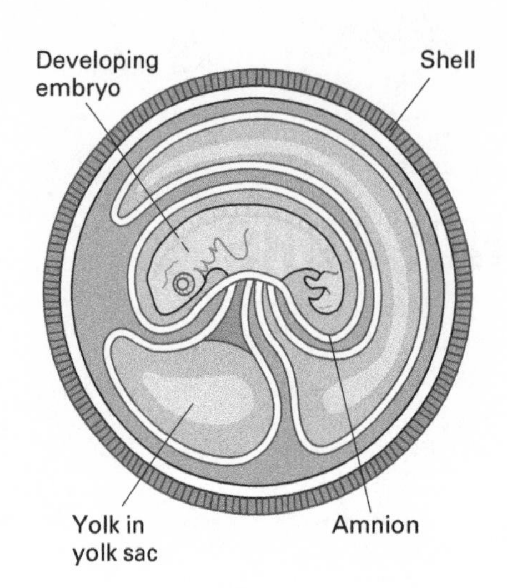


**FIGURE 4.1**
The amniotic egg does not have to be laid in water. It freed reptiles from the dependence on water that characterizes their ancestors, the amphibians. (Drawing by Network Graphics)

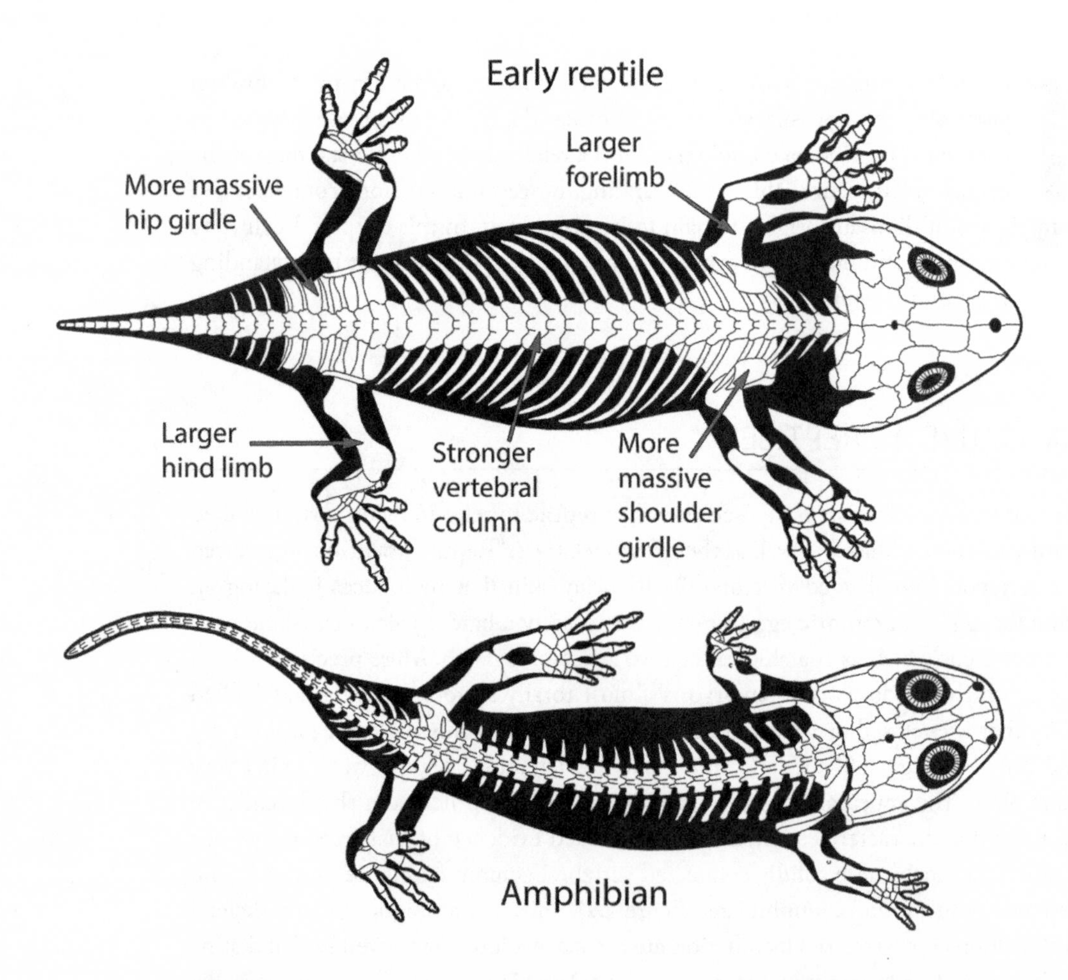


**FIGURE 4.2**
The skeleton of an early reptile differs from that of an amphibian principally in the modification of the limbs and girdles to support the reptile in fully terrestrial locomotion. (© Scott Hartman)

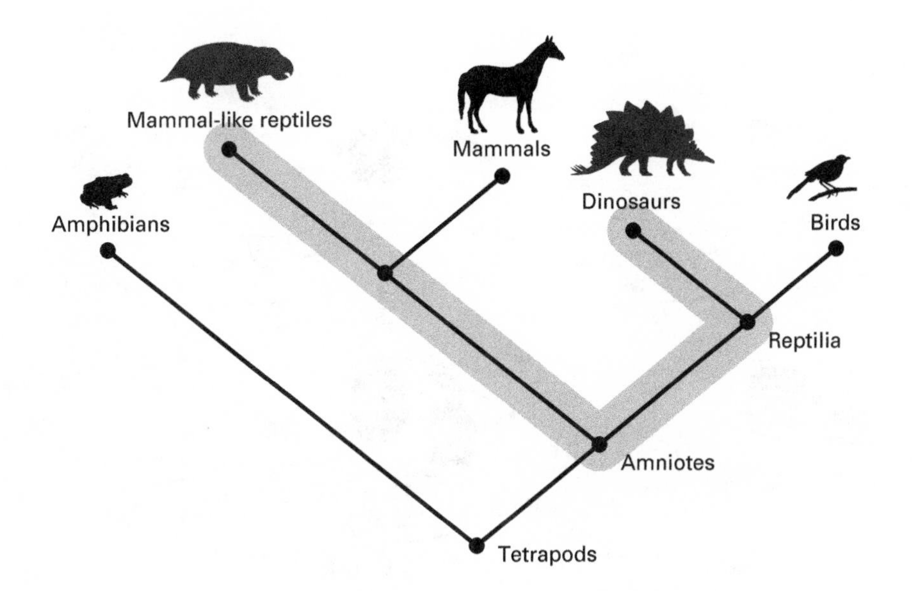


**FIGURE 4.3**
The cladogram of the amniotes indicates that Reptilia is only part of a clade because it does not include the descendants of reptiles: birds and mammals. In cladistic terms, Reptilia is a paraphyletic group. (Drawing by Network Graphics)

we call birds and mammals. This means that the word "reptile" refers to a paraphyletic group, unless we include birds and mammals in that group. **Paraphyletic** groups are clades that do not include all of their terminal branches (figure 4.3). They are only portions of clades and are avoided by most cladists.

There are several ways to deal with this cladistic problem. One way is to abandon the term "reptile" and use the term "amniote" to refer to what were formerly termed reptiles, as well as their descendants, the birds and mammals. Another way, and the one employed here, is to continue, in the interest of easy communication, to use "reptile" as a term to refer to all amniotes except birds and mammals. This use of "reptile" identifies dinosaurs as reptiles. Some paleontologists, however, prefer to recognize birds and dinosaurs as a group separate from reptiles, which they name Dinosauria, Avialae, or Aves (the last name normally applied only to birds). Because there is abundant evidence that dinosaurs laid amniotic eggs, and dinosaur skeletons show extensive modifications for terrestrial locomotion, we shall continue to classify them as reptiles.

## DINOSAURS AS DIAPSIDS

Reptiles have long been classified into four groups based on the pattern of openings in the skull roof behind the orbits (figure 4.4). The idea of a skull roof having openings or holes behind the eyes may strike you as strange. This is because our skull roof behind the eyes seems to be a solid wall of bone surrounding our extremely large brains. We actually have one small opening, behind the eye, but it is not obvious. In contrast, reptilian brains are relatively smaller and are contained in a solid braincase. This braincase is housed within a second wall of bone that contains the openings characteristic of different reptile groups.

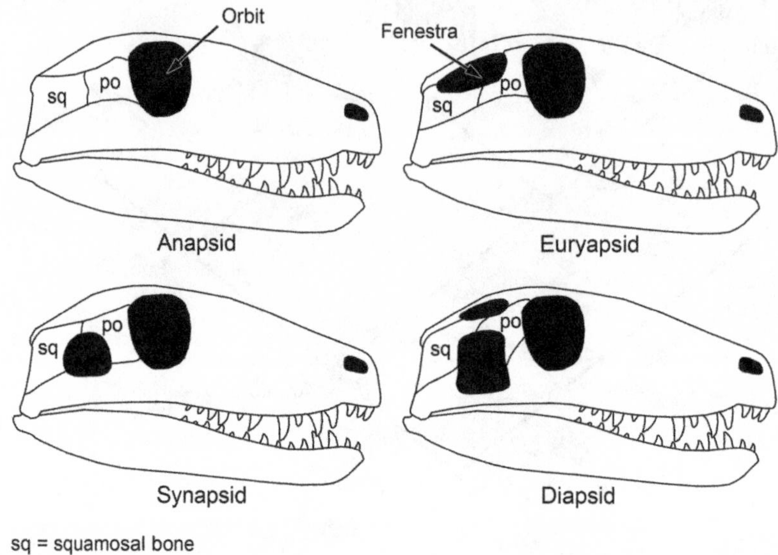


**FIGURE 4.4**
Four groups of reptiles can be recognized based on the number and position of temporal fenestrae. (© Scott Hartman)

The most primitive reptiles had no openings in the wall of bone behind the eyes. These reptiles are called **anapsids** (*an* [without]; *apsis* [opening]), and turtles today have an anapsid skull structure. Other reptiles evolved one or more pairs of openings (see figure 4.4).

The development of these openings, the **temporal fenestrae**, may be attributed to two things. The first is the concentration of mechanical stress in the skull. Experiments have shown that bone grows more thickly in stressed areas of the skull, and between such areas bone is thin or absent. The second is the distribution of the muscle attachments. Muscles can be more strongly anchored to ridges and edges of bone than to flat surfaces. The stresses created by chewing and the attachment of chewing muscles to the skull determined the number and position of the temporal fenestrae.

Dinosaurs have two temporal fenestrae on each side of their skulls and thus belong to the group of reptiles called Diapsida. All living reptiles, except turtles, also are **diapsids**. Euryapsids include two groups of Mesozoic marine reptiles, the very fish-like ichthyosaurs and the long-necked plesiosaurs. Synapsids encompass the mammal-like reptiles, which include the ancestors of mammals, as well as the pelycosaur *Dimetrodon*, a reptile often mistaken for a dinosaur.

## DINOSAURS AS ARCHOSAURS

The diapsid reptiles consist of two great groups, the lepidosaurs and the archosaurs, that diverged from each other during the Permian Period, more than 250 million years ago. Living lizards and snakes, and their ancestors, are **lepidosaurs** and are of no further concern to us in this book.

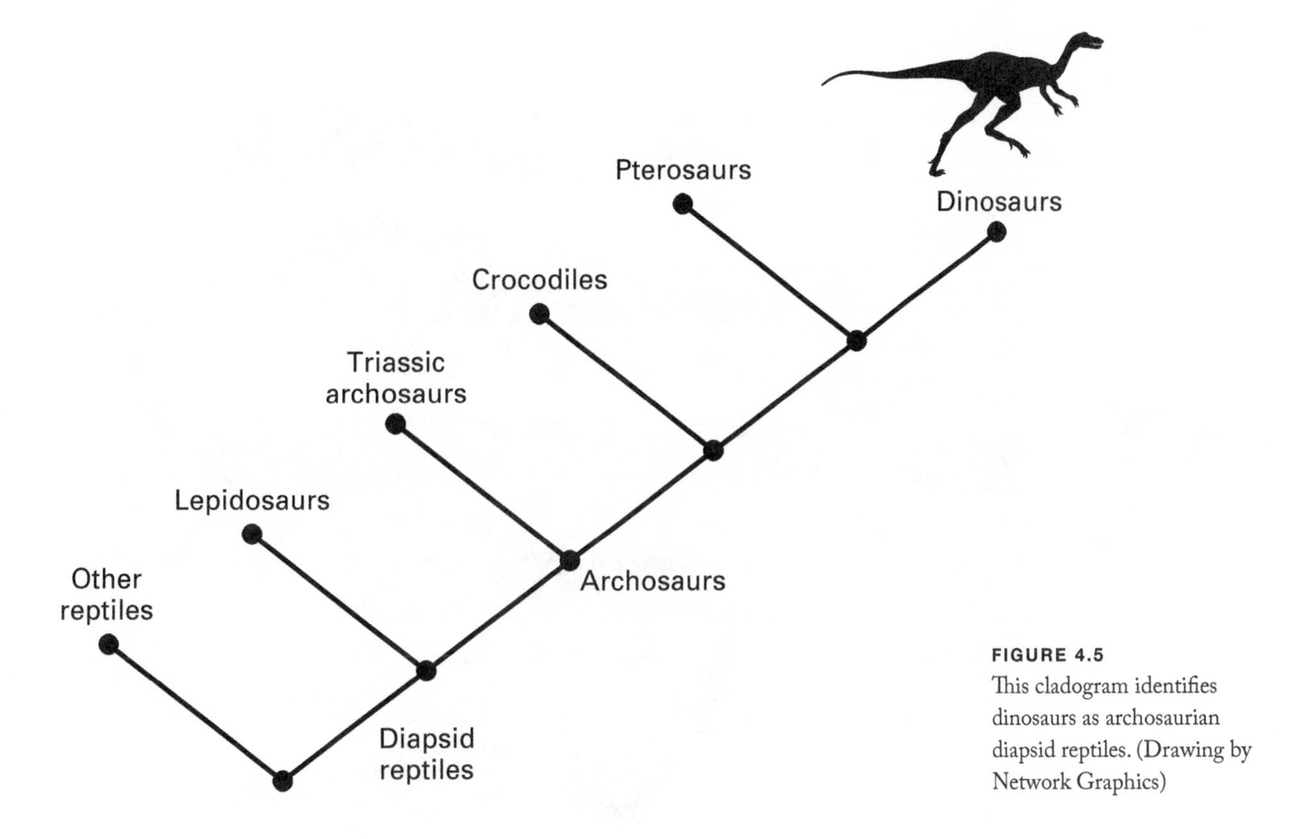


**FIGURE 4.5**
This cladogram identifies dinosaurs as archosaurian diapsid reptiles. (Drawing by Network Graphics)

**Archosaurs** include the dinosaurs, as well as diverse Triassic taxa that include the immediate ancestors of dinosaurs and two groups of close dinosaur relatives, the **pterosaurs** (flying reptiles) and the crocodiles (figure 4.5). Key evolutionary novelties among archosaurs that distinguish most of them from lepidosaurs include having no teeth on the palate and features of the limb (especially ankle) skeleton that indicate a semi-upright or upright limb posture.

Among the archosaurs, crocodilians first appeared in the Late Triassic and survive today. Pterosaurs are often included with the dinosaurs in popular books. But, they are a distinct group of archosaurs with a fascinating and complicated evolutionary history that took place at the same time as the dinosaurs, during the Late Triassic through to the Late Cretaceous (figure 4.6). The ancestry of dinosaurs, which are specialized archosaurs, must be sought among other Triassic archosaurs.

## THE ARCHOSAURIAN ANCESTRY OF DINOSAURS

Archosauria includes crocodiles, dinosaurs, and pterosaurs, as well as a variety of very diverse and successful Triassic taxa (see figure 4.6). Interpreting archosaur phylogeny and dinosaur ancestry has relied heavily on the analysis of ankle structures. There are two reasons for this. First, quite simply, the bones of the ankle are very compact and dense and thus are usually well preserved and common as fossils. Second, and more

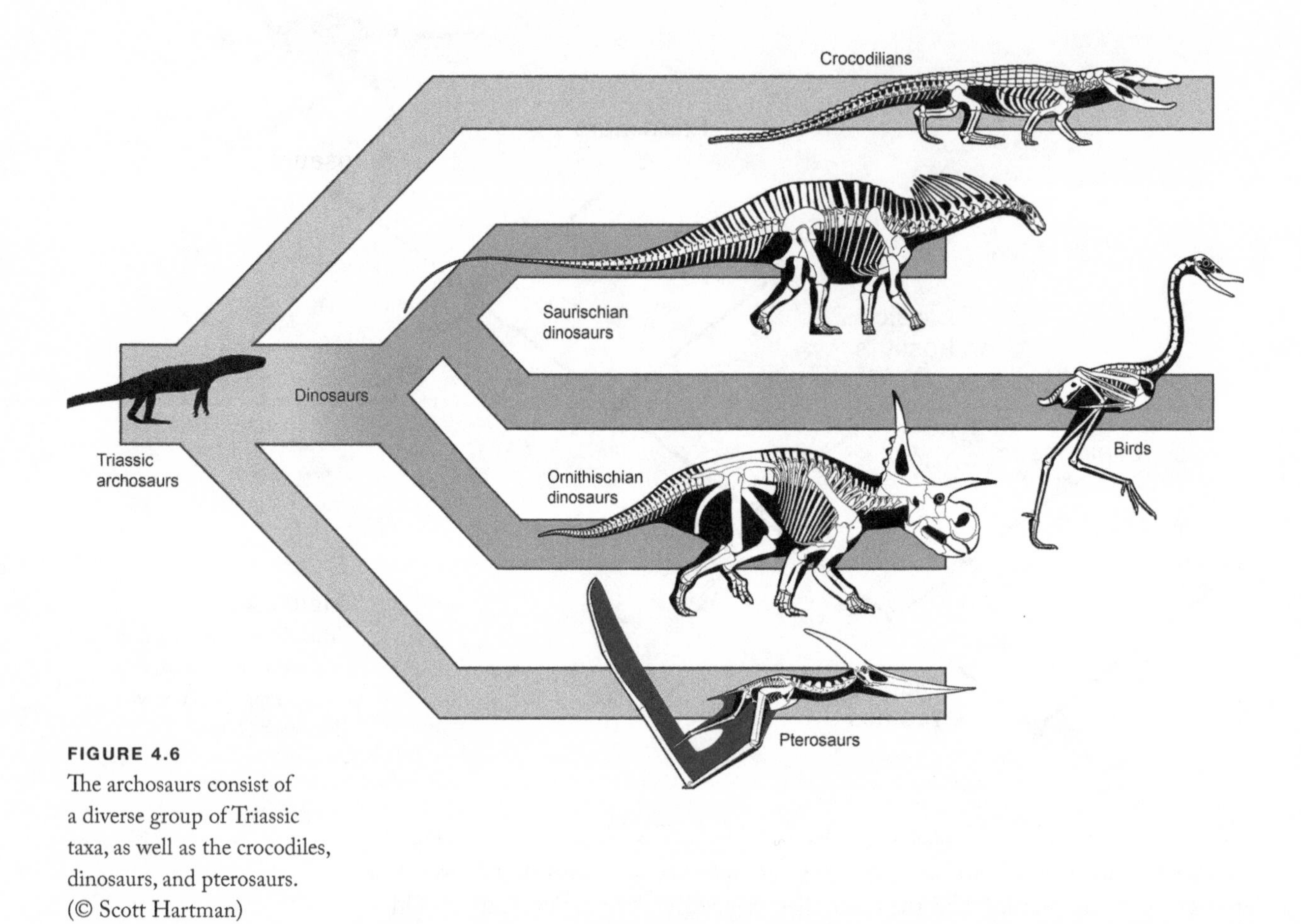


**FIGURE 4.6**
The archosaurs consist of a diverse group of Triassic taxa, as well as the crocodiles, dinosaurs, and pterosaurs. (© Scott Hartman)

important, the structure of archosaur ankles provides us with insight into how these animals walked. This is significant because the origin of dinosaurs is, in one important sense, the evolution of archosaurs with an **upright posture** from those with a semi-upright (sprawling) posture.

Skeletal features indicative of an upright posture thus are among those that paleontologists identify as evolutionary novelties of the dinosaurs (figure 4.7). A special type of ankle structure, called the **advanced mesotarsal (AM) ankle**, is one of these evolutionary novelties (figure 4.8). In the AM ankle, the astragalus is much larger than the calcaneum, and both bones are rigidly attached to each other and to the tibia. Because of this, the AM ankle has a single hinge between the calcaneum and astragalus and the rest of the foot. This hinge allows little twisting of the ankle when walking. Instead, the foot below the AM ankles swings backward and forward in a straight line to enable speedy, upright walking and running.

The AM ankle of dinosaurs contrasts with the **crocodile-normal (CN) ankles** of most other archosaurs (see figure 4.8). These ankles have joints that allow them to twist while walking. This twisting is normal when walking with a sprawling posture, because it helps the entire foot to stabilize slowly as it touches the ground. Most archosaurs, including the heavily armored, plant-eating **aetosaurs** and the crocodile-like **phytosaurs** (figure 4.9), walked with a sprawling posture.

**FIGURE 4.7**
The skeleton of Late Triassic *Coelophysis* displays several evolutionary novelties of dinosaurs, including three or more sacral vertebrae, an expanded ilium, a shoulder joint that faces backward, a neck-and-ball offset on the proximal femur, and an advanced mesotarsal (AM) ankle. (© Scott Hartman)

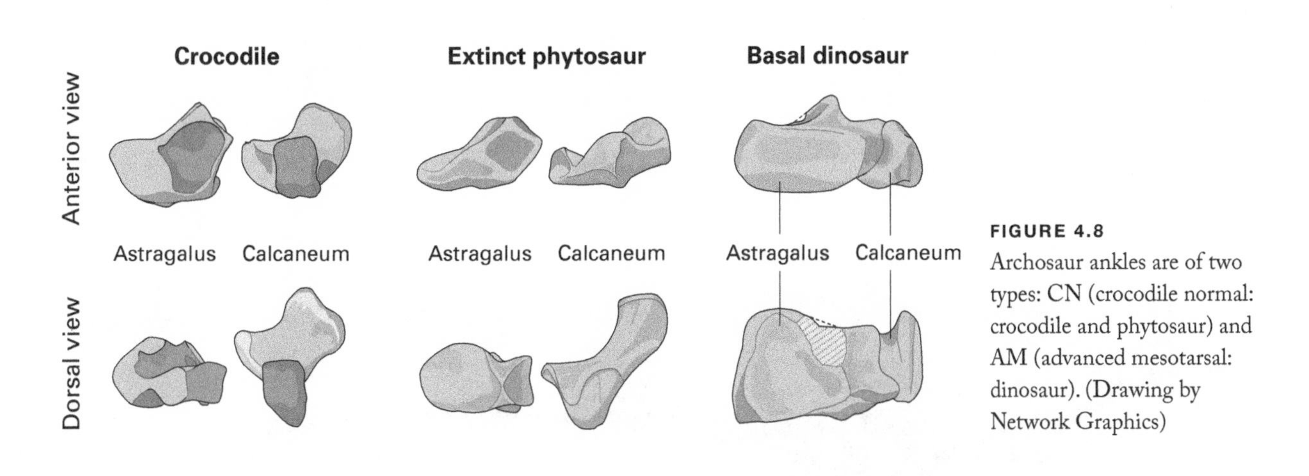


**FIGURE 4.8**
Archosaur ankles are of two types: CN (crocodile normal: crocodile and phytosaur) and AM (advanced mesotarsal: dinosaur). (Drawing by Network Graphics)

The shapes of the dinosaur pelvis and femur also contrast with the shapes of those bones in other archosaurs. These shapes indicate that dinosaurs held the hind limbs upright and directly under the body.

To identify a possible ancestor—or an animal closely related to that ancestor—paleontologists look to small, predatory archosaurs with limb structures that indicate they walked with a nearly upright posture. This would have been a posture intermediate between the sprawling posture of many other Triassic archosaurs, such as the phytosaurs and the aetosaurs (see figure 4.9), and the fully upright posture of the dinosaurs.

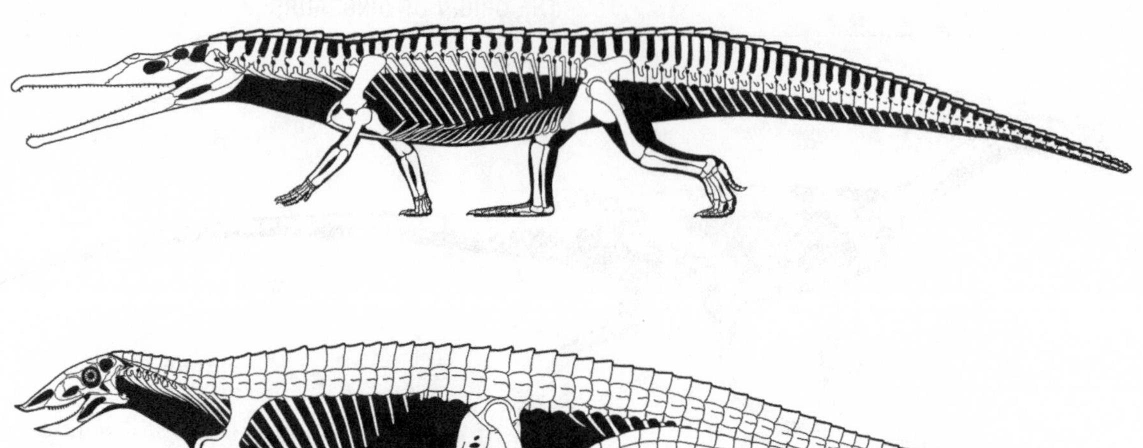

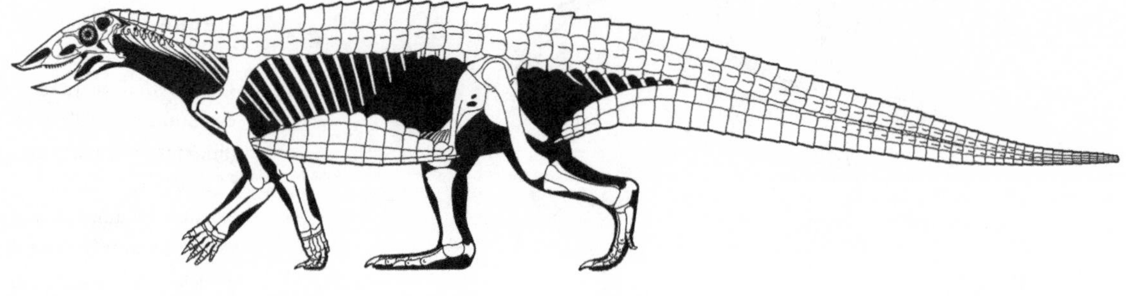

**FIGURE 4.9**
Phytosaurs (*above*) and aetosaurs (*below*) were archosaurs whose crocodile normal (CN) ankle indicates they are not included in the ancestry of dinosaurs. They are more closely related to crocodilians. (© Scott Hartman)

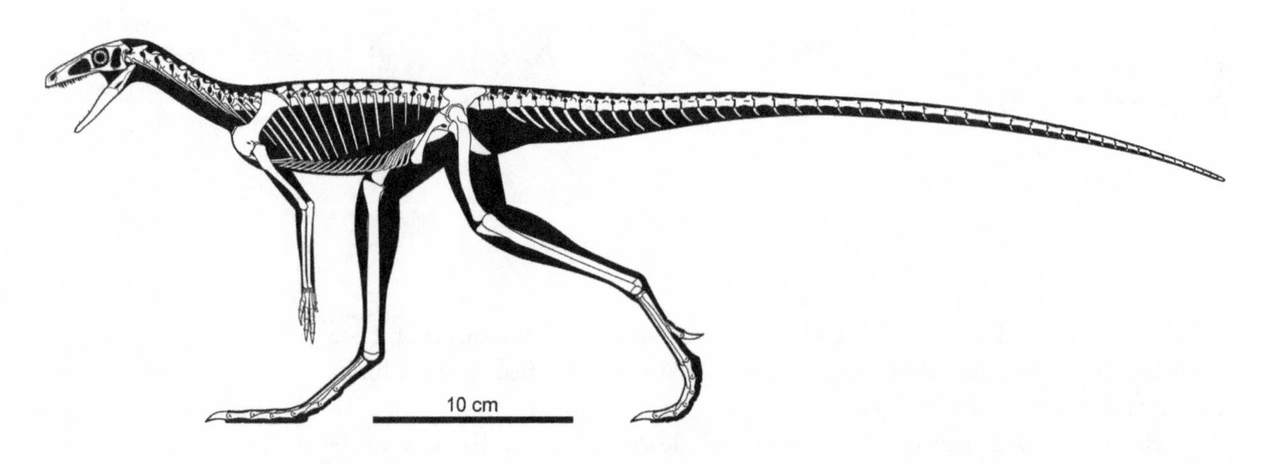


**FIGURE 4.10**
*Marasuchus*, from the Middle Triassic of Argentina, well approximates the dinosaur ancestor. (© Scott Hartman)

Triassic archosaurs with these modifications are well represented by ***Marasuchus*** (figure 4.10). This small (about 40 centimeters long), facultatively bipedal predator is known from complete skeletal material from the Middle Triassic of Argentina. Indeed, the anatomy and geological age of *Marasuchus* are as close to those of a suitable ancestor of dinosaurs as in any known Triassic archosaur.

## THE PHYLOGENY OF DINOSAURS

Dinosaurs appear suddenly in the Late Triassic fossil record (box 4.1). These diapsid archosaur reptiles were distinguished primarily by skeletal features indicating an upright posture. Their ancestry lay among achosaurs similar to *Marasuchus*.

In the more than 150 years since dinosaurs were first recognized, many dinosaur phylogenies have been proposed. Recent studies of dinosaur phylogeny employ the methods of cladistics, as discussed in chapter 2, and this book also follows a cladistic phylogeny (figure 4.11).

For many years, paleontologists viewed dinosaurs as a polyphyletic group of two or more distinct clades that evolved from different groups of archosaurs. But, now it is generally agreed that dinosaurs are a monophyletic, single clade united by evolutionary novelties that include three or more sacral vertebrae, a shoulder joint that faces backward, three or fewer phalanges in the fourth digit of the hand, a distinct neck-and-ball offset from the shaft of the femur to articulate with the open hip socket, and an expanded ilium (see figure 4.7). These and other features distinguish all dinosaurs from their closest relatives among the Archosauria.

Dinosaurs were long ago divided into two groups by British paleontologist Harry G. Seeley. The two groups, **Saurischia** (lizard hips) and **Ornithischia** (bird hips), were

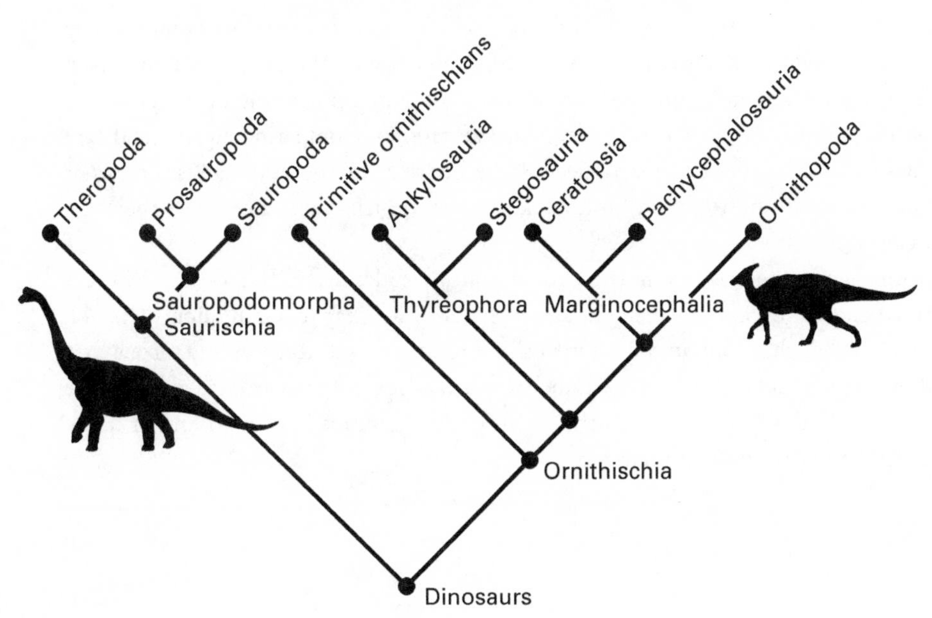


**FIGURE 4.11**
This phylogeny of the major groups of dinosaurs is the one employed in this book. (Drawing by Network Graphics)

## Box 4.1

### Which Dinosaur Is the Oldest?

Dinosaurs of Late Triassic age are known from the United States, Canada, Brazil, Argentina, Morocco, South Africa, Lesotho, Madagascar, Great Britain, and Germany. But, the idea has long persisted that of the Late Triassic dinosaurs, the oldest is either ***Staurikosaurus*** from Brazil or *Herrerasaurus* and *Eoraptor* (box figure 4.1) from Argentina. A recent re-evaluation of the ages of these dinosaurs, however, has indicated that they are of the same age and that equally old dinosaurs are also known from the western United States, Europe, and India. This means that dinosaurs appeared almost simultaneously over a broad geographic area during the Late Triassic.

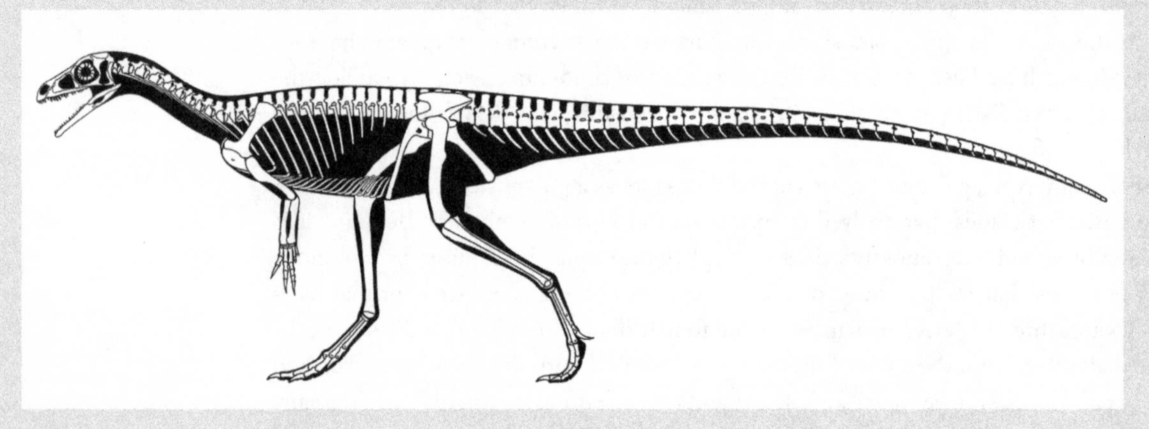

**BOX FIGURE 4.1**
*Eoraptor*, from the Upper Triassic of Argentina, is one of the oldest dinosaurs.
(© Scott Hartman)

Late Triassic tetrapod faunas are of two types: aquatic and terrestrial. The aquatic faunas are dominated by fossils of large amphibians and phytosaurs, a group of crocodile-like archosaurs. The terrestrial faunas lack phytosaurs and amphibians and instead are dominated by rhynchosaurs, a group of plant-eating primitive diapsids, and traversodontids, a group of plant-eating mammal-like reptiles. Comparing aquatic and terrestrial Late Triassic faunas with each other in order to decide whether they are the same age is somewhat difficult. Fortunately, enough animals are found in both aquatic and terrestrial faunas to make their age relationships reasonably clear.

We can thus identify three aquatic faunas—from the United States, Scotland, and India—and two terrestrial faunas—from Brazil and Argentina—as being of about the same age, 225 to 230 million years old. These are the oldest Triassic faunas that contain dinosaur fossils. Thus, there is a nearly simultaneous first record of dinosaur fossils across a broad expanse of the Late Triassic landscape. Furthermore, even at their first appearance, a variety of dinosaurs can be distinguished. This probably means that the origin of dinosaurs actually took place well before 230 million years ago.

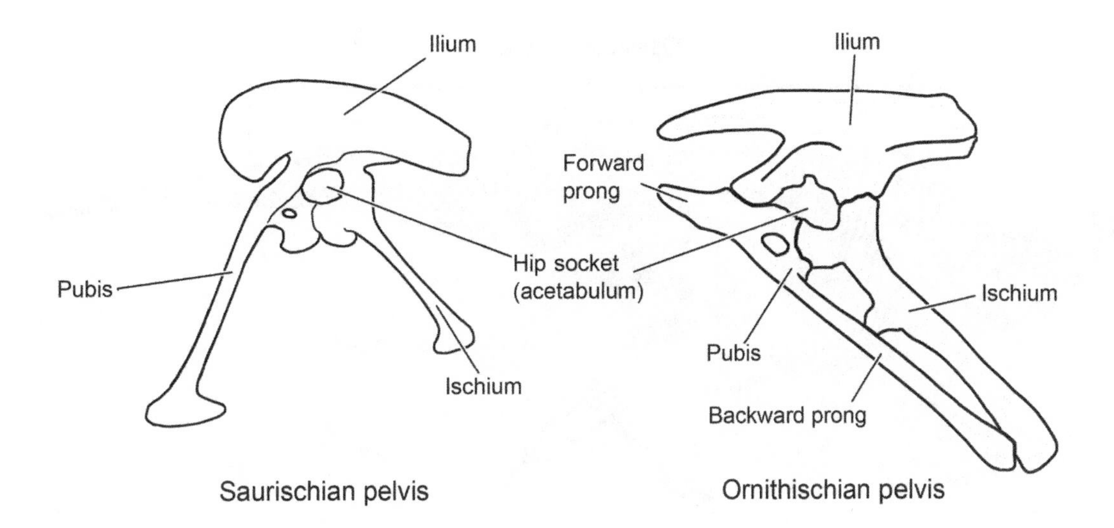


**FIGURE 4.12**
The two major groups of dinosaurs, Saurischia and Ornithischia, are distinguished primarily by differences in pelvic structure. (© Scott Hartman)

distinguished primarily on the basis of their pelvic structure (figure 4.12). In the pelvis of saurischian dinosaurs, the three pelvic bones radiate in different directions from the hip socket (**acetabulum**), as in most other reptiles. In the pelvis of ornithischian dinosaurs, the pubis is parallel to the ischium, a feature seen also in birds.

Despite the resemblance of the dinosaur pelvis to that of either a living lizard or a living bird, it had many distinctive features. These include an opening in the acetabulum below a lip of bone against which the femur pressed and an elongated pubis and ischium that hung down between the legs.

Recent cladistic analysis has upheld a twofold division of all but a few dinosaurs (such as *Herrerasaurus*) into two clades: Saurischia and Ornithischia (see figure 4.11). Indeed, a number of other evolutionary novelties distinguish these two dinosaurian clades. These include the presence of a toothless predentary bone in the lower jaw and ossified (bone-like) tendons in the back and tail, both distinctive features of ornithischians (figure 4.13).

Saurischia itself consists of two clades: **Theropoda** and **Sauropodomorpha**. The evolutionary novelties of these clades are discussed in chapters 5 and 6, respectively. ***Herrerasaurus*** and some other Late Triassic dinosaurs were neither saurischians nor ornithischians because they lacked the evolutionary novelties that unite the saurischians and ornithischians into a clade.

The most primitive ornithischians lacked the evolutionary novelties of other ornithischians, especially the spout-shaped front end of the lower jaw and the well-defined fenestra in front of the eye. **Thyreophoran** (shield-bearing) dinosaurs are united principally by distinctive body armor (see chapter 8), whereas the incipient frill (backward-projecting shelf of bone) present on the skulls of both **ceratopsians** and **pachycephalosaurs** is considered to be the principal reason for their classification as **Marginocephalia** (see chapter 9). The evolutionary novelties of the **Ornithopoda** are discussed in chapter 7.

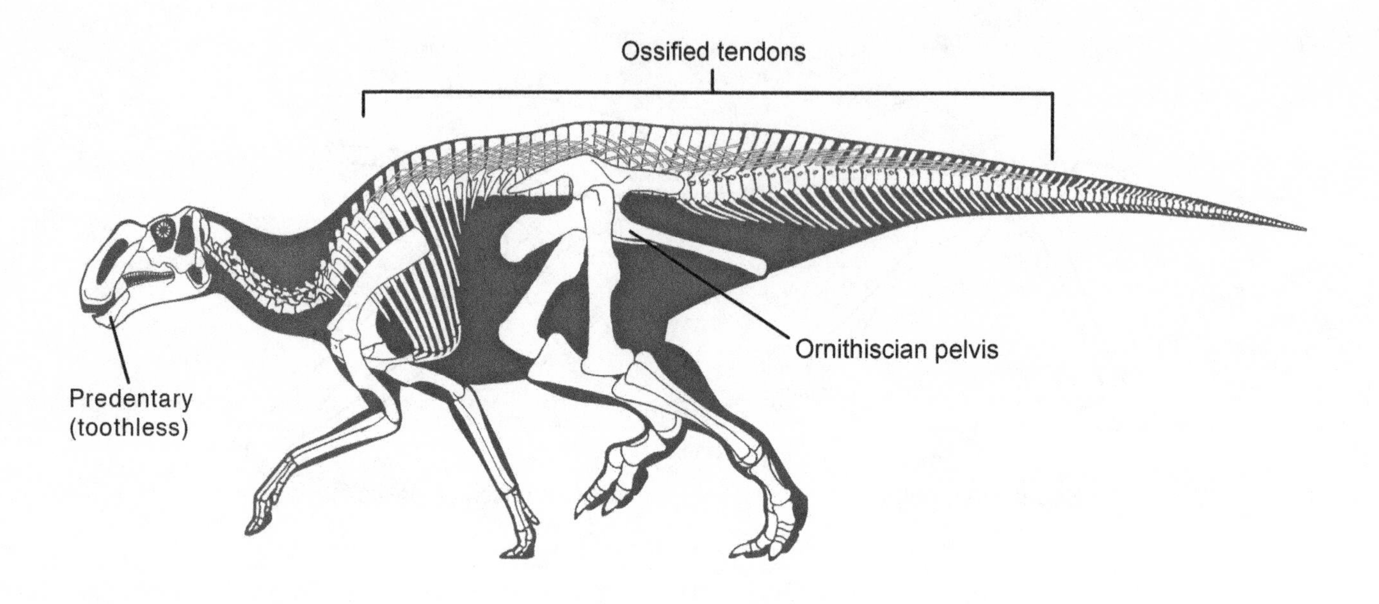


**FIGURE 4.13**
The predentary bone and ossified tendons along the vertebral column of ornithischians are two features, in addition to pelvic structure, that distinguish them from saurischians. (© Scott Hartman)

The phylogeny of the ornithischian groups identifies the close relationships between **stegosaurs** and **ankylosaurs** and between ceratopsians and pachycephalosaurs. Disagreement exists as to how these groups relate to ornithopods and as to the composition of the ornithopods. The phylogeny employed here (see figure 4.11) recognizes ornithopods as close relatives of marginocephalians because of several shared evolutionary novelties, including a gap between teeth in the premaxillary and the maxillary, five or fewer premaxillary teeth, and the absence of a bony lip above the acetabulum.

A classification of the major groups of dinosaurs can be derived from the dinosaur phylogeny presented in figure 4.14. What rank in the Linnaean hierarchy each group is assigned is somewhat arbitrary, but the process of assigning the major groups at the branch tips of the cladogram—theropods, sauropodomorphs, and so on—to orders is followed in this book.

More interesting, perhaps, than turning the cladogram into a classification is to read it as a road map of the important milestones in dinosaur evolution. Thus, from an as yet undiscovered ancestor (see box 4.1), the primitive herrerasaur clade(s) diverged from the clade of other dinosaurs by the Late Triassic. Also by the end of the Triassic, the saurischian and ornithischian clades had diverged from an unknown common ancestor. And, during the Late Triassic, the two major saurischian clades—theropods and sauropodomorphs—must also have diverged. Clearly, much happened in terms of the origin and evolution of dinosaurs during the Late Triassic and Early Jurassic (figure 4.15).

Major events in ornithischian evolution seem to have taken place a bit later, with basal thyreophorans and ornithopods not appearing until the Early or Middle Jurassic. The ankylosaur–stegosaur split in the thyreophorans took place by Middle Jurassic, and marginocephalians appear to not have diverged from ornithopods until the Middle Jurassic.

- Kingdom Animalia
  - Phylum Chordata
    - Subphylum Vertebrata
      - Class Reptilia
        - Subclass Diapsida
          - Infraclass Archosauromorpha
            - Superorder Saurischia
              - Order Theropoda
              - Order Sauropodomorha
                - Suborder Prosauropoda
                - Suborder Sauropoda
            - Superorder Ornithischia
              - Order Ankylosauria
              - Order Stegosauria
              - Order Ceratopsia
              - Order Pachycephalosauria
              - Order Ornithopoda

**FIGURE 4.14**
This classification of dinosaurs is based on the phylogeny in figure 4.11. (Drawing by Network Graphics)

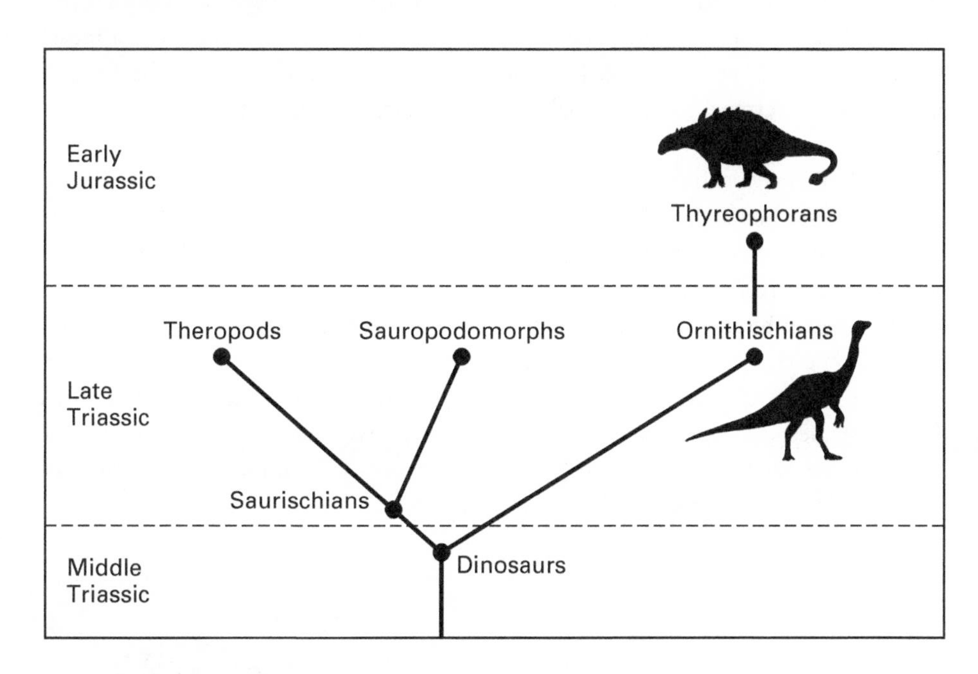


**FIGURE 4.15**
Much divergence took place in the early evolution of dinosaurs during the Late Triassic and Early Jurassic. (Drawing by Network Graphics)

## THE OLDEST DINOSAURS

Some paleontologists assign Late Triassic dinosaurs to the Theropoda (see chapter 5), but others argue that some of these would-be theropods lacked the evolutionary novelties that are diagnostic of the Dinosauria. In other words, they are dinosaurs that are not clearly saurischians or ornithischians. Also, there are some very dinosaur-like archosaurs that are not assigned to the dinosaurs but are instead called dinosauromorphs.

**Dinosauromorphs** were very dinosaur-like Triassic archosaurs. However, they lacked the key evolutionary novelties of dinosaurs. Still, not all paleontologists agree on whether a taxon such as *Silesaurus* is a dinosaur or a dinosauromorph. Such disagreements show the limits of cladistic phylogenetic analysis when there is much convergence and mosaic evolution, as there evidently was during the origin of the dinosaurs. Because of these processes, the dinosauromorphs and some of the early dinosaurs showed a combinations of features, some dinosaurian, others nondinosaurian, even recognizable from footprints (box 4.2), that make it difficult to decide which should be included in the Dinosauria.

If we look at the fossil record of Late Triassic dinosaurs, it captures what appears to be a significant diversification of prosauropods, a lesser diversity of theropods, and very few ornithischians. Two dinosaurs—*Herrerasaurus* and *Lesothosaurus*—are well enough known to give us some idea of the diversity of the most primitive dinosaurs. Triassic prosauropods are discussed in chapter 6.

*Herrerasaurus* is best known from two skulls and several skeletons from the Upper Triassic of Argentina (figure 4.16). This dinosaur was about 4.5 meters long with very long, slender hind limbs, apparently a biped, and a fast runner. The head was relatively large and had numerous blade-like teeth, suggesting that *Herrerasaurus* was a meat-eater. However, *Herrerasaurus* lacks some key theropod features, such as having five or more sacral vertebrae, so many paleontologists exclude it from the Theropoda (and from the Saurischia).

Closely related primitive dinosaurs include *Eoraptor*, from the Upper Triassic of Argentina (see box 4.1), and *Staurikosaurus*, from the Upper Triassic of Brazil. Some paleontologists refer to these dinosaurs as the herrerasaurs, an early clade, or grade, of the most primitive, saurischian-like dinosaurs.

Triassic ornithischians are known from little more than isolated teeth, jaws, and bones (box 4.3). *Pisanosaurus*, from the Upper Triassic of Argentina, is known from part of a skull and skeleton and was a small, 1.3-meter-long primitive dinosaurian herbivore. Because Late Triassic ornithischians have such a poor fossil record, we will examine *Lesothosaurus* (formerly called *Fabrosaurus*), from the Lower Jurassic of southern Africa, as a typical early ornithischian.

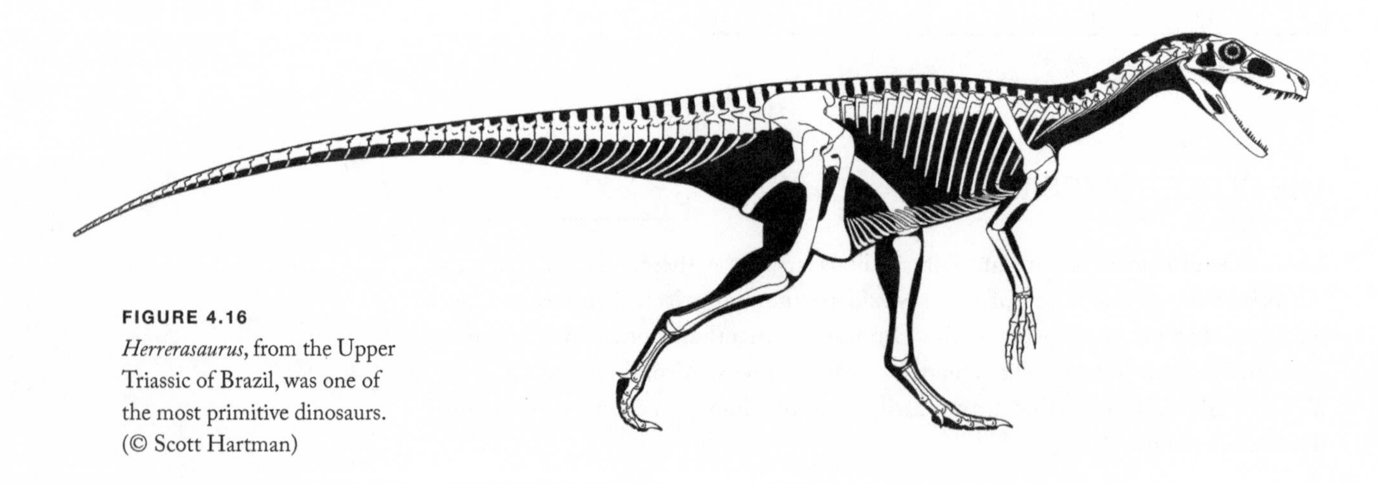

**FIGURE 4.16**
*Herrerasaurus*, from the Upper Triassic of Brazil, was one of the most primitive dinosaurs. (© Scott Hartman)

## Box 4.2

### The Oldest Dinosaur Footprints

The oldest dinosaur body fossils (bones and teeth) are widely acknowledged to be of Late Triassic age, between 225 and 230 million years old. However, several records of older, Middle Triassic footprints have been claimed to be dinosaur tracks. These are usually three-toed tracks that look very much like Late Triassic (or younger) theropod footprints. Many of these tracks, however, have been shown to be three-toed undertracks of the five-toed tracks of non-dinosaurian archosaurs. But, a few three-toed tracks of Middle Triassic age remain that may have been made by dinosaurs, or by dinosauromorphs. Nevertheless, most paleontologists remain reluctant to accept them as the oldest dinosaur fossils.

A relatively recent claim based on footprints is of Early Triassic dinosauromorph footprints. But, this claim also does not stand up to critical scrutiny. These supposed Early Triassic dinosauromorph tracks are from the Lower Triassic of Poland and were named *Protodactylus* (box figure 4.2). Based on tracks and trackways, the supposed dinosauromorph features of *Protodactylus* are that it was made by an animal walking nearly upright, that it had a hind foot in which the metatarsals were bunched, and that digit lengths matched the foot skeleton of some dinosauromorphs.

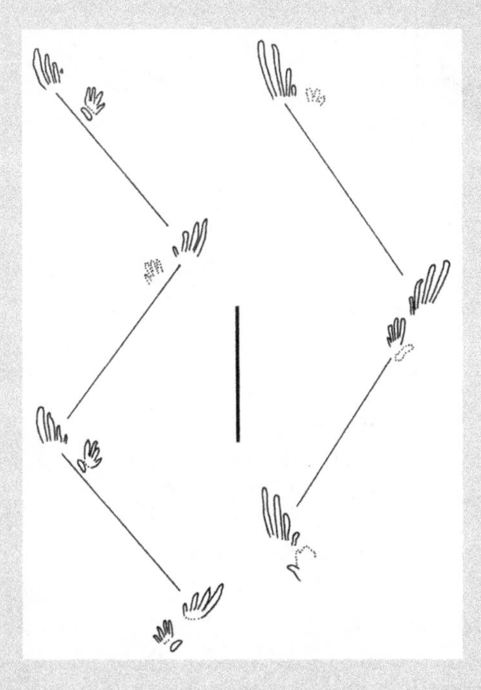

**BOX FIGURE 4.2**
Footprints called *Protodactylus* (*left*), from the Lower Triassic of Poland, are very similar to lepidosauromorph tracks (*right*) called *Rhynchosauroides*.

However, a critical re-evaluation of *Protodactylus* has shown that it is very similar to *Rhynchosauroides*, another kind of Triassic footprint long agreed to have been made by lepidosaurs (see box figure 4.2). Note, for example, the similarity in relative digit length, curvature, and trackway pattern of *Protodactylus* and *Rhynchosauroides*. Indeed, *Protodactylus* and *Rhynchosauroides* are found in the same Lower Triassic rocks in Poland, so there is even doubt that the tracks deserve two different generic names. Some students of fossil footprints think that some tracks called *Rhynchosauroides* were made by lizard-like archosauromorphs, but no *Rhynchosauroides* track can be matched to a dinosauromorph foot. The identification of Early Triassic dinosauromorph tracks thus can be rejected.

## Box 4.3

### Isolated Teeth Versus Skulls and Skeletons

The Late Triassic fossil record of the earliest dinosaurs includes relatively few complete skeletons but many isolated bones and teeth. Recent attempts to interpret these fragmentary fossils has led to some disagreement among paleontologists.

Some paleontologists believe that isolated teeth provide a good basis for distinguishing different types of early dinosaurs. According to these paleontologists, at least 20 different dinosaur tooth types are known from the Late Triassic, and each deserves recognition as representing a distinct dinosaur genus. One tooth type to be recognized is *Revueltosaurus*, a name coined for the supposed ornithischian teeth from the Upper Triassic of New Mexico and Arizona (box figure 4.3). However, subsequent discovery of a skull and skeleton with *Revueltosaurus* teeth in it indicated that these are not dinosaur teeth, but instead the teeth of a crocodile-like archosaur. Clearly, there had been evolutionary convergence in tooth types between such archosaurs and ornithischian dinosaurs.

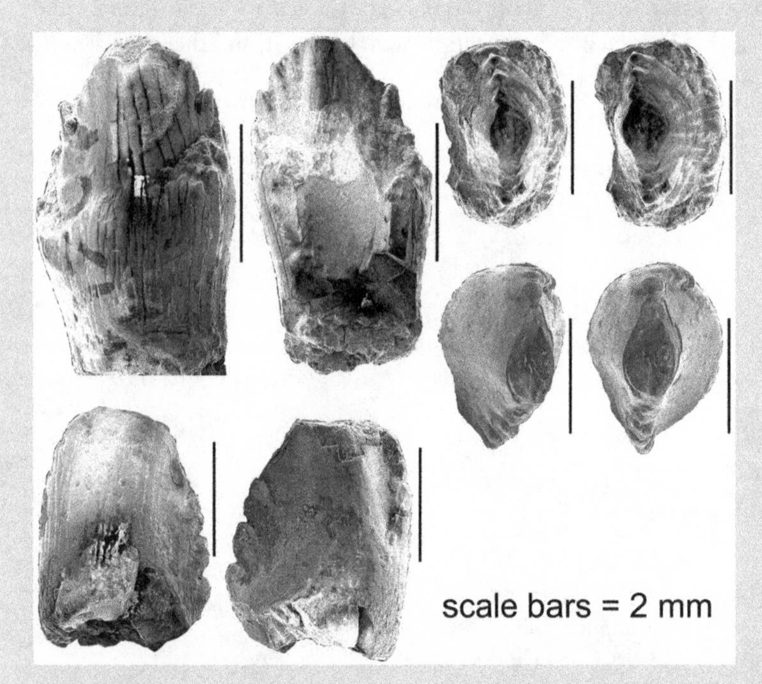


**BOX FIGURE 4.3**
These teeth from the Upper Triassic of New Mexico have been named *Revueltosaurus*. (Courtesy A. B. Heckert)

An alternative point of view, held by other paleontologists, is that teeth alone are insufficient to distinguish Late Triassic dinosaurs. These paleontologists argue that tooth shape varies within a single dinosaur jaw, and that we don't have enough complete Late Triassic dinosaur jaws to understand that variation. It is thus very risky to attempt to identify distinct dinosaurs by their teeth alone, since several distinct tooth types might come from a single dinosaur jaw.

This argument might seem so strong that nobody would try to identify distinct types of Late Triassic dinosaurs based on isolated teeth. But, in fact we don't have enough Late Triassic dinosaur jaws to understand tooth variation, and what jaws we have do not show sufficient variation in tooth shape in a single jaw to encompass many of the types of isolated Late Triassic dinosaur teeth. So, the argument between paleontologists who use isolated teeth to identify dinosaurs and their critics is at a standstill pending the discovery of more complete jaws of Late Triassic dinosaurs.

**FIGURE 4.17**
*Lesothosaurus*, from the Lower Jurassic of Lesotho, was a primitive ornithischian. (© Scott Hartman)

*Lesothosaurus* was a small dinosaur (1.5 meters long) known from several skulls and skeletons (figure 4.17). This dinosaur was characterized by a short, high skull; very large eyes; and numerous slender, leaf-shaped cheek teeth that clearly were those of a plant-eater. A pocket on the side of the skull in front of the eye may have contained a salt gland.

The neck of *Lesothosaurus* was long and flexible, the back long, the tail very long, and the slender hind limbs much longer than the forelimbs. The limbs were those of an agile biped that probably plucked vegetation with its front teeth and hands. *Lesothosaurus* had the proportions of primitive ornithopods (see chapter 7), so some paleontologists include it among the ornithopods. But, *Lesothosaurus* lacked the evolutionary novelties diagnostic of ornithopods, although it did have some features (predentary bone, pelvis, and ossified tendons) characteristic of ornithischians. Here, *Lesothosaurus* is regarded as a primitive ornithischian.

## Summary

1. "Reptile" is a term that refers to tetrapods that lay an amniotic egg, have limb skeletons modified for fully terrestrial locomotion, and are ancestral to birds and mammals.
2. Dinosaurs are reptiles because they laid amniotic eggs and had limb skeletons adapted to terrestrial locomotion.
3. Reptiles are classified into four groups based on the number and position of openings (temporal fenestrae) in the skull roof behind the orbits.
4. Dinosaurs are diapsids because they had two temporal fenestrae on each side of their skull.
5. Diapsids consist of two groups: the lepidosaurs (lizards, snakes, and their ancestors) and the archosaurs (an array of Triassic taxa, crocodilians, pterosaurs, and dinosaurs).
6. The ancestry of dinosaurs lies among the Triassic archosaurs. Key to deciphering this ancestry is ankle structure and other skeletal features that indicate the evolution of upright dinosaurs from semi-upright archosaurs.

7. Dinosaurs appeared suddenly during the Late Triassic over a geographically wide area. They were the descendants of *Marasuchus*-like archosaurs.
8. The earliest dinosaurs include theropods, *Herrerasaurus* (a close relative of the Saurischia), and very primitive ornithischians, such as *Lesothosaurus*.

## Key Terms

acetabulum
advanced mesotarsal (AM) ankle
aetosaur
amniotic egg
anapsid
ankylosaur
archosaur
ceratopsian
crocodile-normal (CN) ankle
crocodilian
diapsid
dinosauromorph
*Herrerasaurus*
lepidosaur
*Lesothosaurus*
*Marasuchus*
Marginocephalia
Ornithischia
Ornithopoda
pachycephalosaur
paraphyletic
phytosaur
pterosaur
reptile
Saurischia
Sauropodomorpha
*Staurikosaurus*
stegosaur
temporal fenestrae
Theropoda
Thyreophora
upright posture

## Review Questions

1. Explain why dinosaurs can be called a group of archosaurian diapsid reptiles.
2. Why are cladists not satisfied with groups of animals like reptiles?
3. How do the ankle structures of archosaurs provide a key to their phylogenetic relationships?
4. What is a dinosaur?
5. What features distinguish saurischian from ornithischian dinosaurs?
6. Summarize briefly the "when," "where," and "from what" of dinosaur origins.
7. How do two of the earliest dinosaurs, *Herrerasausus* and *Lesothosaurus*, differ from each other?

## Further Reading

Benton, M. J. 2012. Origin and early evolution of dinosaurs, pp. 331–345, in M. K. Brett-Surman, T. R. Holtz, Jr., and J. O. Farlow, eds., *The Complete Dinosaur*. 2nd ed. Bloomington: Indiana University Press. (A concise review of dinosaur origins)

Brusatte, S. L., G. Niedźwiedzki, and R. J. Butler. 2011. Footprints pull origin and diversification of dinosaur stem lineage deep into Early Triassic. *Proceedings of the Royal Society B* 278:1107–1113. (Claims that Early Triassic *Protorodactylus* is a dinosauromorph footprint)

Cruickshank, A. R. I., and M. J. Benton. 1985. Archosaur ankles and the relationships of the thecodontian and dinosaurian reptiles. *Nature* 317:715–717. (A technical analysis of the origin of the advanced mesotarsal ankle of dinosaurs)

Klein, H., and G. Niedźwiedzki. 2012. Revision of the Lower Triassic tetrapod ichnofauna from Wióry, Holy Cross Mountains, Poland. *Bulletin of the New Mexico Museum of Natural History and Science* 56:1–62. (Refutes the claim that Early Triassic *Protorodactylus* is a dinosauromorph footprint)

Langer, M. C. 2014. The origins of Dinosauria: Much ado about nothing. *Palaeontology* 2014:1–14. (Editorial review of recent ideas about dinosaur origins)

Langer, M. C., M. D. Ezcurra, J. S. Bittencourt, and F. E. Novas. 2009. The origin and early evolution of dinosaurs. *Biological Reviews* 84:1–56. (Comprehensive technical review of dinosaur origins)

Nesbitt, S. J. 2011. The early evolution of archosaurs: Relationships and the origin of major clades. *Bulletin of the American Museum of Natural History* 352:1–292. (Recent cladistic analysis of archosaur phylogeny)

Nesbitt, S. J., J. B. Desojo, and R. B. Irmis, eds. 2013. Anatomy, phylogeny and palaeobiology of early archosaurs and their kin. *Geological Society, London, Special Publications* 379:1–624 (A collection of 24 technical articles on Triassic archosaurs)

Sereno, P. C., R. N. Martinez, and O. A. Alcober. 2012. Osteology of *Eoraptor lunensis* (Dinosauria, Sauropodomorpha). *Journal of Vertebrate Palaeontology* 32(suppl 1):83–179. (Discusses the skeletal anatomy of *Eoraptor*)

## Find a Dinosaur!

The most complete early dinosaur known is the Late Triassic theropod *Coelophysis*. This is because hundreds of complete skeletons of this dinosaur have been collected from a bonebed at Ghost Ranch in northern New Mexico. These skeletons tell paleontologists about growth, sexual dimorphism, and the behavior (including cannibalism) of this strikingly bird-like dinosaur. To find *Coelophysis* fossils on display, you can visit the Ruth Hall Museum of Paleontology at Ghost Ranch. Nearby, in Albuquerque, a block of rock with 21 skeletons of *Coelophysis* is exhibited at the New Mexico Museum of Natural History and Science. (Incidentally, *Coelophysis* is the official state fossil of New Mexico.) In New York, the American Museum of Natural History also has skeletons of *Coelophysis* on display—their field team discovered the Ghost Ranch bonebed in the 1940s. What you will see at Ghost Ranch, in Albuquerque, and in New York are skeletons embedded in rock. To see freestanding *Coelophysis* replicas, go to the Denver Museum of Nature & Science in Colorado, where the early dinosaur comes alive as an active and agile predator.

# 5

# THEROPODS

MOST predatory (meat-eating) dinosaurs belong to one group, Theropoda (beast foot) (box 5.1). The theropod dinosaurs spanned the entire duration of dinosaurs, from the Late Triassic until the end of the Cretaceous. They include animals that ranged in size from the chicken-size *Compsognathus* to the most famous land-living predator of all time, *Tyrannosaurus rex*. Throughout most of their long history, theropods were a diverse and successful group of dinosaurs.

In this chapter, I examine the main types of theropods and discuss their evolution. But, before doing this, we need to consider some of the complexity associated with the fossil record of theropods and how it affects our view of the evolution of these dinosaurs.

## THE PHYLOGENY OF THEROPODS

Theropods are the most diverse group of dinosaurs, and recent discoveries have dramatically changed how we view their composition. Clearly, these dinosaurs had a very complex evolution from the Late Triassic through the Late Cretaceous. Indeed, no area of dinosaur phylogeny is more fluid or more debated. At least half a dozen phylogenetic analyses of all the theropods, or of major segments of the Theropoda, can be considered current. The phylogeny used here embodies much of what can be considered consensus (figure 5.1).

The oldest widely accepted scheme of theropod phylogeny recognized two clades. According to this scheme, the very large theropods with short necks and small forelimbs, the **carnosaurs** (meat lizards), diverged early from the smaller, long-necked theropods with large forelimbs, the coelurosaurs (hollow-tailed lizards). This view of theropod phylogeny, however, has been superseded by much more complex phylogenies needed to accommodate the new array of theropods discovered during the

## Box 5.1

### What's in a Name?

The Latinized names that paleontologists give to extinct animals usually make some sort of sense. They typically either describe some feature or attribute of the animal, make reference to the place where its fossils were found, or honor another paleontologist (see box 2.1). Some names are very colorful, like *Tyrannosaurus rex*, the "king of the tyrant lizards." Other names are rather pedestrian, like *Allosaurus*, the "different lizard."

The name applied to all meat-eating dinosaurs is **Theropoda**, from the Greek *therios*, meaning "beast," and *podos*, meaning "foot." Although *therios* literally means "beast," it usually refers to mammals, so the name Theropoda also means "mammal foot." Yale paleontologist Othniel Charles Marsh coined the name Theropoda for the meat-eating dinosaurs in 1881. Ten years earlier, Marsh had coined the name Ornithopoda (bird foot), and in 1878 he introduced the name Sauropoda (lizard foot). Indeed, 1882 was a milestone in the history of dinosaur studies, for in that year Marsh published a comprehensive classification of dinosaurs. He divided them into theropods, ornithopods, sauropods, and stegosaurs (plated lizards). Marsh also considered the Hallopoda (leaping foot) to be possible dinosaurs, but they are now thought to be crocodiles.

Clearly, Marsh relied heavily on foot structure as a guide to classifying dinosaurs. In the feet of the dinosaurs he called sauropods, Marsh saw resemblances to the feet of living lizards, hence the name "lizard foot."

But, what similarities in the feet of meat-eating dinosaurs and mammals did Marsh see? And, given the very bird-like feet of these dinosaurs, why didn't Marsh name them Ornithopoda? Furthermore, why did Marsh assign the name Ornithopoda to dinosaurs with feet that were more mammal-like than bird-like (box figure 5.1)?

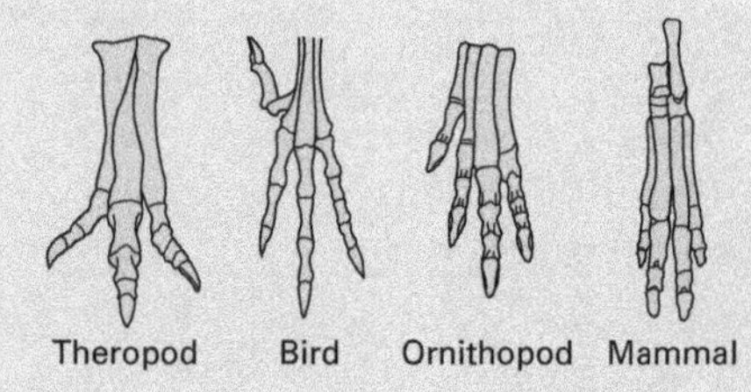


**BOX FIGURE 5.1**
Theropod feet were bird-like, whereas ornithopod feet were mammal-like. (Drawing by Network Graphics)

No clear answer to these questions can be found in Marsh's writings, which do, however, make it clear that he knew the meanings of the words "theropod" and "ornithopod." The best guess is that Marsh erred when he introduced the names "ornithopod," and later, "theropod," applying them in reverse order from what had been his original intention. The result is that the bird-footed dinosaurs are not called ornithopods, but, rather, nonsensically, theropods. Conversely, the name "ornithopod" is applied to dinosaurs with rather mammal-like feet.

Marsh was a very famous paleontologist whose knowledge of dinosaurs was immense. So, his names for two major groups of dinosaurs, Theropoda and Ornithopoda, were accepted and used by paleontologists worldwide. They continue to be used to this day, despite the fact that neither name quite makes sense.

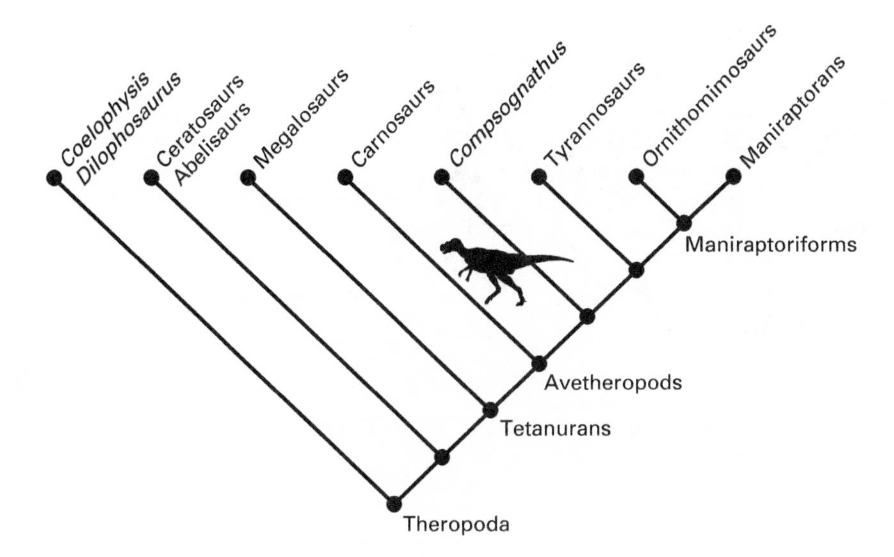


**FIGURE 5.1**
The phylogeny of theropods currently accepted by most paleontologists shows a complex evolutionary history of this diverse group of dinosaurs. (Drawing by Network Graphics)

past two decades (see figure 5.1). Thus, primitive theropods, including the ceratosaurs (horned lizards), represent multiple branches in the evolution of theropods. The clade that led to the carnosaurs and coelurosaurs is referred to as the tetanuran theropods. The ancestry of birds lies among the tetanurans and is discussed in chapter 15.

Although the picture of theropod phylogeny presented here (see figure 5.1) seems neat and clear, ongoing debate about the phylogeny of theropods makes it one of the least-agreed-upon phylogenies among the dinosaurs. In part, this is because different paleontologists place greater emphasis on different features as significant evolutionary novelties in theropod evolution. These disagreements cannot be readily resolved at present. So, suffice it to say that several alternative views of theropod phylogeny are plausible, and the one presented here represents a consensus of most paleontologists.

## WHAT IS A THEROPOD?

Because theropods comprise most of the meat-eating dinosaurs, it might seem sufficient to identify theropods by the skeletal features, especially the teeth, that were involved in acquiring and processing animal food. It is important to remember, however, that many of the most primitive dinosaurs (such as *Staurikosaurus* and *Herrerasaurus*) were also meat-eaters (see chapter 4). Indeed, meat-eating is the primitive condition for all non-dinosaurian archosaurs. Thus, to identify theropods specifically, we must distinguish them from the primitive meat-eating dinosaurs and from other meat-eating archosaurs.

Most of the novelties that distinguish theropods reflect more specialized structures for meat-eating than were present in the primitive, meat-eating dinosaurs. They also reflect the fact that theropods were highly specialized runners and predators. These novelties include broad exposure of the lachrymal bone on the skull roof; a joint

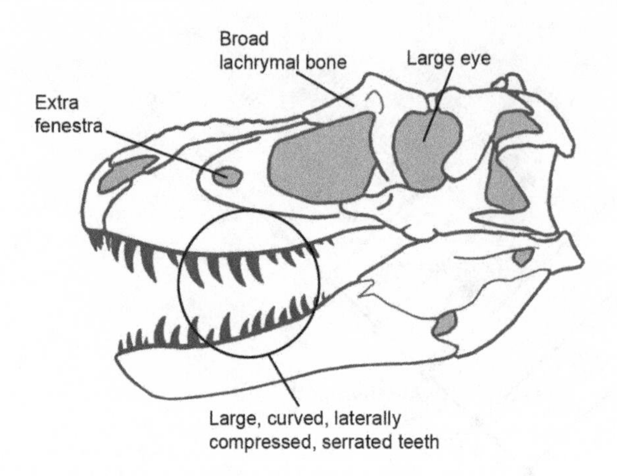


**FIGURE 5.2**
This tyrannosaurid skull shows typical features of theropods. (© Scott Hartman)

(hinge) within the jaw; a sacrum comprising at least five vertebrae; clavicles fused into a wishbone (furcula); four or fewer digits on the hand; and a hind foot that was compact, narrow, and long with three functional digits and a first digit separated from the rest of the foot (figures 5.2 and 5.3). These novelties and others distinguish theropods from other dinosaurs. Combining these novelties with the primitive features inherited from their nontheropod ancestors gives us an overall picture of theropods as remarkable predators.

The teeth of many theropods were relatively large, curved, compressed, and serrated. The mouth bore numerous such teeth that were held in the jaw by a ligament (see figure 5.2). The relatively large eyes suggest that theropods located their prey visually. Theropods had the largest and most sophisticated brains of any known dinosaurs (see chapter 14). Indeed, the brains of the more advanced theropods were very bird-like, suggesting that at least some theropods may have been very sophisticated behaviorally.

The slicing teeth of theropods were set in a lightly built skull in which many of the bones were only loosely joined to each other. Even the bones of the theropod lower jaw were not tightly sutured to each other (see figure 5.2). These loose joints probably

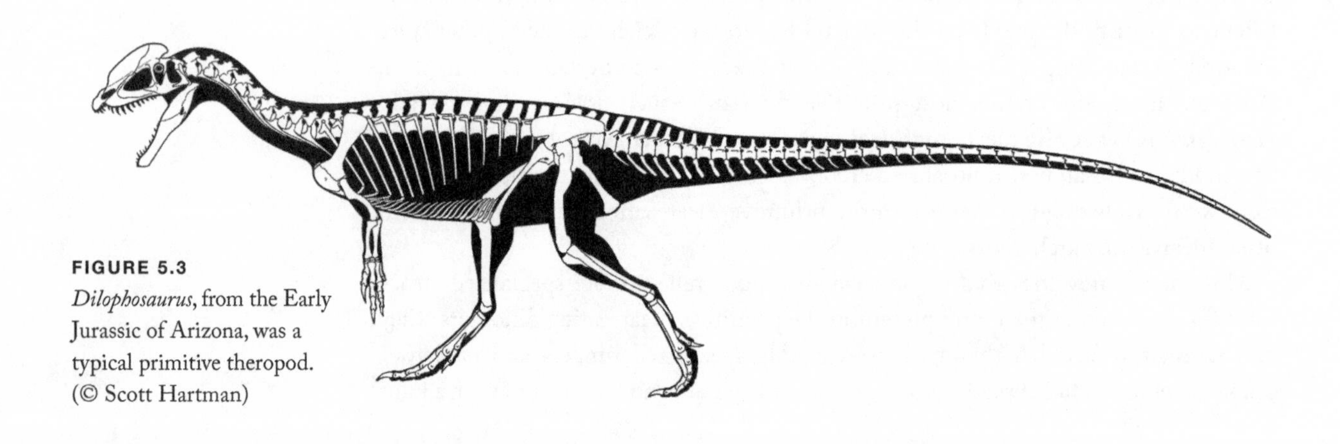

**FIGURE 5.3**
*Dilophosaurus*, from the Early Jurassic of Arizona, was a typical primitive theropod. (© Scott Hartman)

helped the skull compensate for the shocks it encountered when the theropod grabbed and chewed its prey, whereas the rigid lower jaw helped grab and crush the prey.

The theropod skull was attached to the neck at a highly mobile joint. This joint allowed rapid and precise movement of the head, and its presence further supports the idea that theropods were highly sophisticated visual hunters. Today, birds have the same type of head–neck joint, as did their theropod ancestors.

We also see many distinctive features designed for predation in the postcranial skeletons of theropods. Thus, the hands of most theropods had only three digits, one of which (the first digit) was offset from the other two (see figure 5.3). This digit, like our thumb, may have allowed the theropod hand to grasp. On all three digits of the theropod hand, the next-to-last phalanges were elongated, to extend the reach, and the tips of the digits were long, laterally compressed, and bore curved claws.

Theropod hands clearly were not used for walking. So, theropods were obligate bipeds, and their vertebral columns and hind limbs were modified accordingly. At least five vertebrae made up the theropod sacrum to provide greater rigidity at the attachment point of the powerful hind limbs and backbone. The pelvis was very large and provided considerable space for the attachment of the large hip muscles. The femur was distinctly bowed downward, and the hind foot was elongate and symmetrical around the middle digit. Indeed, the theropod hind foot was the very compact foot of a fast runner and was strikingly bird-like (see box 5.1).

In many theropods, the tail changed about midway from being flexible proximally to being stiff distally. A similar type of tail is still present in living birds. The rigid distal portion could have been used by theropods as a stabilizer while running and as a counterweight to the body.

A final typical theropod feature was hollow bones. This lightened the theropod skeleton and also increased the strength of the bones to resist bending. Hollow bones evolved in theropods as an aid to rapid, terrestrial locomotion, but later aided their descendants, the birds, in flight.

## PRIMITIVE THEROPODS

Theropods were already important terrestrial predators in the Late Triassic and became the dominant predators of the land during the Jurassic and Cretaceous. Theropods can be divided into three groups: primitive taxa, such as the coelophysoids and dilophosaurids; the ceratosaurs; and the tetanurans. The primitive theropods were relatively small bipedal hunters, ranging from the 1-meter-long *Procompsognathus* to the 3-meter-long *Coelophysis* (box 5.2).

One of the best known primitive theropods is ***Dilophosaurus***, from the Lower Jurassic of Arizona (see figure 5.3). This large (6-meter-long) dinosaur displayed many typical theropod features, such as a large head full of long, blade-like teeth; a large antorbital fenestra; and a thick, powerful lower jaw. The neck and tail were long, and the forelimb was much shorter than the hind limb. *Dilophosaurus* had many fused bones in the hind limb. It also had a distinct gap between the premaxillary and maxillary bones in the skull, and the hand had four fingers, the first three of which bore claws.

Box 5.2

## A Flock of *Coelophysis*

Very few theropod dinosaurs are known from mass death assemblages that represent a sample of a theropod population. Late Triassic **Coelophysis** is an exception. In 1947, an expedition of the American Museum of Natural History discovered a *Coelophysis* bonebed in the Upper Triassic rocks of northern New Mexico. At least a few hundred complete skeletons of *Coelophysis* have been excavated from the bonebed, which is still not completely mined out. Taphonomic analysis indicates that a population of the small theropod was washed into a pond, where the dinosaurs likely drowned, and then were rapidly buried by a second flood event. This population can thus be seen as a large sample of a flock of *Coelophysis*.

Careful analysis of the *Coelophysis* skeletons provides rare insight into aspects of the biology of an early theropod dinosaur:

1. *Coelophysis* was a visually oriented diurnal hunter that was an obligate biped and a fast, agile runner (box figure 5.2).
2. The fact that so many dinosaurs were together at death suggests gregarious behavior; they likely aggregated to reproduce.
3. Two kinds of *Coelophysis* have been found in the bonebed: robust dinosaurs with short skulls, short necks, and large forelimbs, and gracile dinosaurs with long skulls, long necks, and smaller forelimbs. These are likely sexual dimorphs, and the gracile dinosaur is probably the female.
4. Cannibalism of the young is evidenced by bones and coprolites in some of the adult body cavities.
5. The distribution of size classes suggests rapid growth with sexual maturity at about two to three years and that the largest dinosaurs were about seven years old at death.
6. Based on the size classes, there are about 20 yearlings per large adult, so juvenile mortality must have been high.

The *Coelophysis* bonebed thus gives paleontologists remarkable insight into the lifestyle and population biology of an early theropod dinosaur. Not surprisingly, *Coelophysis* seems to have been very bird-like in its biology.

**BOX FIGURE 5.2**
The skull of *Coelophysis*, with its large eyes, is that of a diurnal, visual hunter.

The two thin, bony crests on the skull of *Dilophosaurus* are the basis of its name, which means "two-crested lizard."

At the beginning of the Jurassic, *Dilophosaurus* was the first relatively large dinosaurian predator. Its presence marked the ascent of the theropods to the status of apex predators in Jurassic–Cretaceous terrestrial communities.

## Ceratosaurs

Ceratosaurs were a significant group of primitive theropods of the Late Jurassic through to the Late Cretaceous. More than 40 species can be recognized from fossils found in North America, Europe, India, South America, Africa, and Madagascar. Key evolutionary novelties of the Ceratosauria include having six or more sacral vertebrae; very deep coracoids; small forelimbs; and very weak, perhaps even nonfunctional, hands. Ceratosaurs encompass quite a range of theropods, including *Limusaurus* from the Upper Jurassic of China, with its toothless skull, which was probably the first noncarnivorous theropod.

The 6- to 8-meter-long ***Ceratosaurus***, from the Upper Jurassic of North America and Europe, is a characteristic, relatively large ceratosaur. However, most of the diversity of ceratosaurs was found in abelisauroids, a Cretaceous group known largely from the Gondwana continents. Abelisauroids are characterized by an enlarged external mandibular fenestra, arched sacral vertebrae, a round head on their humerus, and distinctively doubled grooves on the distal phalanges of their hind feet, among other features. *Aucasaurus*, from the Late Cretaceous of Argentina, well represents the abelisaurs (figure 5.4). Note its extremely small forearms, a characteristic seen in many abelisaurs, thought to have been useless vestigial structures.

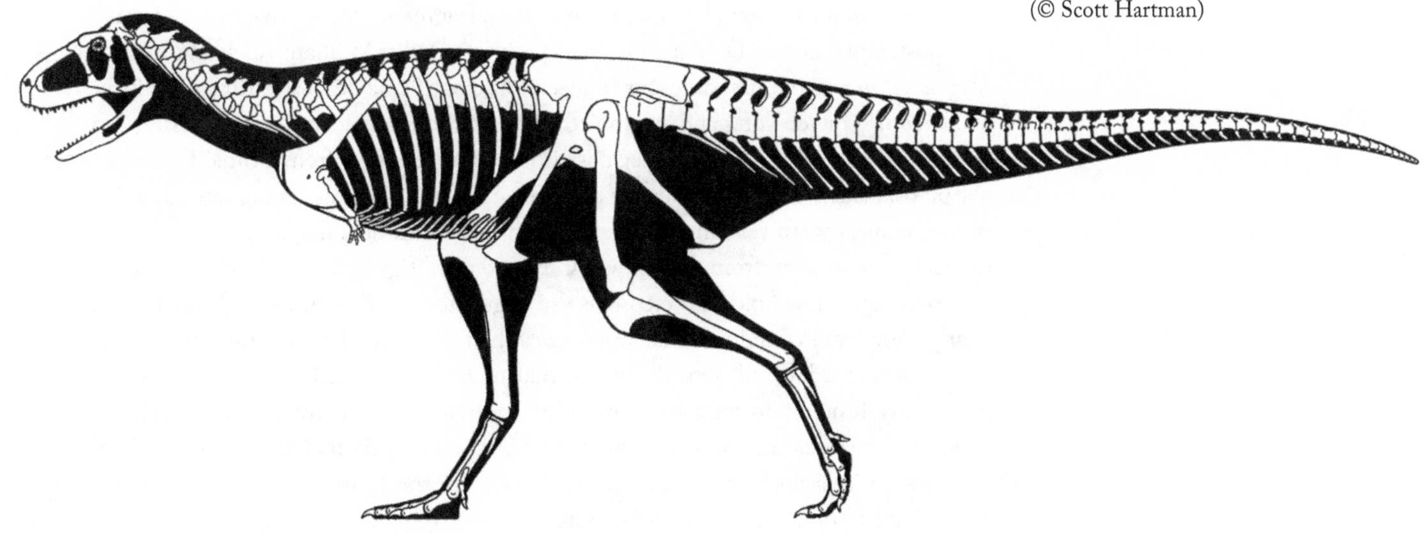

**FIGURE 5.4**
*Aucasaurus*, from the Upper Cretaceous of Argentina, with its tiny forelimbs, is a characteristic abelisaur. (© Scott Hartman)

Some ceratosaur skeletons show marked sexual dimorphism (differences between presumed males and females). Thus, there seems to have been a "male" form with a longer skull and neck, thicker and more robust limbs, and very large muscle-attachment sites. The "female" form, in contrast, had a juvenile-like skeleton with a shorter skull and neck and slender limbs. Further, there were crests on the skulls of some ceratosaurs. These were made of very thin bone and presumably functioned in display, probably by males.

## TETANURANS

The strange name **Tetanurae** (fused tails) is applied by paleontologists to most of the theropods (see figure 5.1). These theropods were much more bird-like than the primitive theropods. Indeed, most paleontologists have concluded that the ancestry of birds lies among the tetanuran theropods. Evolutionary novelties of the tetanurans include relatively large heads, a reduction in the dentition so that all teeth were in front of the orbit, and interlocking vertebrae that stiffened the tail.

Three great groups of tetanuran theropods can be recognized: the megalosaurs, the carnosaurs, and the coelurosaurs. The carnosaurs and coelurosaurs are united in a group called the Avetheropoda, because it is from this group that birds originated.

### Megalosauroids

Megalosaurs were large theropods of the Middle Jurassic through to the Late Cretaceous. They were characterized by long skulls with elongated snouts that had small or no crests on the skull roof. This contrasts with the skull of carnosaurs, such as *Allosaurus*, in which the skull was tall with a short and blunt snout and had horns or crests on the skull roof. The powerful forelimb of megalosaurs also was a distinctive feature.

*Megalosaurus*, from the Lower and Middle Jurassic of Great Britain and France, is the archetypal megalosaur. It is known from skull fragments, lower jaws, teeth, and various postcranial bones. Described by the British naturalist William Buckland in 1824, *Megalosaurus* was the first dinosaur to receive a scientific name. It is estimated to have been about 9 meters long and resembled *Allosaurus* in many features. A better known and characteristic Late Jurassic megalosaur is *Monolophosaurus* from China (figure 5.5).

Spinosaurids were megalosaurs of the Cretaceous with very elongate, crocodile-like snouts, conical teeth (lacking serrations,) and tall dorsal neural spines that formed a sailback. *Spinosaurus*, from the Lower Cretaceous of Egypt, is characteristic and, at 14 meters long, now holds the record as the largest theropod (figure 5.6). Known from nearly complete skeletal material, *Spinosaurus* had 1.8-meter-long neural spines on its back that formed a "sail" reminiscent of that of *Dimetrodon*, a Permian reptile often mistakenly thought to be a dinosaur. Gut contents and the conical teeth indicate it was a fish-eater that swam to hunt. Other spinosaurids include *Baryonyx*, from the Lower Cretaceous of Europe; *Suchomimus*, from the Lower Cretaceous of North Africa; and *Irritator*, from the Lower Cretaceous of Brazil.

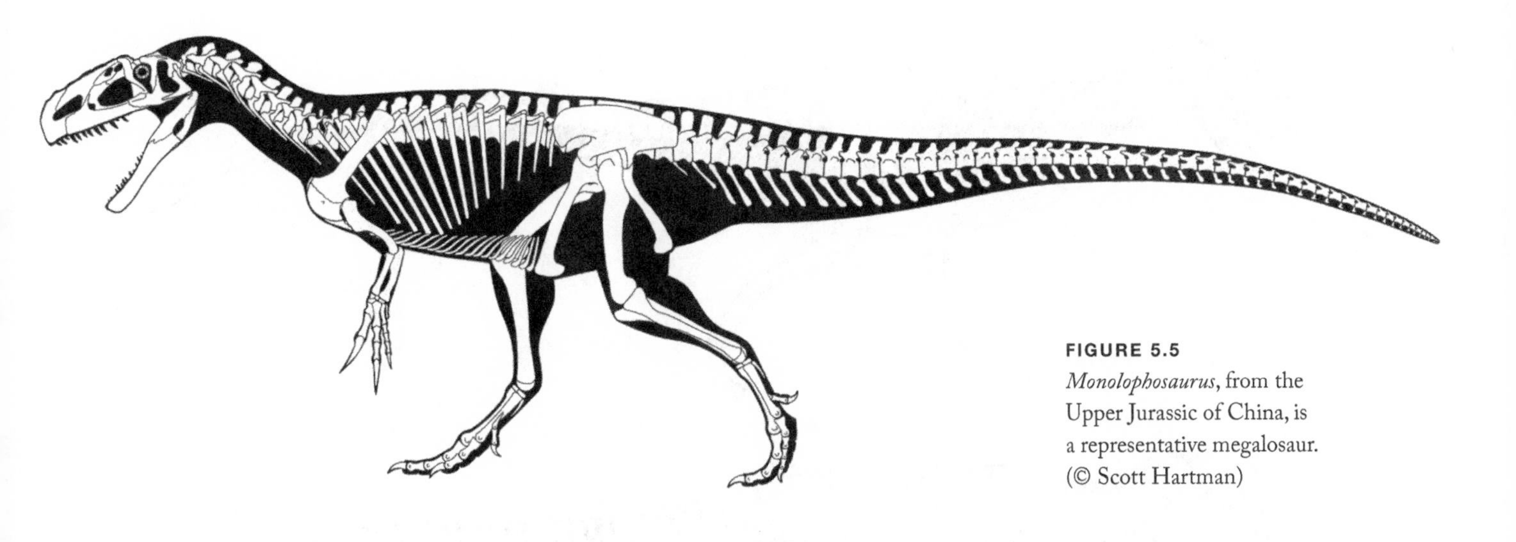

**FIGURE 5.5**
*Monolophosaurus*, from the Upper Jurassic of China, is a representative megalosaur. (© Scott Hartman)

**FIGURE 5.6**
*Spinosaurus* was the largest theropod—a huge, sail-backed fish-eater. (© James McKinnon)

**FIGURE 5.7**
Late Jurassic *Allosaurus* (*above*) and Late Cretaceous *Giganotosauros* (*below*) were typical carnosaurs. (© Scott Hartman)

## AVETHEROPODS

More advanced than megalosaurs, the avetheropods had an additional fenestra in their maxilla, complicated air sac chambers in their vertebrae, and a reduced fourth digit in the manus ("hand"). The dominant large theropods of the Middle Jurassic to Early Cretaceous were the **carnosaurs**, a group of primitive avetheropods. Carnosaurs are readily distinguished from other avetheropods by having an extra opening in the maxillary and very large nasal bones, among other features.

*Allosaurus* is the best known carnosaur (figure 5.7). A unique feature of *Allosaurus* was its lightly constructed skull, which had a distinctive, roughened ridge just above and in front of the orbit. *Allosaurus* is best known from the remains of at least 44 individuals collected at the Cleveland-Lloyd Dinosaur Quarry in eastern Utah (figure 5.8).

Cretaceous carnosaurs are carcharodontosaurs. They include *Giganotosaurus* (see figure 5.7), from the Upper Cretaceous of Argentina, a carnosaur almost as large as *Spinosaurus*.

**FIGURE 5.8**

This map shows some of the scattered bones at the Cleveland-Lloyd Dinosaur Quarry in Utah, where the remains of at least 44 individuals of *Allosaurus* were collected. (From J. H. Madsen, Jr. 1976. *Allosaurus fragilis*: A revised osteology. *Utah Geological and Mineralogical Survey Bulletin* 109:1–163. Copyright © 1976 James H. Madsen, Jr. Reprinted by permission)

## Coelurosaurs

The most bird-like tetanurans were the **coelurosaurs**, and they include the ancestors of birds. Their key evolutionary novelties include enlarged brains (at least twice the size of other theropods of comparable body size), a tridactyl hand with long second and third digits, long and narrow metatarsals, and boat-shaped chevron bones. Primitive coleurosaurs had what have been called "protofeathers," so some sort of feathery body covering may also diagnose the group.

Coelurosaurs were mainly a Cretaceous group, best known from Asia and western North America. Several different phylogenies of coelurosaurs have been proposed, so there is a variety of classifications of the many theropods assigned to Coelurosauria. Here, we divide the coelurosaurs into three groups: the primitive coelurosaurs, represented by *Compsognathus*; the tyrannosauroids; and the Maniraptoriformes.

### *The Genus* Compsognathus

*Compsognathus* (delicate jaw) is a small dinosaur best known from two virtually complete skeletons from the Upper Jurassic of western Europe (figure 5.9). The better of the two specimens is exquisitely preserved on lithographic limestone collected near Solnhofen, Bavaria, in southern Germany. In fact, this skeleton of *Compsognathus* was one of the first complete dinosaur skeletons ever described.

This skeleton represents a small theropod only 70 centimeters long. Its small size, large eyes, and relatively very large head suggest it belonged to an immature theropod. Indeed, the second skeleton of *Compsognathus*, from France, is about 50 percent larger than the German specimen.

*Compsognathus* had the typical proportions of a small tetanuran. It was lightly built and had a long neck, long tail, and short forelimbs that were only 37 percent of the length of the large hind limbs. It also displayed typical tetanuran evolutionary novelties, such as all teeth in the skull and lower jaws being in front of the orbit and a hand lacking a fourth finger.

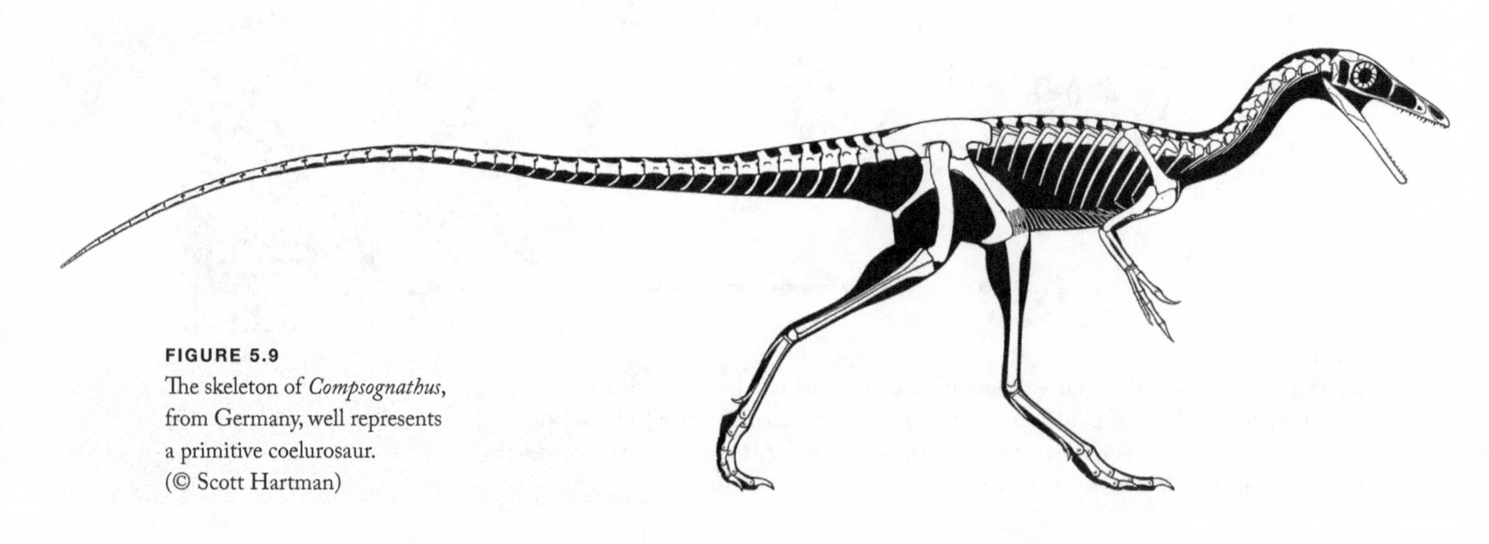

**FIGURE 5.9**
The skeleton of *Compsognathus*, from Germany, well represents a primitive coelurosaur. (© Scott Hartman)

Perhaps one of the most interesting features of *Compsognathus* is that the German specimen preserves its last meal. Inside the abdomen of this skeleton is most of the skeleton of an extinct lizard, *Bavarisaurus*. This small skeleton was originally identified as an embryo of *Compsognathus* and was presented as evidence that the small theropod gave live birth. Closer examination of the skeleton in the *Compsognathus* abdomen, however, now convinces paleontologists that just before this dinosaur died, it ran down and ate a small lizard.

## *Tyrannosauroids*

Tyrannosauroids are known primarily from fossils collected in Upper Cretaceous deposits of Asia and western North America, but their fossil record does extend back to the Middle Jurassic. The best known genera are ***Albertosaurus, Daspletosaurus***, and ***Tyrannosaurus*** from North America and ***Tarbosaurus*** from Asia. Evolutionary novelties of tyrannosauroids include fused nasal bones, incisor-like premaxillary teeth, and a very slender third metacarpal.

The archetypal **tyrannosaurid** is *Tyrannosaurus*, one of the largest (13 meters long, weighing an estimated 6 tons) land-living meat-eaters of all time (figure 5.10). This most famous theropod had a 1.3-meter-long head that bore serrated, blade-like teeth, each as much as 20 centimeters long. Its massive head was supported by a short, thick neck, and its short, thick abdomen projected forward from a massive pelvis supported by huge, pillar-like limbs. The long, heavy tail acted to counterbalance the body when running. The tiny forelimbs make a curious impression; their function is not clear (box 5.3). Other Late Cretaceous tyrannosauroids were in many ways just slightly abbreviated, more primitive versions of *Tyrannosaurus*.

During the Late Cretaceous in Asia and western North America, tyrannosauroids were moderately diverse and successful predators. They were the last large-bodied theropods to evolve and include some of the last dinosaurs.

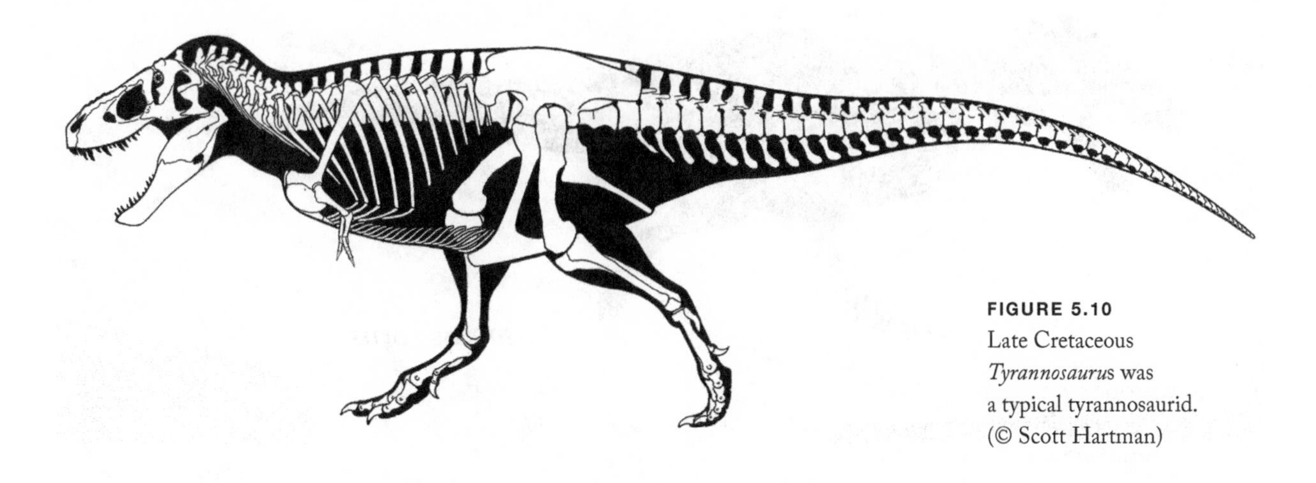

**FIGURE 5.10**
Late Cretaceous *Tyrannosaurus* was a typical tyrannosaurid. (© Scott Hartman)

Box 5.3

## Tiny *Tyrannosaurus* Forelimbs

The tiny forelimbs of *Tyrannosaurus rex* have long puzzled paleontologists (box figure 5.3). What did the dinosaur do with those puny arms? Most answers to that question have focused on what the forelimbs could not do—they could not reach the dinosaur's mouth, and they could not have been used to push the dinosaur up if it laid down. Indeed, this has led many paleontologists to conclude that the forelimbs of *Tyrannosaurus* served no useful function. They were simply vestigial structures on their way to being completely lost. Indeed, the forelimbs of some abelisaurs are even proportionately smaller than those of *Tyrannosaurus*. They, too, are seen as functionally useless vestiges.

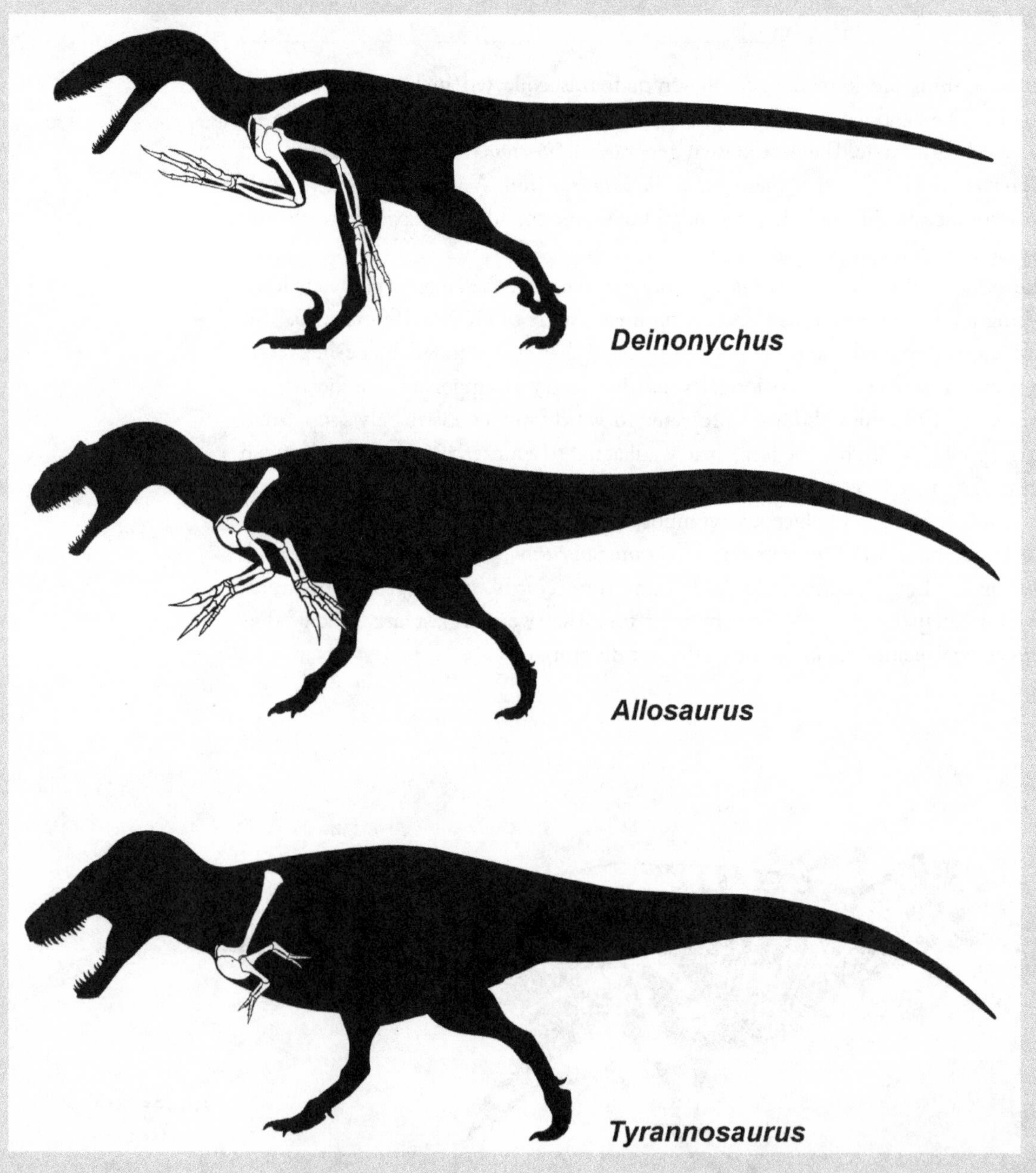


**BOX FIGURE 5.3**
Tiny *Tyrannosaurus* forelimbs are typically considered useless by paleontologists.
(© Scott Hartman)

There have been, however, a couple of positive suggestions about possible forelimb function by *Tyrannosaurus rex*. One is that the tiny arms helped to balance the extremely large head and prevent the dinosaur from tipping over. And, a more compelling analysis demonstrates that the tiny arms of *Tyrannosaurus* were actually strong and powerful—certainly more so than human arms. Thus, reconstructions of the forelimb musculature and strength calculations indicate that they could be moved quickly and resist strong forces. This suggests that they may have been used to clutch prey. Perhaps these forelimbs were vestigial but still were useful to what was one of the great dinosaurian predators.

## *Maniraptoriforms*

The derived coleurosaurs are a mainly Cretaceous group formally called the Maniraptoriformes. They were distinguished from other coelurosaurs by many features, including even larger brains, small skulls relative to body size, and an increased number of relatively smaller teeth. Many of the maniraptoriforms were not carnivorous and instead were either omnivores or herbivores. Here, we divide the maniraptoriforms into the ornithomimosaurs and the maniraptorans.

**Ornithomimosaurs** Well-known and truly abundant primitive maniraptoriforms are the "bird-mimic lizards," the ornithomimosaurids. Best known from the Upper Cretaceous of Asia and western North America, the **ornithomimosaurs** strongly resemble modern flightless birds, such as emus and ostriches, in having small heads, long necks, and long legs (figure 5.11). But their long arms with clawed hands and their long, bony tails are the two most obvious features that distinguish them from birds.

**FIGURE 5.11**
Ornithomimosaurids were very similar in overall body shape and proportions to living ostriches. (Drawing by Network Graphics)

Evolutionary novelties that distinguish ornithomimosaurs from other maniraptoriforms include the lightly built skull with very large orbits and shallow snout, the lack of (or near lack of) teeth (presumably a horny bill covered much of the jaws), and the long neck, which accounts for about 40 percent of the length of the vertebral column in front of the sacrum. ***Struthiomimus***, from the Upper Cretaceous of western Canada, was a typical ornithomimosaur and possessed these evolutionary novelties (figure 5.12). Other features of *Struthiomimus* include the very flexible joints in this dinosaur's long neck, which conferred great mobility. The ventral part of the abdomen was protected by belly ribs (gastralia), and deep scars on the spines of the dorsal vertebrae suggest the presence of powerful ligaments along the back. Such ligaments and the gastralia would have produced a very rigid abdomen that was held forward and nearly horizontal as the dinosaur ran. Large attachment sites for muscles on the distal caudal vertebrae would have helped to hold the tail rigid as a counterbalance.

The forelimbs of *Struthiomimus* were extremely long and slender, about 50 percent the length of the hind limbs. The long, slender hands had three functional fingers that bore long, pointed, but only slightly curved claws. The long, slender hind limbs had three functional toes as well, but they bore flattened claws.

The anatomy of *Struthiomimus* and some other ornithomimosaurs strongly suggests that they were fast ground runners, perhaps as swift as living, similarly proportioned ostriches, which achieve top speeds of 50 kilometers per hour. Their diet, however, is less clearly interpreted from their toothless, beak-like mouths. Most paleontologists view ornithomimosaurs as omnivores that had a mixed diet of plant and animal food.

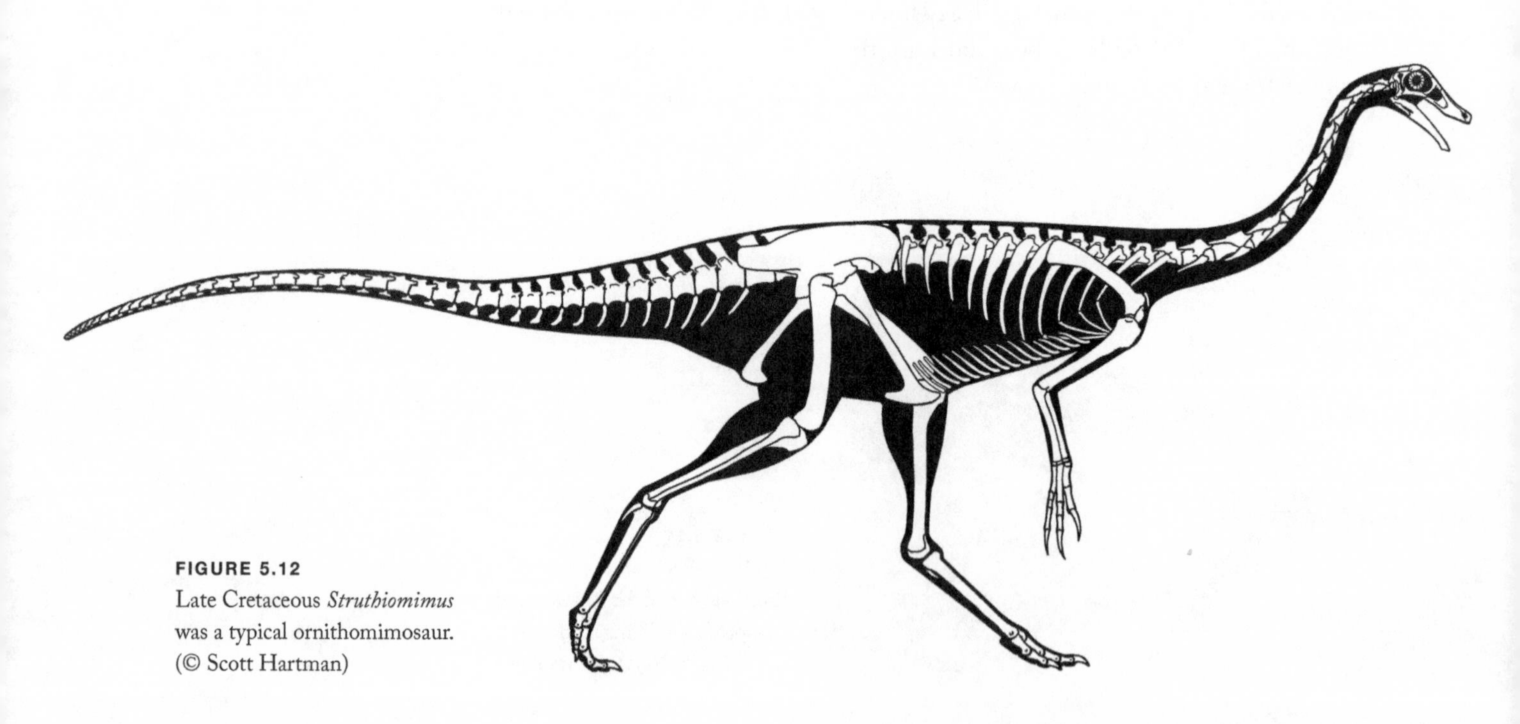

**FIGURE 5.12**
Late Cretaceous *Struthiomimus* was a typical ornithomimosaur. (© Scott Hartman)

**Maniraptorans** The remaining coelurosaurs are placed in the Maniraptora (hand grabbers). They range in age from Middle Jurassic to Late Cretaceous and are the single most diverse group of dinosaurs. Their even larger brains, large bony sternum, and elongate forelimbs are evolutionary novelties that distinguish them from ornithomimosaurs.

Among the best known maniraptorans are the **oviraptorosaurs**, from the Cretaceous of Asia and western North America. Unique features of most oviraptorosaurs include their toothless jaws (Early Cretaceous forms still had teeth); short snouts in short, boxy skulls; very large fenestrae in the mandible; very deep and strong lower jaws; and a crest of sponge-like bone on the tip of the snout (possibly for display).

*Khaan*, from the Upper Cretaceous of Mongolia, was a characteristic oviraptorosaur (figure 5.13). This 2-meter-long coelurosaur had large claws on the hands and slender hind limbs. *Khaan* had a skull and jaws designed to crush tough food items. At first, paleontologists thought that a close relative of *Khaan*, *Oviraptor*, ate the eggs of other dinosaurs; hence its name, which means "egg stealer" (box 5.4). But, now it is believed that oviraptorosaurs primarily ate the freshwater clams that are common fossils in the lake-margin deposits where most oviraptorosaur fossils are found.

A second diverse group of maniraptorans is the deinonychosaurs, popularly known as the "raptors." The sickle-shaped retractable claw on pedal digit II is their most distinctive evolutionary novelty. They comprise two families: the **Dromaeosauridae** and the **Troodontidae**. Dromaeosaurids are known from the Cretaceous of Asia, North America, Europe, Africa, and Madagascar. ***Deinonychus*** (terrible claw), from the Lower Cretaceous of Wyoming and Montana, is one of the best known and characteristic dromaeosaurids (figure 5.14). This rather large (3 to 3.3 meters long) dromaeosaurid

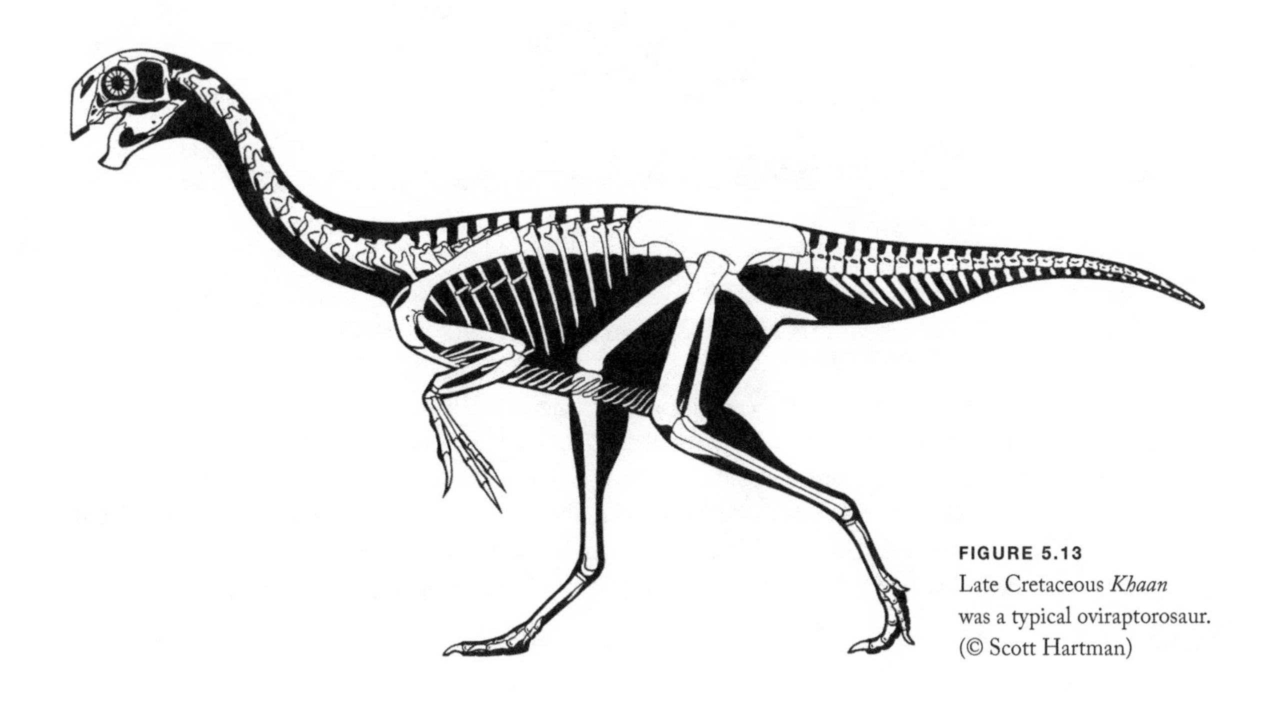

**FIGURE 5.13**
Late Cretaceous *Khaan* was a typical oviraptorosaur. (© Scott Hartman)

Box 5.4

## A Dinosaur Maligned

Discovered in Mongolia in 1923 by the Central Asiatic Expeditions of the American Museum of Natural History (AMNH), ***Oviraptor*** is a small maniraptoran theropod. Found near several nests of dinosaur eggs that were thought to have been laid by the ceratopsian *Protoceratops*, *Oviraptor* was perceived to be an egg-eater. Hence, its name, conferred by AMNH Chief Paleontologist Henry Fairfield Osborn, which in Latin means "egg thief" or "egg taker."

Indeed, the original skull of *Oviraptor* was found just inches away from what was thought to be a nest of *Protoceratops* eggs. Further, Osborn saw the toothless beak of the small theropod as a crushing mechanism to break eggs. This induced Osborn to imagine the dinosaur as a likely egg-eater that plundered *Protoceratops* nests. He thus coined the colorful name *Oviraptor* but cautioned that this "may entirely mislead us as to the feeding habits and belie its character." Indeed, Osborn went on to conclude that the dinosaur was "herbivorous or omnivorous." One has to wonder why, then, he named it *Oviraptor*!

Decades later, in the 1990s, a new AMNH expedition to the Gobi Desert found more of the same kinds of dinosaur egg, but with *Oviraptor* embryos inside them. And, they found an *Oviraptor* skeleton sitting on top of a nest of eggs (box figure 5.4). This dinosaur had evidently been buried in a sandstorm as it brooded more than 15 eggs in its nest. Clearly, *Oviraptor* was not likely to have been eating its own eggs, so the name of the dinosaur mistakes its diet, at the very least.

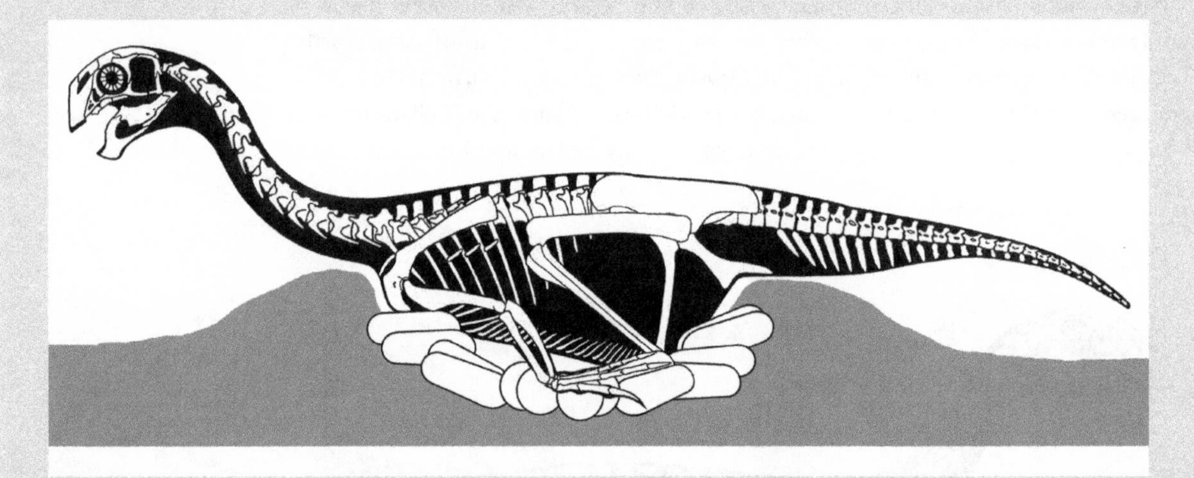

**BOX FIGURE 5.4**
The skeleton of the theropod *Khaan* brooding its nest of eggs. (© Scott Hartman)

So, what did *Oviraptor* eat? This has been discussed at length, because the peculiar toothless beak of the dinosaur suggests it may not have been just a meat-eater. Still, a lizard skeleton in the stomach region of a close relative of *Oviraptor* indicates some predation. Thus, an omnivorous diet seems most likely for *Oviraptor*.

Perhaps most significant is the bird-like behavior inferred for *Oviraptor* from the discovery of the skeleton on the nest. The eggs in the nest were arranged, which suggests the dinosaur manipulated them into a pattern, as do many modern birds. And, the dinosaur sat on the nest to incubate the eggs, also a very avian behavior.

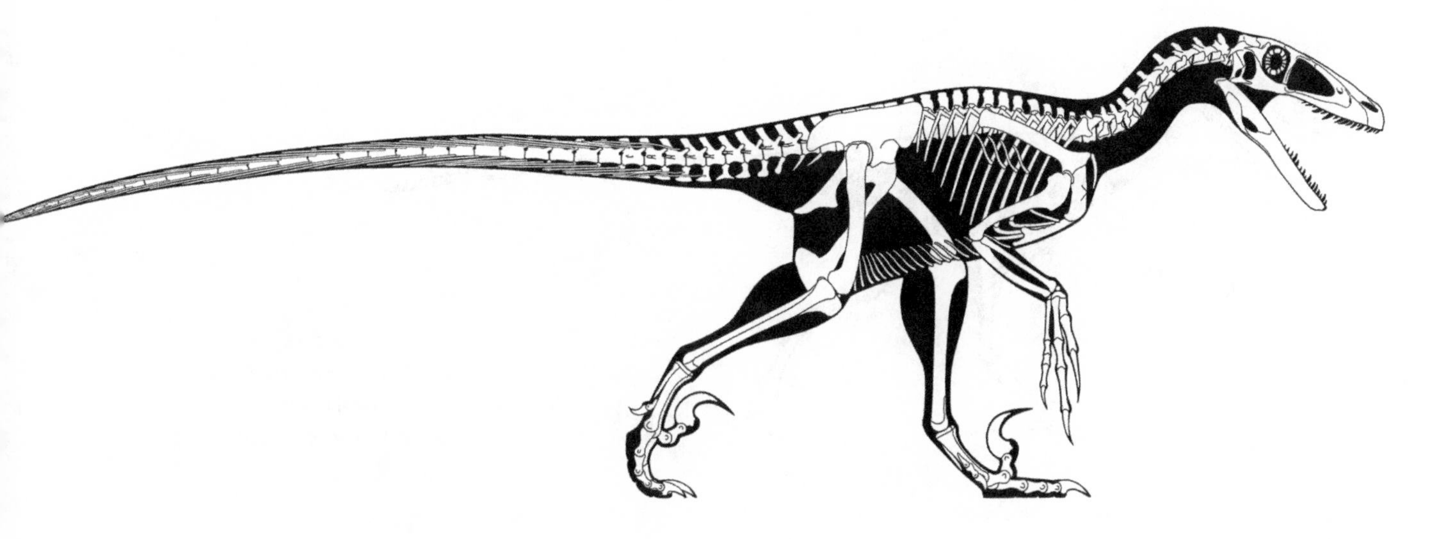

**FIGURE 5.14**
Early Cretaceous *Deinonychus* was a typical dromaeosaurid. (© Scott Hartman)

displays characteristic features of this group of dinosaurs. Note especially the large skull full of serrated teeth, the backward-directed pubis, the stiffened tail, and the huge claws on the second toes of the hind feet. Aside from these specializations, much of the anatomy of *Deinonychus* was typically theropod.

A quick examination of the skeleton of *Deinonychus* convincingly demonstrates it was a habitual biped. How the dinosaur might have used the huge, curved claws on its hind feet, however, is puzzling. The only way to use these claws would have been to have one or both hind feet off the ground, stabbing the prey. To do so would have required great agility and balance, aided in part by the stiff counterbalancing tail. The skeleton of *Deinonychus* was clearly that of a very active bipedal predator.

The second group of deinonychosaurs is the troodontids. These coelurosaurs, best known from ***Saurornithoides*** (includes *Troodon* of some paleontologists) from the Upper Cretaceous of North America and Asia, are closely related to dromaeosaurids (figure 5.15). Characteristic features of the troodontids include a long skull with a narrow snout, a large braincase (troodontids have the largest brains relative to body size of any dinosaur), and an inflated bony casing for the middle ear. The teeth of troodontids were small and numerous (as many as 35 in the lower jaw), and many were serrated along their posterior edges. Troodontids had ankle bones fused to each other and to the tibia and small claws on the second digits of their hind feet.

Other features that distinguish *Saurornithoides* and other troodontids include their extremely large eyes and possible stereoscopic vision. Like dromaeosaurids, troodontids were agile bipedal predators. The earliest known troodontids are from the Lower Cretaceous of Asia, but they are best (though still poorly) known from the Upper Cretaceous of Asia and western North America.

A very unusual group of maniraptorans is the Therizinosauria (scythe reptiles), from the Cretaceous of Asia and North America. They had small skulls, long necks, huge claws, and short legs. Nearly the size of *Tyrannosaurus*, *Nothronychus* had 1-meter-long claws on its hands (figure 5.16).

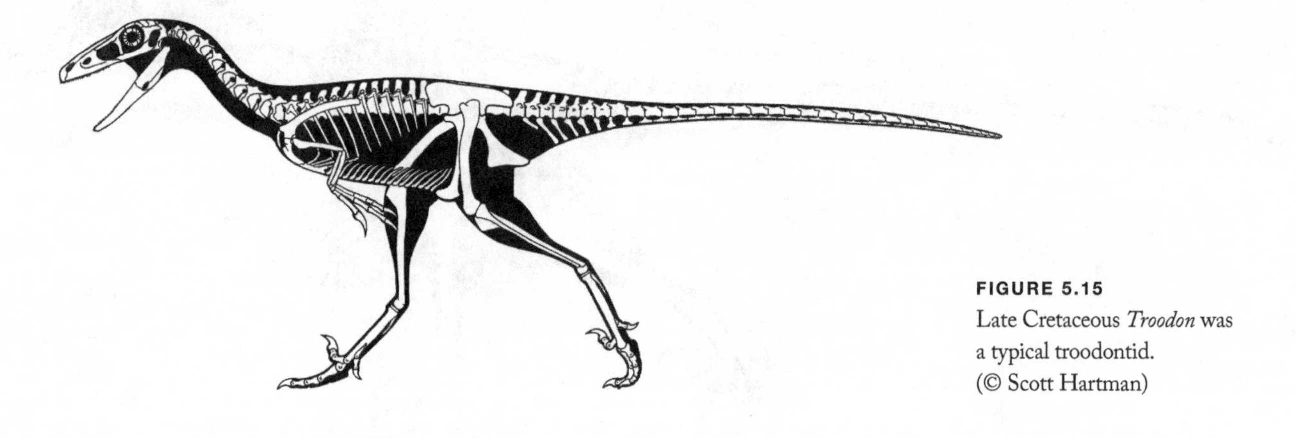

**FIGURE 5.15**
Late Cretaceous *Troodon* was a typical troodontid. (© Scott Hartman)

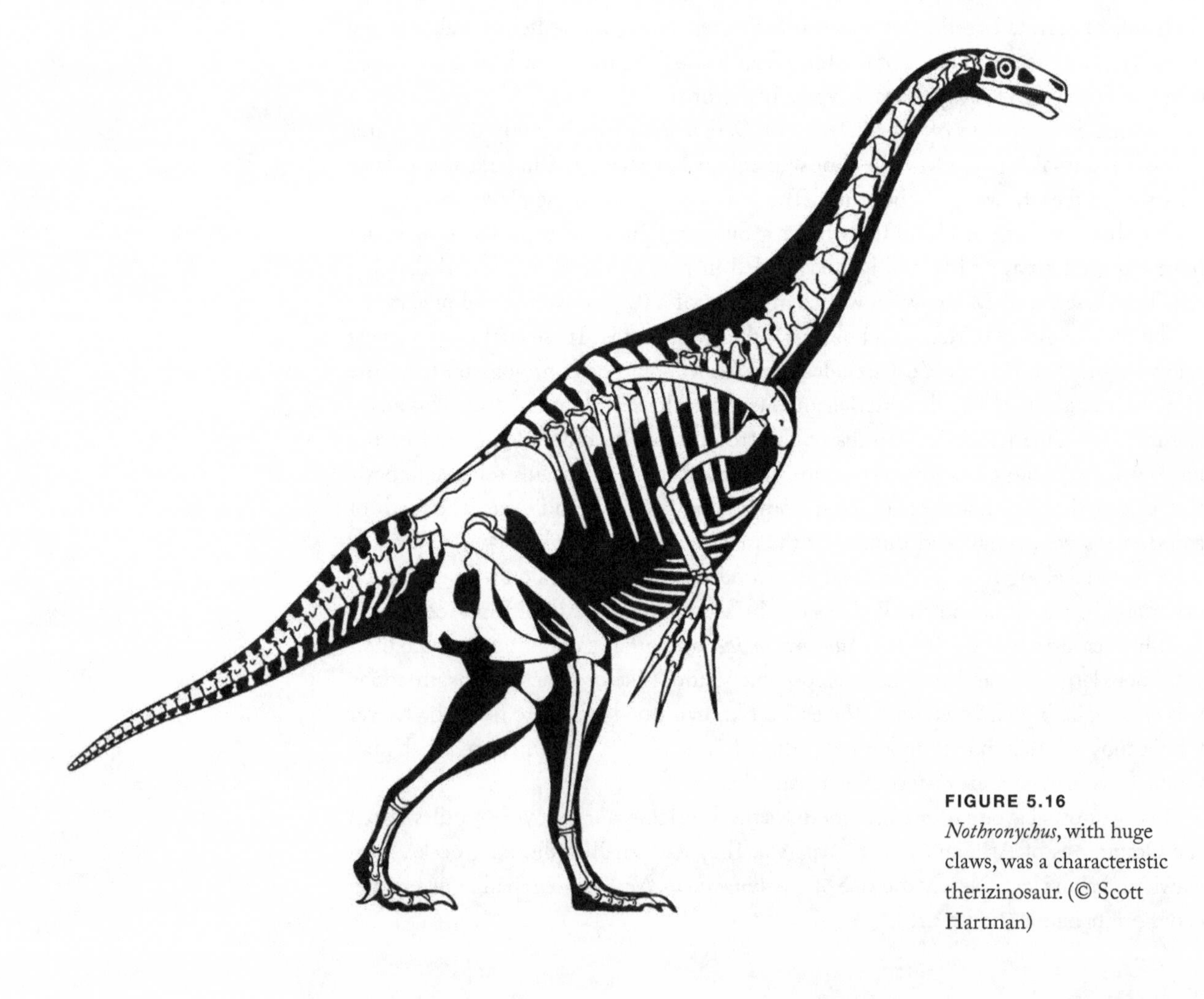

**FIGURE 5.16**
*Nothronychus*, with huge claws, was a characteristic therizinosaur. (© Scott Hartman)

## THEROPOD EVOLUTION

Theropod dinosaurs appeared as highly specialized, running predators almost at the outset of dinosaur evolution. Early theropods of the Late Triassic were important terrestrial predators up to 3 meters long. By the Early Jurassic, 6-meter-long theropods were the apex predators of terrestrial communities. For the remainder of the Jurassic, and through the Cretaceous, theropods remained the dominant terrestrial carnivores.

The evolutionary split between ceratosaurs and tentanurans took place during the Early Jurassic. Although ceratosaurs were relatively rare during the Jurassic, they include the abelisauroids, which were the dominant predatory dinosaurs of the southern continents and Europe during the Cretaceous.

Tetanurans comprise three major groups: the megalosauroids, the carnosaurs, and the maniraptoriforms. Jurassic megalosaurs were large predators, but the Cretaceous megalosaurs, notably the spinosauroids, were a group of fish-eaters that include the largest known theropods. Carnosaurs were large theropods of the Middle Jurassic and Cretaceous and include the gigantic carcharodontosaurs of the Cretaceous of South America and Africa.

Coelurosaurs first appeared as small carnivorous forms during the Middle Jurassic. However, by the Cretaceous, there were herbivorous and omnivorous coelurosaurs as well. Small, *Compsognathus*-like theropods were primitive coelurosaurs. Tyranosauroids and maniraptoriforms were more advanced. The maniraptoriforms were the most diverse dinosaurs and are best known from Cretaceous records in Asia and North America. They include the oviraptorosaurs, ornithomimosaurs, therizinosaurs, and deinonychosaurs. It is among the deinonychosaurs, more specifically the dromaeosaurids, that most paleontologists locate the ancestry of birds.

By the Late Cretaceous, abelisaurids were the top theropod predators everywhere but Asia and North America. In these areas, tyrannosaurids dominated. They were among the last of the dinosaurs.

## Summary

1. Most predatory dinosaurs were theropods.
2. Theropod phylogeny is complex and not fully agreed upon.
3. Nevertheless, theropods can be distinguished from other dinosaurs by many skeletal features, most of which identify them as bird-like, bipedal cursors. Indeed, theropods include the ancestors of birds.
4. The most primitive theropods were the Late Triassic–Late Jurassic coelophysoids and the Jurassic–Cretaceous ceratosaurs.
5. The very bird-like tetanurans were advanced theropods.
6. The megalosaurs of the Jurassic–Cretaceous were large tetanurans and include the largest of all theropods, *Spinosaurus*.
7. Carnosaurs were large theropods of the Jurassic–Cretaceous.
8. The tyrannosaurids, some of the largest meat-eating land animals of all time, were advanced, extremely large carnosaurs of the Late Cretaceous.

9. The coelurosaurs were advanced tetanurans with small skulls and long forelimbs, among other features. Late Jurassic *Compsognathus* is a characteristic primitive coelurosaur.
10. Maniraptoriforms were advanced coelurosaurs that include the ornithomimosaurs and manirpatorans.
11. The maniraptorans encompass many taxa, including the bizarre therizinosaurs, the oviraptorosaurs, and the deinonychosaurs ("raptors").
12. Bird ancestry lies among the deinonychosaurs.

## Key Terms

*Albertosaurus*
*Allosaurus*
carnosaur
*Ceratosaurus*
*Coelophysis*
coelurosaur
*Daspletosaurus*
*Deinonychus*
*Dilophosaurus*
Dromaeosauridae
Khaan
Ornithomimosaur
*Oviraptor*
oviraptorosaur
*Saurornithoides*
*Struthiomimus*
*Tarbosaurus*
Tetanurae
Theropoda
Troodontidae
*Tyrannosaurus*

## Review Questions

1. What features are diagnostic of theropods, and what types of behavior can we infer from these features?
2. What relationship exists between the theropod fossil record and our understanding of theropod phylogeny?
3. What are the bird-like features of theropods?
4. How are avetheropods distinguished from other theropods, and what is the significance of their distinctive features?
5. Draw an evolutionary tree of the coelurosaurs. Where would you place the origin of birds on this tree?
6. Compare and contrast the anatomy and behavior of *Struthiomimus* and *Deinonychus*.

## Further Reading

Carrano, M. T., R. B. J. Benson, and S. D. Sampson. 2012. The phylogeny of Tetanurae (Dinosauria: Theropoda). *Journal of Systematic Palaeontology* 10:211–300. (Detailed cladistic analysis of tetanuran relationships)

Holtz, T. R., Jr. 2012. Theropods, pp. 347–378, in M. K. Brett-Surman, T. R. Holtz, Jr., and J. O. Farlow, eds., *The Complete Dinosaur*. 2nd ed. Bloomington: Indiana University Press. (Concise review of the theropods)

Larson, P., and K. Carpenter, eds. 2008. *Tyrannosaurus rex, the Tyrant King*. Bloomington: Indiana University Press. (Collection of 21 technical articles on *Tyrannosaurus rex*)

Lee, Y., et al. 2014. Resolving the long-standing enigmas of a giant ornithomimosaur *Deinocheirus mirificus*. *Nature* 515:257–260. (Describes a nearly complete skeleton of this therizinosaur)

Madsen, J. H., Jr. 1976. *Allosaurus fragilis*: A revised osteology. *Utah Geological and Mineralogical Survey Bulletin* 109:1–163. (A detailed, technical, and extensive bone-by-bone description of *Allosaurus*)

Ostrom, J. H. 1969. Osteology of *Deinonychus antirrhopus*, an unusual theropod from the Lower Cretaceous of Montana. *Peabody Museum of Natural History, Yale University Bulletin* 30:1–165. (A comprehensive, technical description of the anatomy, behavior, and relationships of *Deinonychus*)

Ostrom, J. H. 1978. The osteology of *Compsognathus longipes* Wagner. *Zitteliana* [Abhandlungen der Bayerischen Staatssammlung für Paläontologie und historische Geologie] 4:73–118. (A comprehensive, technical description of the anatomy, behavior, and relationships of *Compsognathus*)

Parrish, J. M., et al., eds. 2013. *Tyrannosaurid Paleobiology*. Bloomington: Indiana University Press. (Collection of 15 technical articles on all aspects of the biology of tyrannosaurs)

Rinehart, L. F., et al. 2009. The paleobiology of *Coelophysis bauri* (Cope) from the Upper Triassic (Apachean) Whitaker quarry, New Mexico, with an analysis of a single quarry block. *Bulletin of the New Mexico Museum of Natural History and Science* 45:1–260. (Comprehensive study of the biology of *Coelophysis*)

## Find a Dinosaur!

You can find a real fossil of *Tyrannosaurus rex* on display at several natural history museums in the United States and Canada. The type specimen of *T. rex*, for which the name was coined, is exhibited at the Carnegie Museum of Natural History (Pittsburgh, Pennsylvania). The American Museum of Natural History (New York) also has a skeleton on display. Legendary fossil collector Barnum Brown (1873–1963) collected these *T. rex* fossils in the early twentieth century, and they were the first to be described. In Chicago, the Field Museum of Natural History displays the *T. rex* skeleton nicknamed "Sue," after its discoverer, Sue Hendrickson. A similarly nicknamed skeleton, "Stan," can be seen at the museum of the Black Hills Institute of Geological Research (Hill City, South Dakota). Sue and Stan are remarkably complete skeletons of *T. rex*, more complete than the skeletons displayed in Pittsburgh and New York. Other skeletons and skulls of *T. rex* can be seen in the exhibit halls of the Smithsonian's National Museum of Natural History (Washington, D.C.), at the Royal Tyrrell Museum of Palaeontology (Drumheller, Alberta), and at the Natural History Museum of Los Angeles County (Los Angeles, California). Most of the *T. rex* on display in other natural history museums are replicas (casts) made from original fossils.

# 6

# SAUROPODOMORPHS

FEW animals are more awe inspiring than the largest land animals of all time, the **sauropod** dinosaurs. Their anatomy, classification, lifestyle, and evolution are the focus of this chapter, which begins with a consideration of their closest relatives, the prosauropod dinosaurs.

The prosauropods and the sauropods together constitute a group of dinosaurs with the ungainly name **Sauropodomorpha**. They are distinguished from other dinosaurs by having heads that were very small relative to their bodies, spatulate (spoon-shaped) teeth, at least 10 elongated vertebrae in relatively long necks, short feet, and very large claws on the first digits of their forefeet. Sauropodomorphs spanned the entire age of dinosaurs, from the Late Triassic until the end of the Cretaceous. They were one of the most successful groups of plant-eating dinosaurs as well as the result of nature's most amazing experiment in animal gigantism.

## PROSAUROPODS

**Prosauropods** represent one of the first evolutionary diversifications of plant-eating dinosaurs. Known from fossils found on all continents except Antarctica (figure 6.1), they range in age from Late Triassic to Early Jurassic and so include some of the oldest dinosaurs. Prosauropods were sauropod-like in general build, though much smaller and more slender.

Prosauropods were once considered ancestral to sauropods (hence their name). Indeed, some paleontologists believed the "missing link" between the two groups was provided by prosauropods such as ***Vulcanodon***, from the Lower Jurassic of southern Africa, and ***Riojasaurus***, from the Upper Triassic of Argentina (figure 6.2). But all prosauropods, including these two types, had some skeletal features, such as the small size of the fifth digit of the hind foot, the existence of which would require a highly

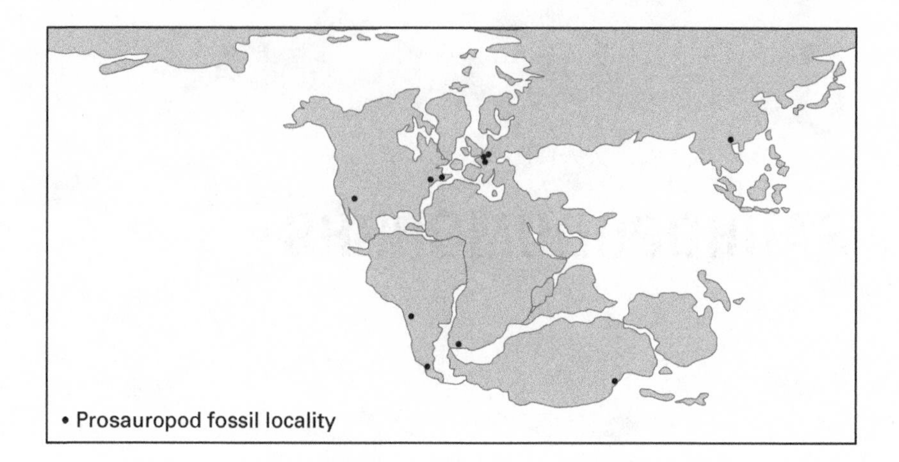


**FIGURE 6.1**
This map of the occurrences of some well-known prosauropods indicates their worldwide distribution during the Early Jurassic. (Drawing by Network Graphics)

unlikely reversal in evolution were prosauropods truly the ancestors of sauropods. Sauropods had large fifth digits on the hind foot, as did, presumably, the ancestors of prosauropods. Therefore, paleontologists now favor the view that prosauropods were the closest relatives, but not the ancestors, of sauropods and that both groups derived from a common ancestor, as yet undiscovered (figure 6.3).

Typical prosauropods include the massive, heavily built 17-meter-long *Riojasaurus* from Argentina, the lightly built *Anchisaurus* from the eastern United States, and *Yunnanosaurus* from southern China, with its chisel-shaped teeth reminiscent of those of some later sauropods. *Plateosaurus* is a particularly well-known and characteristic prosauropod from the Late Triassic of western Europe.

## The Genus *Plateosaurus*

One of the best-known prosauropods is ***Plateosaurus***, from the Upper Triassic of western Europe (figure 6.4). Mass death assemblages of this large (6- to 8 meters in body length) prosauropod are known from Trössingen, Germany, where the animals may have been killed and their skeletons accumulated by mud slides. The complete

**FIGURE 6.2**
*Riojasaurus*, an 11-meter-long prosauropod from the Upper Triassic of Argentina, may best approximate the ancestry of sauropods. (© Scott Hartman)

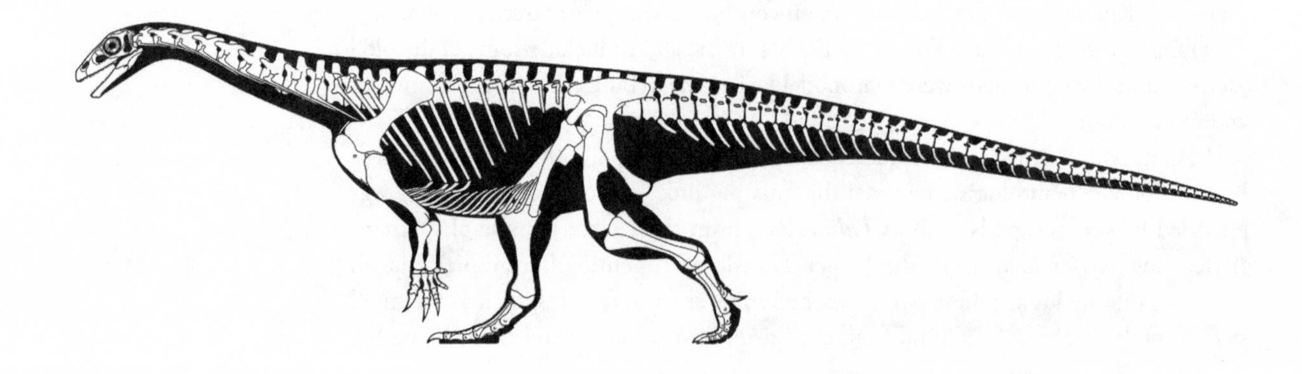

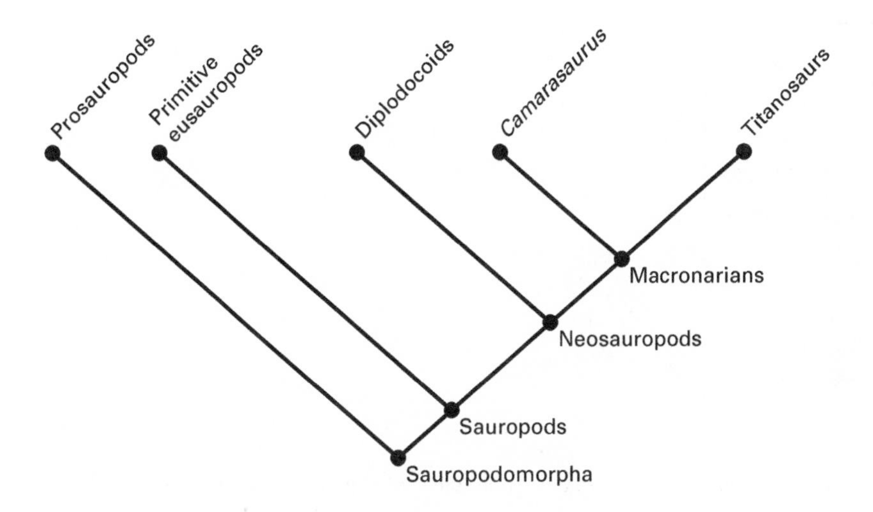


**FIGURE 6.3**
Although some paleontologists see prosauropods as the ancestors of sauropods, the two groups are best regarded as close relatives derived from an as yet undiscovered ancestor. (Drawing by Network Graphics)

skeletons of *Plateosaurus* from Trössingen give us remarkable knowledge of this early prosauropod.

Salient features of the skeleton of *Plateosaurus* include its large size and robustness, the numerous coarsely serrated teeth, and the jaw joint below the level of the teeth. Its small head was attached to a long neck, followed by a long back and tail, in proportions characteristic of prosauropods. The pelvis was typically saurischian and anchored massive hind limbs. The first finger of the hand bore a very large claw, and the fourth and fifth digits of the hind foot were small and must not have borne much weight when the animal walked.

## Prosauropod Lifestyles

Prosauropods are often restored as bipedal animals (see figure 6.4), but their footprints indicate that they also could have walked quadrupedally (figure 6.5). Indeed, the forelimbs of prosauropods were at least two-thirds the length of their hind limbs, as in the later diplodocid sauropods, which are generally considered to have been obligatory quadrupeds. The first digit of the hand was much larger than the others and bore a

**FIGURE 6.4**
The skeleton of 6- to 8-meter-long *Plateosaurus*, from the Upper Triassic of western Europe, shows many characteristic features of prosauropods. (© Scott Hartman)

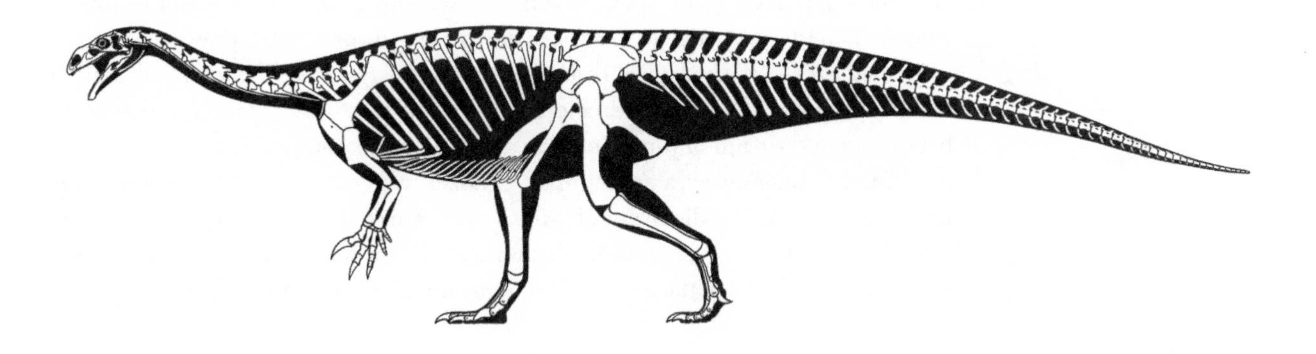

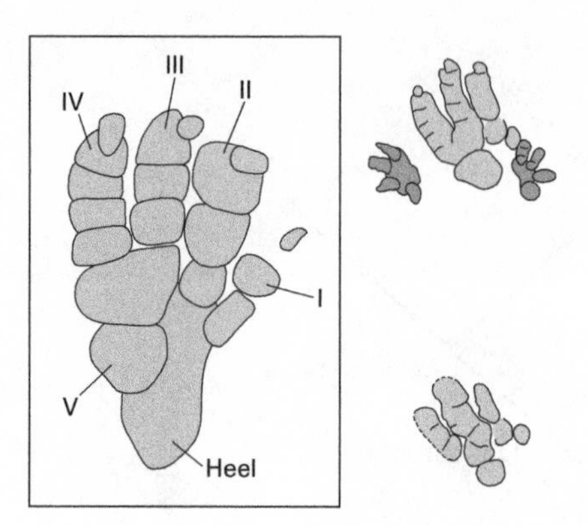


**FIGURE 6.5**
Prosauropod footprints like these indicate that prosauropods did sometimes walk on all fours. In the upper-right portion of the drawing, note both hind foot and forefoot impressions. (Drawing by Network Graphics)

huge claw, and that digit must have been held clear of the ground while walking. There is little doubt that prosauropods could rear up on their hind limbs to reach vegetation in tall trees, which is consistent with the evidence that prosauropods were herbivores. A large number of features suggest that prosauropods ate plants:

1. Their long necks would have extended their vertical feeding range so that they could have cropped vegetation from tall trees, as do living giraffes.
2. The jaw hinge was below the line of the upper tooth row, as in plant-eating ornithischians.
3. The offset of the jaws of prosauropods ranged from small to large. This offset allowed the tooth rows to be almost parallel so that contact would have occurred nearly simultaneously along their entire length when chewing, as in many living plant-eating mammals.
4. The spatulate teeth of prosauropods had about 20 coarse serrations per tooth that were set at an angle of about 45 degrees to the cutting edge. They thus bore an uncanny resemblance to the teeth of the living *Iguana*, a plant-eating lizard.
5. A mass of small stones (gastric mill or gizzard) found in the stomach region of a skeleton of the African prosauropod *Massospondylus* may have been used to crush swallowed vegetation.

## The Genus *Mussaurus*

No review of prosauropods would be complete without mention of ***Mussaurus*** (mouse lizard) of the Upper Triassic, described in the late 1970s from a tiny skeleton and associated eggs from the Upper Triassic of southern Argentina. The total body length of *Mussaurus* was about 20 centimeters. Clearly, this tiny baby dinosaur, which could have stood in the palm of your hand, was a hatchling prosauropod. It may have been a juvenile of *Coloradisaurus*, a prosauropod dinosaur from the same locality known only from adult bones. The discovery of *Mussaurus* not only brought to light one of the world's smallest dinosaurs, but also demonstrated that prosauropods, like sauropods, were egg layers. Indeed, the eggs of *Mussaurus* are the oldest known dinosaur eggs.

## Prosauropod Evolution

Prosauropods were among the first dinosaurs to appear during the Late Triassic. They rapidly achieved a worldwide distribution and were among the largest, and most successful, plant-eating vertebrates of the Late Triassic and Early Jurassic. The great similarity of prosauropods such as *Ammosaurus* from the United States and *Lufengosaurus* from southern China suggests that by the Early Jurassic, environments favorable to prosauropods were present across the world's continents.

The extinction of prosauropods at the end of the Early Jurassic marks the first significant disappearance of a major group of dinosaurs. It also coincided with the diversification of sauropod dinosaurs as well as other large plant-eaters, the stegosaurs and ankylosaurs. So it is tempting to believe that the prosauropods were replaced by these larger herbivores.

# SAUROPODS

These gigantic quadrupedal herbivores include the largest land animals of all time, behemoths that weighed as much as 59 tons (possibly more) and attained body lengths of at least 35 meters. ***Brontosaurus***, correctly called ***Apatosaurus***, is the sauropod most familiar to the general public, followed by ***Diplodocus*** and ***Brachiosaurus***. Paleontologists recognize as many as 75 genera of sauropods, although most of these animals are not known from complete skeletons. Evolutionary novelties of sauropods include their gigantic size, long necks and tails, tiny heads, and dorsally located nostrils, among other features.

The oldest known sauropods are of Late Triassic age, and the group persisted and had a worldwide distribution until the extinction of all the dinosaurs at the end of the Cretaceous (figure 6.6). In terms of general size and diversity, the zenith of sauropod

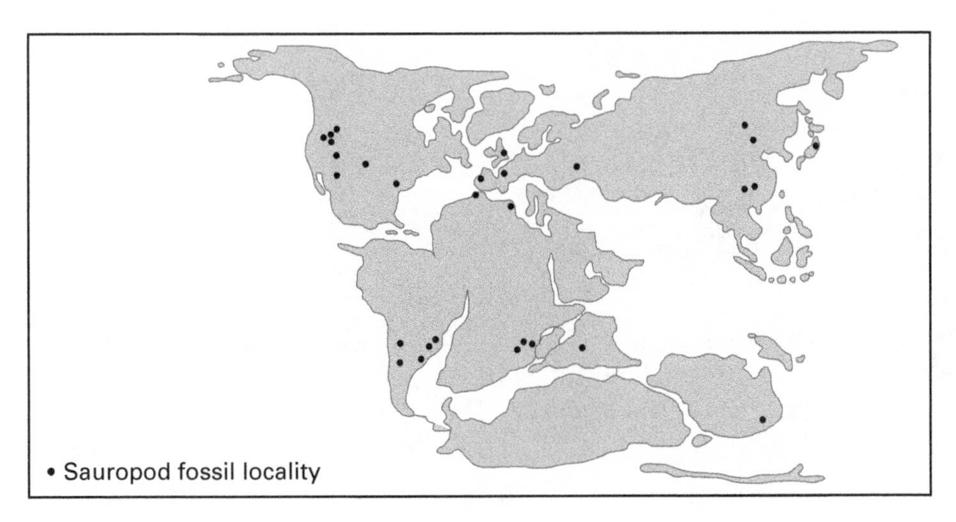


**FIGURE 6.6**
Sauropod dinosaurs were the largest and among the most successful plant-eating vertebrates of the Jurassic and Cretaceous. Their fossils are known from all continents except Antarctica. (Drawing by Network Graphics)

evolution occurred during the Jurassic–Cretaceous transition. At least three very distinct types of sauropods coexisted: the diplodocoids and primitive macronarians, and the Titanosauria (advanced macronarians), which are distinguished by features of their skulls, vertebrae, and limb skeletons (box 6.1). Thus, one current and well-accepted phylogeny of sauropods recognizes an array of primitive eusauropods (including the very long-necked ***Mamenchisaurus***) and two groups of neosauropods: Diplodocoidea and Macronaria. Primitive macronarians include *Brachiosaurus*, but most macronarians are titanosaurs, a mostly Cretaceous group known from fossils found on the former Gondwana continents.

Box 6.1

## Sauropod Vertebrae: Key to Classification

The skulls of sauropods were small and weakly connected to their vertebral columns. Because of this, they were easily detached and lost after death, so it is the norm to find sauropods without skulls (see box 6.2). Indeed, the skulls of many sauropod taxa are unknown. This has forced paleontologists to rely on other aspects of the skeletons of sauropods, especially their vertebrae, to classify them. Fortunately, sauropod vertebrae were intricate structures that varied significantly from sauropod to sauropod. So they may offer a useful basis for the classification of these giants.

The basic sauropod vertebra, like that of most vertebrates, consists of a **centrum** beneath a **neural arch** (box figure 6.1A). The centrum is the body of the vertebra and provides a site of attachment for muscles that support the body. The connected centra together comprised the flexible rod upon which the body of the dinosaur was hung. The centrum also supports the neural arch, a complex structure that encloses the spinal cord and sends out bony struts to which the ribs and other vertebrae are attached. The top of the neural arch is a flange of bone called the neural spine, another site of attachment for tendons, muscles, and ligaments, especially in the shoulder and hip regions.

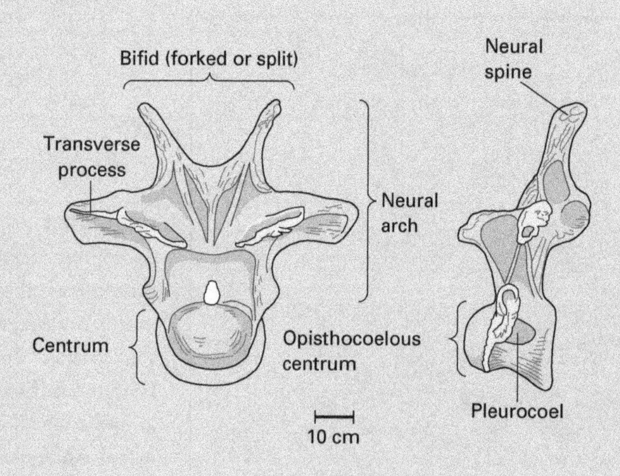


**BOX FIGURE 6.1A**
Sauropod vertebrae, such as this dorsal vertebra of *Apatosaurus*, present a variety of features that can be used in sauropod classification. (Drawing by Network Graphics)

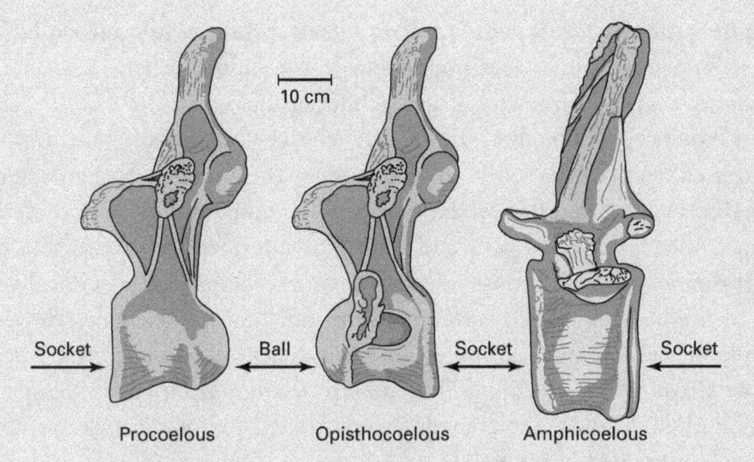


**BOX FIGURE 6.1B**
Sauropod vertebrae are procoelous, opistocoelous, or amphicoelous, depending on the arrangement of the ball-and-socket joints of the centra. (Drawing by Network Graphics)

As stated in the appendix, "A Primer of Dinosaur Anatomy," the vertebral column of all dinosaurs, including sauropods (and most other vertebrates), can be divided into four regions: cervical (neck), dorsal (back), sacral (hip), and caudal (tail). In addition, the shapes of the surfaces at which the centra meet each other (articulate) have descriptive names. In sauropods, the centra meet at ball-and-socket joints (box figure 6.1B). In a single centrum, if the socket is anterior and the ball is posterior, the vertebra is termed **procoelous**, from the Greek words *pro*, meaning "before," and *koilus* (*coelos*), meaning "hollow" or "cavity." If the situation is reversed—socket posterior and ball anterior—the vertebra is termed **opisthocoelous** (*opisthen* is Greek for "behind"). Vertebrae bounded by two cavities are called **amphicoelous** (*amphi* is Greek for "double"). Indeed, one sauropod genus, ***Amphicoelias***, takes its name from the amphicoelous nature of some of its vertebral centra.

The centra of many sauropod dinosaurs, particularly in the cervical region, were a lightly built framework of delicate struts and buttresses bounding deep cavities, especially along the sides of the centra (see box figure 6.1A). Such cavities are called **pleurocoels** (*pleura* is Greek for "side"; thus, "side cavities"). Through evolution, the solid, spool-like vertebrae of sauropod ancestors were greatly modified by removing bone so that the vertebrae became much lighter. The development of pleurocoels may thus have helped lighten an otherwise inordinately heavy structure in a very large animal already fighting a tremendous battle against gravity during its day-to-day activities. Some paleontologists have also suggested that the pleurocoelous vertebrae may have been invested with air sacs that functioned in a fashion similar to the air sacs that modern birds have in their bones, which help to lighten them and increase the efficiency of their respiration. Another suggestion is that the pleurocoels were sites of glycogen (glucose, a natural sugar) storage. These ideas are difficult to evaluate from just the structure of the vertebrae alone, however. What is certain is that the vertebrae were much lightened by the presence of pleurocoels.

The caudal vertebrae of sauropods had **chevron** bones on their undersides that surrounded and protected blood vessels for the tail. The shape of these chevrons; the shape and length of the neural spines; and whether various vertebrae are procoelous, opisthocoelous, or amphicoelous are important features used to classify sauropods. Thus, for example, the titanosaurians are identified in part on the basis of their procoelous anterior caudal vertebrae, which differ from the caudals of other sauropods.

## Primitive Eusauropods

The most primitive sauropods are considered primitive **eusauropods**. Among the best-known primitive eusauropods are the Chinese genera *Shunosaurus* and *Datousaurus*, described from the Zigong dinosaur quarry of the Middle Jurassic in Sichuan Province (see box 10.2). These were relatively small sauropods; *Datousaurus*, the longer of the two, was "only" 10 meters long. Their skulls were rather similar to that of *Camarasaurus* but had longer muzzles. The numerous teeth were slender but had small, spoon-shaped crowns and were thus intermediate in structure between those of diplodocoids and those of *Camarasaurus*. There were 12 cervical vertebrae and 13 in the back. The neck vertebrae were short and lacked much development of the pleurocoels. The neural spines were not divided, and there were three bones in the wrist and two in the ankle. The humerus-to-femur ratio was about 0.66, the same as in diplodocoids. The chevrons were forked, and in *Shunosaurus*, there was even a club at the end of the tail.

## Diplodocoids

**Diplodocoids** were Late Jurassic sauropods known mostly from North America, Europe, and Africa. They are most easily distinguished by their skulls, which were long and slender with elongate muzzles (figure 6.7). The jaws bore slender, peg-like teeth confined to the front of the mouth. The nostrils were on top of the skull, in front of and above the orbits.

Diplodocoid bodies were long and relatively lightly built, and diplodocoids were among the longest, though not necessarily the heaviest, dinosaurs (figure 6.8). A 26-meter-long *Diplodocus* would have weighed about 12 tons, considerably less than the shorter, 40-ton *Brachiosaurus*. The necks of diplodocoids were extremely long, with an increased number of vertebrae at the expense of those in the back. This was foreshadowed by the neck of the primitive eusauropod *Mamenchisaurus*, from the Cretaceous of China, which had 19 vertebrae but only 12 in the back (figure 6.9).

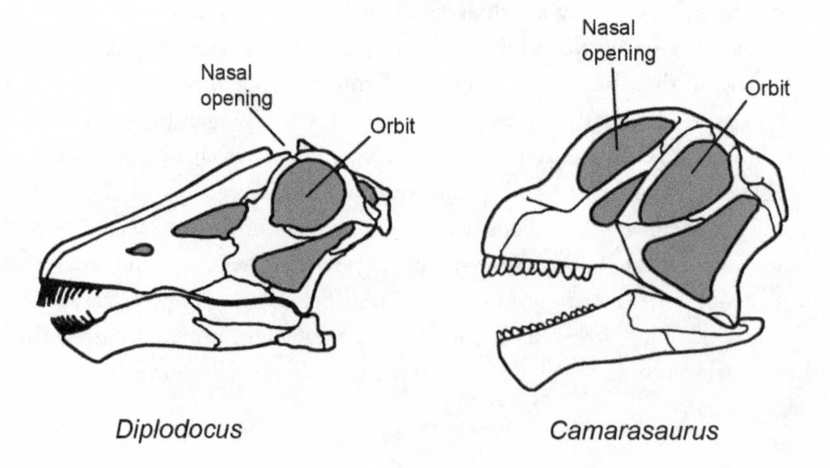


**FIGURE 6.7**
Diplodocoids and primitive macronarians, such as *Camarasaurus*, can most easily be distinguished by features of their skulls. (© Scott Hartman)

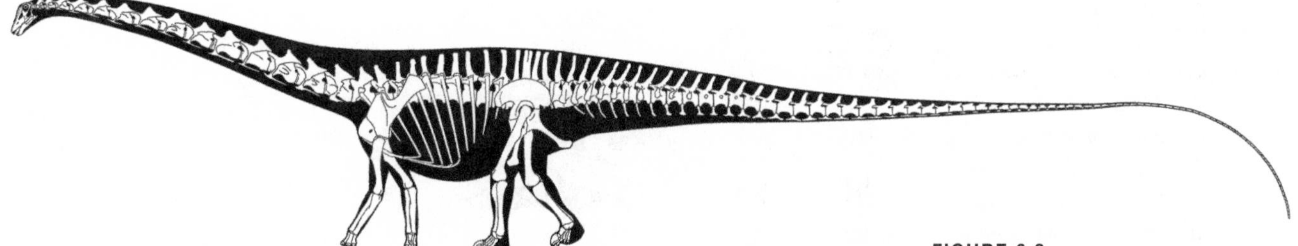

**FIGURE 6.8**
An adult *Diplodocus* was a 27-meter-long, lightly built sauropod, characteristic of the diplodocoids. (© Scott Hartman)

The neck vertebrae of diplodocoids had unusually short ribs, and their neural spines were marked by a deep, V-shaped groove, as were those of the anterior dorsal vertebrae. At least one pulley-like muscle or tendon may have occupied this groove as part of the neck- and head-lifting musculature. The spines of the vertebrae in the hip region were high, and the chevrons of the tail were much modified from the type of chevron characteristic of more primitive sauropods. Instead of projecting downward as simple spines, the chevrons of diplodocoids were modified with fore and aft expansions into parallel rods (figure 6.10). The tails of diplodocoids were very long and made up of at least 80 vertebrae. The last 30 to 40 of these vertebrae were little more than elongate rods that formed what has been called a "whiplash."

The limbs of diplodocoids were relatively slender, and their rib cages were deep and narrow. The forelimbs were fairly short for sauropods, with a typical humerus-to-femur ratio of 0.66. The number of bones in the wrist and ankle were much reduced: only one or two in the wrist and one (astragalus) in the ankle. Apparently, both the

**FIGURE 6.9**
The development of an extremely long neck culminated in *Mamenchisaurus*, from the Upper Jurassic of China. (© Scott Hartman)

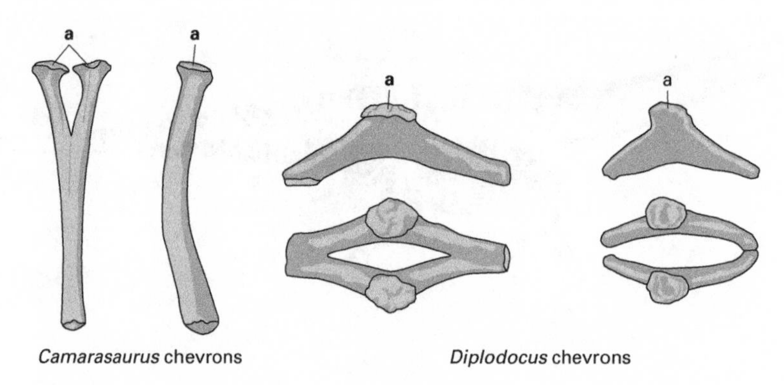


**FIGURE 6.10**
The normal chevrons of sauropod dinosaurs were shaped like "tuning forks" of bone directed downward. Those of diplodocids were modified to have fore-and-aft expansions that, in some posterior caudal vertebrae, became rod-like extensions. (Drawing by Network Graphics)

wrist and the ankle were heavily invested with cartilage, which absorbed the stresses produced by so heavy a creature.

*Diplodocus*, of course, is the typical and best-known diplodocoid (see figure 6.8). Its fossils are known from Upper Jurassic deposits in the western United States, as are those of the closely related diplodocoid, *Barosaurus*. *Barosaurus* has also been reported from eastern Africa. It had much longer neck vertebrae and a shorter tail than did *Diplodocus*.

Perhaps the most famous diplodocoid is *Apatosaurus*, also known as *Brontosaurus* (box 6.2). *Apatosaurus* was one of the bulkiest diplodocoids, with robust limb bones, heavy cervical ribs, and a single bone in its wrist.

## Box 6.2

### The *Brontosaurus* Business

*Brontosaurus* stands out as one of the most popular names for a dinosaur. Yet few grade-school children shrink from pointing out that *Brontosaurus* is not technically the correct name for this sauropod; *Apatosaurus* is. This is because scientists have, since 1903, agreed that *Apatosaurus*, named by Yale paleontologist Othniel Charles Marsh in 1877, and *Brontosaurus*, also named by Marsh, but in 1879, represent the same kind of sauropod dinosaur. When more than one name exists for the same genus, the internationally accepted rules for naming animals, living and extinct, force paleontologists to use the earliest proposed name (see box 2.1). The later name, in this case *Brontosaurus*, is then branded a synonym and abandoned.

Marsh's research also lies at the crux of another aspect of the *Brontosaurus* business, the question of what type of head this sauropod had. In 1883, Marsh produced the first reconstruction of a sauropod dinosaur, that of *Brontosaurus excelsus* (box figure 6.2). In so doing, he relied primarily on a fairly complete skeleton from a dinosaur quarry at Como Bluff, Wyoming. But, this skeleton lacked a head. To complete his reconstruction Marsh had to guess which type of head, from among the sauropod heads known to him, belonged to *Brontosaurus*. Unfortunately, he guessed incorrectly when he based the head in the reconstruction on two incomplete detached skulls, one from a dinosaur quarry east of Como Bluff and another from Garden Park, Colorado. So, *Brontosaurus* was presented to the public with a *Camarasaurus*-like skull, one with large, spatulate teeth.

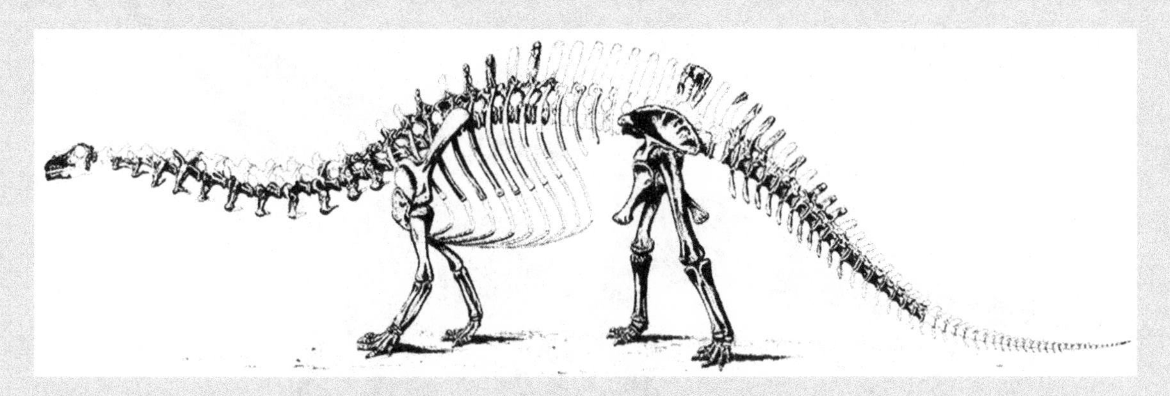

**BOX FIGURE 6.2**
Othniel Charles Marsh's classic reconstruction of the skeleton of *Apatosaurus* included a *Camarasaurus*-like head. Only 75 years later was the correct, *Diplodocus*-like skull placed on the body of *Apatosaurus*. (Courtesy of Yale Peabody Museum of Natural History, New Haven, Connecticut)

More than a decade after Marsh's death, between the years 1909 and 1915, the Carnegie Museum began to excavate the famous dinosaur quarry near Jensen, Utah, now known as Dinosaur National Monument. The most complete *Apatosaurus* known was one of the fruits of their labors, but it was also missing a head connected to the vertebral column. In 1916, this skeleton was named *Apatosaurus louisae* by W. J. Holland, the director of the Carnegie Museum, after Louise Carnegie, wife of the museum's patron, Andrew Carnegie. The *Apatosaurus louisae* skeleton was found lying on top of another smaller, headless *Apatosaurus* skeleton. Nearby lay part of the skeleton of a third sauropod thought to be a *Diplodocus*. Also among those bodies was a *Diplodocus*-like skull. Holland thought this skull was too large to belong to the supposed *Diplodocus* skeleton or to the smaller *Apatosaurus*. In his mind, it made sense to attach it to the skeleton of *Apatosaurus louisae*. Thus, Holland reasoned, Marsh had guessed wrong.

The skeleton of *Apatosaurus louisae* was mounted at the Carnegie Museum, and Holland planned to place the *Diplodocus*-like skull on it. But, Henry Fairfield Osborn, one of the most eminent vertebrate paleontologists of the time, believed that Marsh was right and "convinced" Holland not to do so. Or did he? Holland left the *Apatosaurus louisae* skeleton on display without a head for more than 20 years. Only after Holland's death in 1932 was a head placed on the skeleton—a copy of the *Camarasaurus*-like skull originally chosen by Marsh!

Holland, as the saying goes, must have rolled over in his grave. But, in the late 1970s he was vindicated. Further preparation of the skeletons from the Utah quarry showed that the skeleton Holland thought was a *Diplodocus* was actually another *Apatosaurus*. Three *Apatosaurus* skeletons in the same quarry as a *Diplodocus*-like skull provides very strong evidence that the skull belonged to *Apatosaurus*, too. Indeed, modern studies of sauropods have listed numerous similarities shared by the skeletons of *Diplodocus* and *Apatosaurus*, making it no surprise that their skulls would also be similar. So, the *Camarasaurus*-like skulls that rode the necks of *Apatosaurus* skeletons for nearly 100 years came off and were replaced by *Diplodocus*-like skulls with rod-shaped teeth.

**FIGURE 6.11**
*Camarasaurus*, one of the best known primitive macronarians, would have reached a length of as much as 18 meters. (© Scott Hartman)

## Primitive Macronarians

***Camarasaurus*** is the best known of all sauropods, known from dozens of skulls and many skeletons. It is a characteristic primitive macronarian and thus provides a remarkable contrast to the diplodocoids (figure 6.11). This is most obvious from its skull (see figure 6.7). That of *Camarasaurus* was short and heavy with a blunt snout. The jaws bore relatively large, spatulate teeth along their entire length. The nostrils were large (hence the name **Macronaria**, which means "large nose") and located on the sides of the skull, just in front of the eyes.

The body of *Camarasaurus* was solidly built, but neither exceptionally long nor overly heavy. There were only 12 neck vertebrae (the number found in primitive sauropods), and a U-shaped trough in the neural spines extended along the first four dorsal vertebrae. The spines of the sacral vertebrae were lower and thicker than in diplodocoids. In the neck and back vertebrae, the pleurocoels were very extensive. The tail chevrons were of the standard sauropod type.

*Camarasaurus* had limbs that were stout, and the typical humerus-to-femur ratio was 0.70 or more. So, the forelimbs were relatively longer than in diplodocoids. The wrist and ankle each had two bones, and the forefoot was elongated. Besides *Camarasaurus*, the best-known primitive macronarians are its compatriot *Haplocanthosaurus*, from the Upper Jurassic of the western United States, and *Brachiosaurus*, a behemoth best known from the Upper Jurassic of Tanzania (figure 6.12).

In features of the skull and teeth, *Brachiosaurus* closely resembled *Camarasaurus*. But, in *Brachiosaurus*, the forelimb was relatively long—the humerus-to-femur ratio

**FIGURE 6.12**
*Brachiosaurus* (called *Giraffatitan* by some paleontologists) was one of the heaviest land animals of all time, weighing as much as 50 tons and measuring 22 meters from snout to tail tip. (© Scott Hartman)

was greater than 1.0. Thus, the shoulders of *Brachiosaurus* were positioned higher than its hips. Although there were only 13 neck vertebrae and 11 or 12 vertebrae in the back, those of the neck were quite elongate, resulting in a very long neck. The neural spines were all undivided, and those in the sacral region were low. Although there were about 50 vertebrae in the tail, each vertebra was short, which made for a relatively short tail. The chevrons were simple, and the hand bore a very short claw on its first digit. *Brachiosaurus*, from the Upper Jurassic of the western United States, Tanzania, and Portugal, was not even as long as *Apatosaurus*, but it was at least twice as heavy and could raise its head 12 meters off the ground.

## Titanosaurs

**Titanosauria** encompasses about 50 species of derived macronarians. These sauropods are identified by their procoelous anterior tail vertebrae, widened sacrum, short and robust limbs, and short forefeet, among other features (figure 6.13). Furthermore, the shape of a titanosaur's humerus and femur indicate that these bones were bowed out (not held strictly vertical) while walking.

**FIGURE 6.13**
This composite skeleton of *Alamosaurus*, from the Upper Cretaceous of North America, is of a relatively well-known titanosaur. (© Scott Hartman)

One titanosaur, *Saltasaurus*, from the Cretaceous of Argentina, had body armor (figure 6.14). So it is reasonable to assume that at least some other titanosaurids were armored. Most titanosaurs were of Cretaceous age and inhabited Argentina (*Antarctosaurus* and *Saltasaurus*), India (*Titanosaurus*), and other parts of the Mesozoic southern supercontinent called Gondwana. A rare exception is *Opisthocoelicaudia*, a titanosaur from the Upper Cretaceous of Mongolia.

In North America, the last known sauropod of the Late Cretaceous is ***Alamosaurus***. This fairly large titanosaur is known from just a few teeth (of the diplodocoid type), a shoulder blade and forelimb, some pelvic bones, a femur, a foot, some cervical and dorsal vertebrae, and a tail, yet it is one of the better known titanosaurs. It almost certainly emigrated from South America into North America at about the end of the Cretaceous (box 6.3).

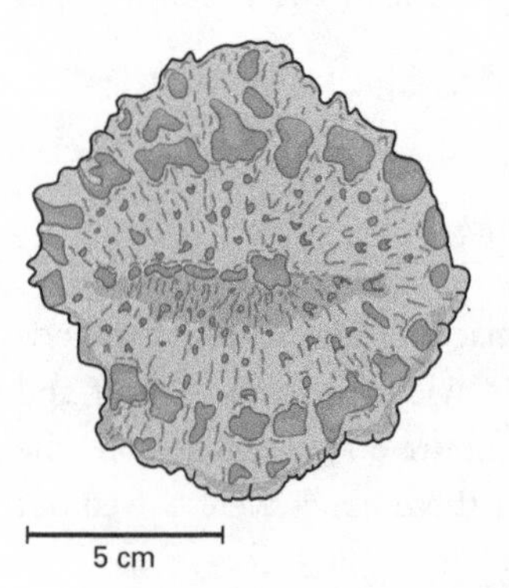


**FIGURE 6.14**
Armor plates like this one adorned the back of the titanosaurian *Saltasaurus* and may have characterized all titanosaurians. (Drawing by Network Graphics)

## Box 6.3

### The Sauropod Hiatus

It is a curious fact that sauropod dinosaurs have a temporally disjunct distribution in the western United States during the Cretaceous. In Lower Cretaceous strata, sauropod bones and footprints are found in Wyoming, Texas, Oklahoma, and Arkansas. Thereafter, sauropod fossils are absent for about 30 million years, reappearing solely in the form of the titanosaurid *Alamosaurus*, in the uppermost Cretaceous deposits of Texas, New Mexico, and Utah. Two explanations have been offered to explain this 30-million-year absence, the **sauropod hiatus**.

The first explanation is that sauropods were not absent in western North America during the hiatus but were actually there all along. The environments the sauropods lived in during the hiatus have either not been preserved in the rock record or have not yet been sampled by paleontologists. So, sauropods may have lived in the western United States during the hiatus, but their remains either were not fossilized or are still to be discovered.

The second explanation posits the extinction of sauropods in western North America at about the end of the Early Cretaceous, about 100 million years ago. Thirty million years later, *Alamosaurus* arrived from elsewhere, probably from South America, where sauropods prospered throughout the Cretaceous.

The balance of evidence favors the second explanation. Particularly important is the fact that there are some very well-known dinosaur faunas in the western United States during the 30-million-year-long hiatus, representing a variety of environments from which no trace of a sauropod fossil is known. Also significant is the fact that the last North American sauropod, *Alamosaurus*, is a titanosaurian closely related to South American Cretaceous sauropods. Although it does seem reasonable to believe that in the western United States, sauropods became extinct at the end of the Early Cretaceous and invaded 30 million years later (box figure 6.3), the reasons for their extinction and subsequent invasion remain unknown.

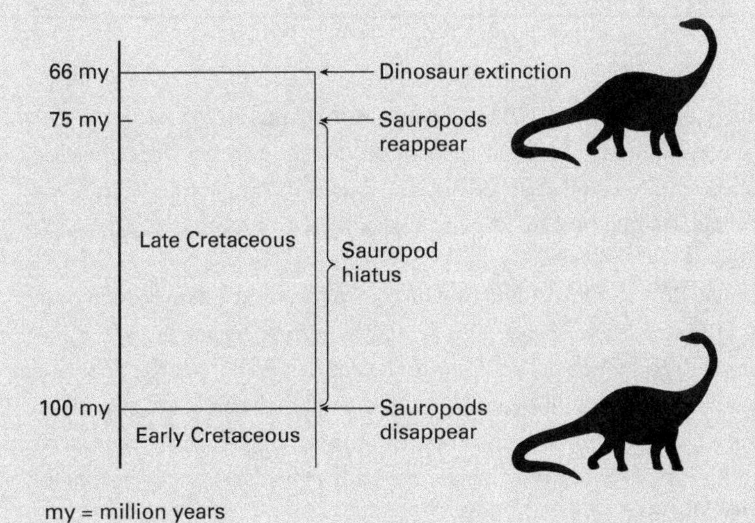


**BOX FIGURE 6.3**
The sauropod hiatus lasted for 30 million years. (Drawing by Network Graphics)

## How Large Was the Largest?

Recent discoveries of huge sauropods have increased the known limits of body size in terrestrial vertebrates and raised the question, which was the largest sauropod? (box 6.4). When using the word "large," however, it is important to distinguish between weight (body mass) and length. As the previous discussion indicates, some of the longest sauropods were relatively lightly built and thus generally lighter than the shorter, more robust sauropods such as *Brachiosaurus*.

Some recent candidates for the award for largest, in terms of weight, are *Antarctosaurus*, a titanosaur from the Cretaceous of Argentina estimated to have weighed up to 69 tons; *Argentinosaurus*, also from the Argentine Cretaceous, estimated at up to 73 tons; *Brachiosaurus*, weighing in at up to 78 tons; *Paralititan* from the Upper Cretaceous of Egypt and *Dreadnoughtus* from the Upper Cretaceous of Argentina, both weighing up to 59 tons; and *Sauroposeidon*, from the Lower Cretaceous of Oklahoma, weighing as much as 60 tons. For perspective, a Boeing 737 airplane weighs about 50 tons, and an African elephant weighs about 5 tons.

Other than *Brachiosaurus*, none of these huge sauropods are known from a complete skeleton, so their weight estimates are based on various extrapolations. Yet, even based on complete skeletal material, published weight estimates of *Brachiosaurus* range from 16 to more than 78 tons, a nearly fivefold difference. This shows how problematic dinosaur weight estimates can be. Different methods can yield very different estimates. About 50 tons seems to be the most reliable upper limit for known humongous sauropods (see box 6.4).

### Box 6.4

### The Largest Sauropod?

In 1878, Edward Drinker Cope described *Amphicoelias fragillimus*, long considered to be the largest sauropod ever. He based the new dinosaur species on parts of a huge vertebra and femur from the Upper Jurassic Morrison Formation in Colorado. The vertebra was only part of the neural arch (box figure 6.4), but Cope reported it to be a whopping 1.5 meters tall, leading him to estimate a total height of more than 1.8 meters for the dorsal vertebra that it represented.

In the 1890s, Cope sold his fossil collection to the American Museum of Natural History, including the bones of *Amphicoelias fragillimus*. However, these bones disappeared by 1900, perhaps discarded because they were very fragile and broken. Fortunately, the lost dinosaur has remained part of the scientific literature. Estimates of its size based on the data and illustration in Cope's original article are of a 58- to 60-meter-long monster weighing as much as 150 tons. For comparison, most Late Jurassic sauropods weighed an estimated 7 to 26 tons. And, a typical living blue whale, generally considered the largest of all animals, is about 24 meters long and weighs in at about 190 tons.

Most paleontologists have accepted the colossal size of *Amphicoelias fragillimus*, even though this makes it a sauropod three to four times larger than any contemporary Late Jurassic sauropod. Furthermore, the size estimates of *Amphicoelias fragillimus* are of a sauropod much heavier than the 75 tons considered to be the maximum weight limit of any terrestrial organism, based on calculations using biomechanical and gravitational constraints.

These considerations encouraged a recent re-evaluation of *Amphicoelias fragillimus* by paleontologists Cary Woodruff and John Foster. They suggest that Cope made a typographical error, reporting the preserved portion of the vertebra as 1.5 meters tall, when it was more likely 1.05 meters tall. That still makes it the largest Late Jurassic sauropod, but one at least one-third smaller than had been estimated.

Of course, without the fossil and without Edward Drinker Cope (who died in 1897), we will never know whether the measurement Cope published was correct or a typo. What we do know is that no sauropod that approaches the size of *Amphicoelias fragillimus* has been discovered since 1878 and that the size estimates based on Cope's publication seem implausible.

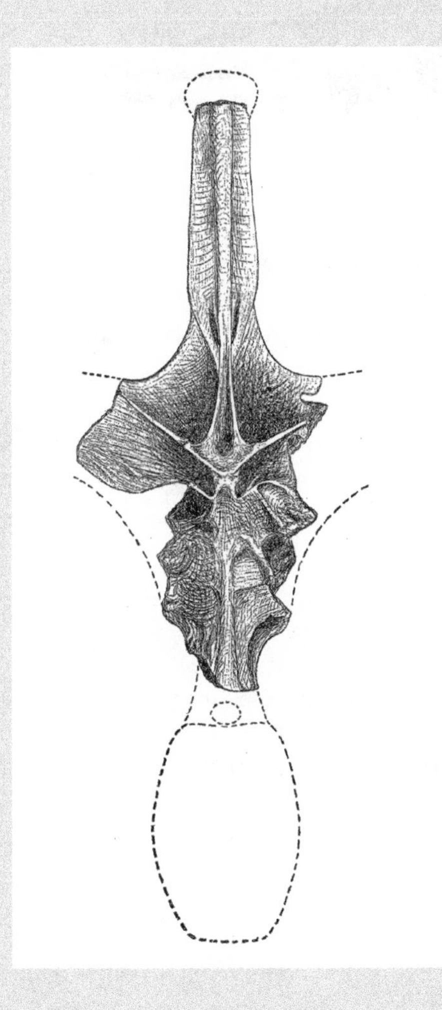

**BOX FIGURE 6.4**
Edward Drinker Cope's original drawing of part of a vertebra of *Amphicoelias fragillimus*. (From E. D. Cope. 1878. A new species of *Amphiocoelias*. *American Naturalist* 12:563–564)

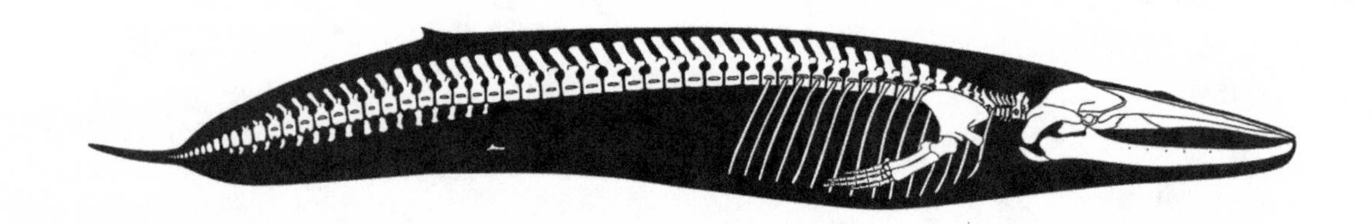

**FIGURE 6.15**
At 50 or more tons, *Brachiosaurus* (or *Giraffatitan, left*), *Puertasaurus* (*center*), and *Argentinosaurus* (*right*) were the heaviest dinosaurs and would have dwarfed a human and a living elephant but still were significantly lighter than a blue whale. (© Scott Hartman)

Length estimates based on incomplete skeletons are also tricky. The longest estimates that may be reliable are 35 meters for *Mamenchisaurus*, 36 meters for *Argentinosaurus*, and 39 meters for *Turiasaurus*, from the Upper Jurassic of Spain. Length estimates as long as 58 meters were made for "*Seismosaurus*," a synonym of *Diplodocus*, but did not stand up to critical scrutiny.

At present, conservative estimates of the upper limits of known sauropod size are about 50 tons for an extremely heavy *Brachiosaurus* and about 40 meters long for a very elongate *Turiasaurus* (figure 6.15).

## Sauropod Lifestyles

The enormous size and the unusual body shape (especially the long neck and minuscule head) of sauropod dinosaurs have led to much speculation about their way of life. Particularly significant topics for discussion are sauropod diet, metabolism, locomotion, habitat preferences, and reproduction, as well as evidence of social behavior among sauropods.

### *Diet*

The two primary types of sauropod teeth, well represented by the peg-like teeth of diplodocoids and the spatulate teeth of *Camarasaurus*, appear to have been well

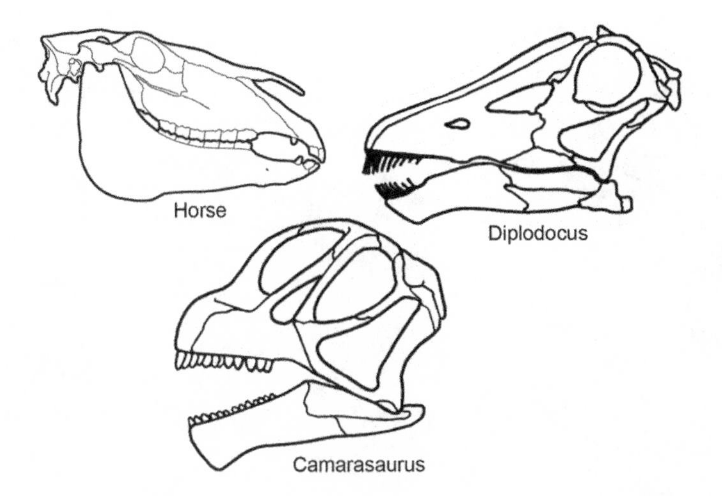


**FIGURE 6.16**
The teeth of sauropods probably functioned like the incisors of living horses and were used to crop vegetation. (© Scott Hartman)

suited to cropping vegetation in much the same way that a living horse uses its incisors (figure 6.16). Some sauropod teeth are found worn, but sauropods clearly had no grinding or shearing teeth for smashing and pulping vegetation as did the duck-billed dinosaurs, with their dental batteries, or living horses, with their premolars and molars. Sauropods must have swallowed vegetation with little chewing and relied on chemical or bacterial mechanisms in the gut to break it down.

Some paleontologists have argued that the rod-like teeth of the diplodocids were used as a sieve to collect algae and invertebrates. This is plausible, though it would necessitate an incredible supply of algae and invertebrates for the diplodocids to feed on. Sauropod cropping of different kinds of plants, quickly swallowed and broken down in the gut chemically or bacterially, seems most plausible. Assuming modern nutritional values for the vegetation of the Mesozoic, it has been estimated that a 29-ton sauropod would have had to eat 50 kilograms of vegetation each day!

## *Metabolism*

The 50-kilogram-per-day calculation is based on the notion that sauropods had metabolic rates comparable to those of living reptiles. Of course, if they had higher metabolic rates, comparable to those of living mammals, they would have had to eat more. There is no conclusive evidence, however, that this was so (see chapter 14), so it is reasonable to conclude that they had an ectothermic ("cold-blooded") reptilian metabolism, with one important difference.

This difference is evident considering the great bulk of sauropods; their massiveness would have made them what physiologists term inertial homeotherms (also called gigantotherms). In other words, sauropods would have maintained an essentially constant body temperature (homeothermy) because of the resistance to heat transference inherent in their large mass (thermal inertia). Thus, by virtue of its size alone, once it reached its ideal body temperature, a sauropod would have retained that temperature much longer than a smaller animal. Of course it would also have taken much longer to reach that temperature in the first place!

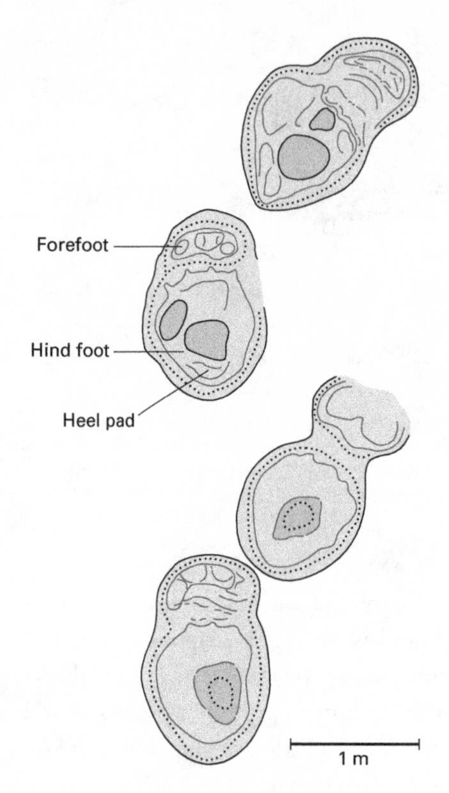


**FIGURE 6.17**
Sauropod footprints like these from Morocco indicate that sauropods walked on all fours and that their heels were supported by thick, fleshy pads, as are the heels of living elephants. (Drawing by Network Graphics)

## *Locomotion and Habitat Preferences*

The nearly equal length of the fore and hind limbs; the thick, solid limb bones; the massive pelvic and shoulder girdles; and the short, broad feet of sauropods indicate that they were obligatory quadrupeds. Sauropod foot structure indicates that they walked on the tips of their front toes with their hind-foot heels resting on large pads, similar to living elephants. Sauropod footprints confirm these conclusions (figure 6.17). The limb structure of sauropods is what anatomists term **graviportal**, meaning designed to bear great weight on land.

The oldest idea about sauropod lifestyle is that they were aquatic. For nearly a century, paleontologists believed that sauropod limbs were not strong enough to support their bulky bodies, that the "weak" teeth of sauropods could be used only to eat soft, aquatic vegetation, and that the nostrils on the top of some sauropod skulls functioned as snorkels, allowing the animal to breathe while most of the head was submerged. The image of a sauropod dinosaur standing in deep water, its head emergent like the periscope of a submarine, is classic (figure 6.18). Indeed, one paleontologist even went so far as to conclude that the differing neck lengths and limb proportions indicated habitation of different depths of water. According to this scheme, *Brachiosaurus* lived in water depths of about 8 meters, *Diplodocus* in depths of 4 to 5 meters, and *Apatosaurus* in shallow depths of 2 to 3 meters.

A modern view of sauropod habitat preferences has resulted from a reappraisal of sauropod anatomy. This reappraisal demonstrates that there is little, if any, evidence

**FIGURE 6.18**
Classic reconstructions of sauropods (drawn by paleontologist Edward Drinker Cope and published in 1897) show them submerged in water. Paleontologists now envision sauropods as tree-top browsers that spent most of their lives on dry land.

to support the idea of aquatic sauropods. Indeed, it has been estimated that water pressure would have prevented expansion and contraction of a sauropod's lungs if the dinosaur were submerged. The massive limbs of sauropods could certainly have supported their huge bodies; the graviportal limb skeletons of sauropods seem to have been designed to do just that. The teeth of sauropods were as robust as a horse's incisors and could easily have cropped tough vegetation. And, the nasal openings on top of some sauropod heads were remarkably like those of living tapirs, land-living mammals that have a short trunk (proboscis) like a miniature version of an elephant's trunk. The nostrils far back on a tapir's skull are an outgrowth of the presence of a proboscis. Perhaps some sauropods, most likely the diplodocoids, had a tapir-like proboscis. The evidence thus seems to support a view of sauropods as land-living treetop browsers analogous to living giraffes and elephants.

## *Reproduction*

Like prosauropods, sauropod dinosaurs were egg layers. Sauropod eggs are well known from the Jurassic of the United States and Portugal and from the Cretaceous of the United States, Argentina, France, Spain, Romania, and India. Titanosaur embryos in eggs are known from the Upper Cretaceous of Argentina. At Aix-en-Provence in France, numerous eggs as much as 25 centimeters long, with bumpy shell textures, have been attributed to the titanosaur ***Hypselosaurus***.

## Social Behavior

Although discussed at greater length in chapter 13, it is appropriate to mention the circumstantial evidence for social behavior in sauropods. Many of the great dinosaur quarries of the Late Jurassic, such as those at Como Bluff, Wyoming, are populated mostly by sauropod skeletons. These mass death assemblages of sauropods may indicate group (gregarious) behavior among these giant reptiles, but some of the dinosaur quarries, such as the one at Dinosaur National Monument, Utah, are populated mostly by river-transported carcasses that accumulated in sand banks. The evidence of group behavior among sauropods provided by the mass death assemblages is thus not totally convincing.

Sauropod trackways also support, to some extent, the notion of gregarious behavior among sauropods. For example, Early Cretaceous footprints from near Glen Rose, Texas (figure 6.19), seem to document a group of sauropods walking together in the

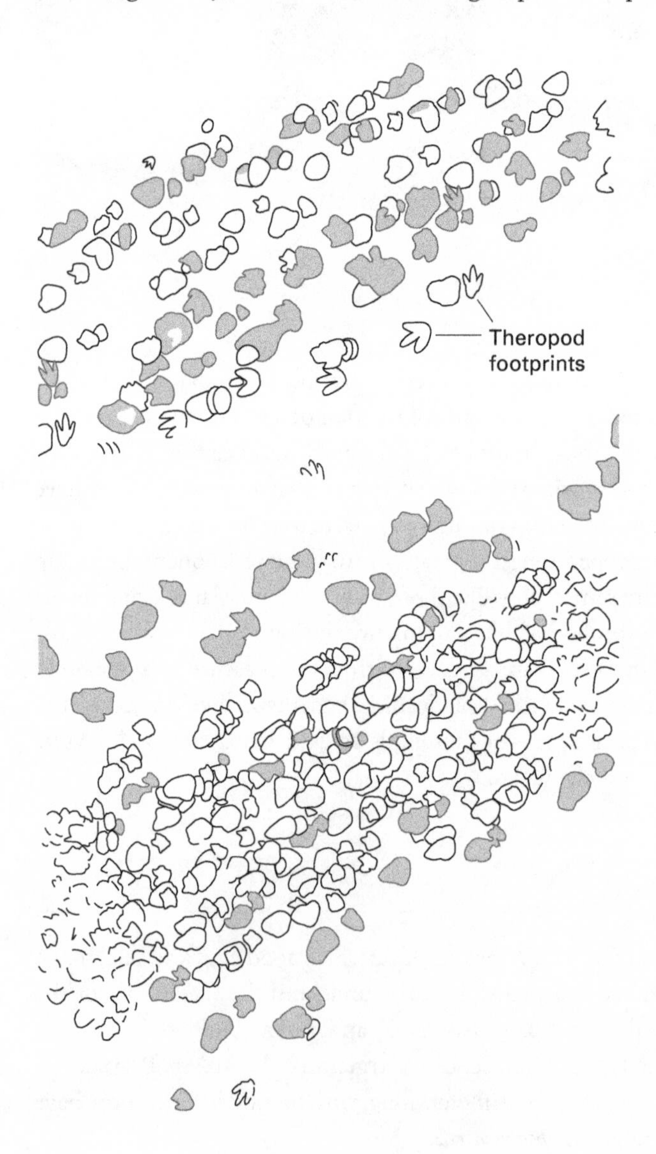


**FIGURE 6.19**
These sauropod trackways, from the Lower Cretaceous of Texas, are taken by some paleontologists as evidence of group behavior among sauropods. (Drawing by Network Graphics)

same direction. But, to call this group a herd, which implies a complex social structure, may be overly generous. Not only is there no unequivocal evidence that all these sauropods walked together at one time, but the distribution of track sizes, which would indicate the presence and number of sauropods of different ages, is scattered, which suggests that no orderly movement of young and old was involved.

The evidence for sauropod group behavior is at best circumstantial. Mass death assemblages and multiple trackways suggest some sort of group behavior among sauropods, but little else can be concluded at present.

## Sauropod Evolution

Sauropod dinosaurs evolved from a prosauropod-like ancestry late in the Triassic. Although no known prosauropod is a strong candidate for sauropod ancestry (unless evolutionary reversals be accepted), *Riojasaurus*, from the Late Triassic of Argentina, or *Vulcanodon*, from the Early Jurassic of southern Africa, may best approximate that ancestor.

The oldest sauropods are of Late Triassic and Early Jurassic age and include *Barapasaurus* from India, *Isanosaurus* from Thailand, and *Vulcanodon* from southern Africa. They and Middle Jurassic sauropods, such as *Datousaurus* and *Shunosaurus* (China), *Cetiosaurus* (England), and *Patagosaurus* (Argentina), were long included in a group of primitive sauropods nominally termed cetiosaurids. This "family" is clearly a grade level (not a clade) in the evolution of sauropods characterized by relatively small (up to 12 meters long), short-necked forms with *Camarasaurus*-like heads and teeth. It is interesting that these large animals appear suddenly in the fossil record with most of the graviportal limb specializations we associate with sauropods. Their geographic distribution suggests a likely origin of sauropods in the southern supercontinent of Gondwana.

The zenith in size and diversity of sauropods was achieved during the Jurassic–Cretaceous transition. The terrestrial giants had a worldwide distribution by this time, and *Diplodocus*, *Brachiosaurus*, and *Camarasaurus* well represent their structural diversity. In terms of size, diversity, and sheer abundance, sauropod dinosaurs were the dominant herbivores on land during the Jurassic–Cretaceous transition.

Sauropod fortunes seem to have waned during the Cretaceous, when there were fewer and mostly smaller animals. It is tempting to suggest that the rise of ornithopod dinosaurs, potential herbivorous competitors of sauropods, and the appearance and evolution of flowering plants during the Cretaceous, may have contributed to or resulted from the decline in the sauropods. Sauropods were plant eaters, but they may not have been well adapted to eating flowering plants. Particularly significant is that most sauropod diversity during the Cretaceous was limited to the southern continents where ornithopods were not very successful. The extinction of sauropods coincided with the extinction of the dinosaurs at the end of the Cretaceous. Though reduced in diversity and numbers, sauropods can truly be said to have met the end with "their heads held high."

## Summary

1. Sauropodomorph dinosaurs comprise two closely related groups, prosauropods and sauropods, distinguished from other dinosaurs by having had small heads, spatulate teeth, long necks, short feet, and large claws on the first digits of their forefeet.
2. No known prosauropod is a suitable ancestor of sauropods, although *Riojasaurus*, from the Upper Triassic of Argentina, may best approximate that ancestry.
3. Prosauropods lived during the Late Triassic and Early Jurassic and so are some of the oldest dinosaurs, representing the first evolutionary diversification of plant-eating dinosaurs.
4. Sauropod dinosaurs first appeared in the Late Triassic, and by the Late Jurassic they were the dominant very large plant-eating dinosaurs.
5. At least four types of sauropods can be recognized in the Jurassic–Cretaceous: primitive eusauropods, diplodocoids, primitive macronarians, and titanosaurs.
6. The heaviest known sauropod was *Brachiosaurus*, weighing up to 50 tons, and the longest sauropod was *Turiasaurus*, with a length of 39 meters.
7. Sauropods were plant eaters, egg layers, and inertial homeotherms. There is no evidence to support the long popular notion of aquatic sauropods. Sauropods had elephant-like limb structures and probably spent most of their time on dry land.
8. By the Jurassic–Cretaceous transition, sauropods had reached their evolutionary zenith in terms of size and diversity.
9. Sauropods were much less successful after the Jurassic but did survive until the end of the Cretaceous.

## Key Terms

*Alamosaurus*
*Amphicoelias*
amphicoelous
*Apatosaurus*
*Brachiosaurus*
*Brontosaurus*
*Camarasaurus*
centrum
chevron
diplodocoid
*Diplodocus*
eusauropod
graviportal
*Hypselosaurus*
Macronaria
*Mamenchisaurus*
*Mussaurus*
neural arch
opisthocoelous
*Plateosaurus*
pleurocoel
procoelous
prosauropod
*Riojasaurus*
sauropod
sauropod hiatus
Sauropodomorpha
*Seismosaurus*
Titanosauria
*Vulcanodon*

## Review Questions

1. Why do most paleontologists exclude any known prosauropod from the ancestry of sauropods?
2. What characteristic features of prosauropods are exemplified by *Plateosaurus*?

3. What anatomical evidence supports the notion that prosauropods and sauropods were herbivores?
4. Distinguish diplodocoid from macronarian sauropods.
5. What other groups of sauropods are recognized by paleontologists, and what are their salient characteristics?
6. Describe the vertebrae of sauropods, and explain how they are relevant to sauropod classification.
7. How was the confusion about the name and the skull of *Brontosaurus* resolved?
8. Why did paleontologists believe sauropods were aquatic, and why do they no longer hold this view?
9. Review the evolutionary history of the sauropodomorph dinosaurs.

## Further Reading

Berman, D. S., and J. S. McIntosh. 1978. Skull and relationships of the Upper Jurassic sauropod *Apatosaurus* (Reptilia, Saurischia). *Bulletin of the Carnegie Museum of Natural History* 8:1–35. (An in-depth technical description of the skull of *Apatosaurus* and related sauropods as well as an authoritative review of the "*Brontosaurus* business")

Coombs, W. P., Jr. 1975. Sauropod habits and habitats. *Palaeogeography, Palaeoclimatology, Paleoecology* 17:1–33. (A thorough review of the evidence for aquatic habits among sauropods and the key article debunking this notion)

Curry Rogers, K. A., and J. A. Wilson. 2005. *The Sauropods: Evolution and Paleobiology*. Berkeley: University of California Press. (A collection of 11 articles on diverse sauropod research)

Desmond, A. J. 1976. *The Hot-Blooded Dinosaurs*. New York: Dial. (Chapter 5, pp. 100–133, recounts the history of ideas on sauropod behavior, habitat preferences, and the rise and demise of the idea of aquatic sauropod)

Fowler, D. W., and R. M. Sullivan. 2011. The first giant titanosaurian sauropod from the Upper Cretaceous of North America. *Acta Palaeontologica Polonica* 56:685–690. (Giant bones of *Alamosaurus* suggest a sauropod as large as *Argentinosaurus*)

Klein, N., K. Remes, C. T. Gee, and P. M. Sander, eds. 2011. *Biology of the Sauropod Dinosaurs: Understanding the Life of Giants*. Bloomington: Indiana University Press. (A collection of 18 articles on diverse aspects of sauropod biology)

Lavocara, K. J., et al. 2014. A gigantic, exceptionally complete titanosaurian sauropod dinosaur from southern Patagonia, Argentina. *Nature Scientific Reports* 4:6196. (Description of a new behemoth, *Dreadnoughtus*)

Lucas, S. G., and A. P. Hunt. 1989. *Alamosaurus* and the sauropod hiatus in the Cretaceous of the North American Western Interior. *Geological Society of America Special Papers* 238:75–85. (Develops and evaluates explanations for the sauropod hiatus)

Ostrom, J. H., and J. S. McIntosh. 1966. *Marsh's Dinosaurs: The Collections from Como Bluff.* New Haven, Conn.: Yale University Press. (A thorough review of the collections from one of the world's great Jurassic–Cretaceous dinosaur quarries, including publication of 65 lithographs prepared for Othniel Charles Marsh of *Diplodocus*, *Apatosaurus*, and *Camarasaurus*, among others)

Rainforth, E. C. 2003. Revision and re-evaluation of the Early Jurassic dinosaurian ichnogenus *Otozoum*. *Palaeontology* 46:803–808. (Study of early prosauropod footprints)

Sander, P. M., et al. 2011. Biology of the sauropod dinosaurs: The evolution of gigantism. *Biological Reviews* 86:117–155. (Detailed review of the biology of sauropod gigantism)

Wilson, J. A., and K. A. Curry Rogers. 2012. Sauropoda, pp. 445–481, in M. K. Brett-Surman, T. R. Holtz, Jr., and J. O. Farlow, eds., *The Complete Dinosaur*. 2nd ed. Bloomington: Indiana University Press. (Concise review of the sauropods)

Woodruff, D. C., and J. R. Foster. 2014. The fragile legacy of *Amphicoelias fragillimus* (Dinosauria: Sauropoda; Morrison Formation, latest Jurassic). *Volumina Jurassica* 12:211–220. (Argues that the huge size estimates for *Amphicoelias fragillimus* may have been based on a typographical error)

Yates, A. M. 2012. Basal sauropodomorpha, pp. 425–433, in M. K. Brett-Surman, T. R. Holtz, Jr., and J. O. Farlow, eds., *The Complete Dinosaur*. 2nd ed. Bloomington: Indiana University Press. (Concise review of the prosauropods)

## Find a Dinosaur!

Where can you find a giant sauropod dinosaur? Perhaps the most spectacular sauropod fossil in public view is the skeleton of *Brachiosaurus* (some call it *Giraffatitan*) on display in the Museum für Naturkunde (Berlin, Germany). Collected in the early twentieth century from Upper Jurassic rocks in eastern Africa, this monster towers almost 13 meters above the museum floor and is advertised as the tallest dinosaur skeleton on display anywhere. Closer to home, the American Museum of Natural History (New York) exhibits skeletons of *Apatosaurus* and *Barosaurus*. In Washington, D.C., the Smithsonian's National Museum of Natural History has *Camarasaurus* and *Diplodocus* skeletons on display. The Carnegie Museum of Natural History (Pittsburgh, Pennsylvania) has the renowned original skeleton of *Diplodocus carnegii*, named after the museum's benefactor. And, don't forget the infamous skeleton of "*Brontosaurus*" (with the correct head) that is the centerpiece of the Great Hall of Dinosaurs at the Yale Peabody Museum of Natural History (New Haven, Connecticut). In California, the Natural History Museum of Los Angeles County has the long-necked Chinese sauropod *Mamenchisaurus* in its exhibit hall. And, way down south, in Buenos Aires, Argentina, you can see a mounted skeleton of *Argentinosaurus* (not all of it real), touted by some as the largest of all sauropods, at the Museo Argentino de Ciencias Naturales Bernadino Rivadavia. Closer to home, you can see a replica of the *Argentinosaurus* skeleton on display at the Fernbank Museum of Natural History (Atlanta, Georgia).

# 7

# ORNITHOPODS

ORNITHOPODS (bird feet) were bipedal ornithischian dinosaurs that lacked body armor (see box 5.1). They first appeared during the Middle Jurassic and were among the last dinosaurs to become extinct at the end of the Cretaceous. Their fossils are found on all continents. Ornithopods were thus one of the longest-lived dinosaur groups, and they were also one of the most diverse.

Several distinctive clades of ornithopods have been recognized (figure 7.1), from the primitive ornithopods to the enormous iguanodontians and their close relatives, the hadrosaurids (commonly referred to as the duck-billed dinosaurs). Many features of the skull were evolutionary novelties of the ornithopods, including the cranial kinesis (joints within the skull), the ventrally offset premaxillary tooth row (offset with respect to the maxillary tooth row), the closed or reduced external mandibular fenestra, the lower jaw joint location well below the level of the maxillary and lower tooth rows, and the slender process on the ischium. In this chapter, I review the anatomy and evolution of the ornithopods.

## HETERODONTOSAURS

Paleontologists disagree about the phylogenetic position of the heterodontosaurs. They are variously considered very primitive ornithopods, primitive ornithischians, or related to marginocephalians. Although discussed here with ornithopods, we regard them as primitive ornithischians.

Heterodontosaurs are best known from the Lower Jurassic of southern Africa. These small (about 1 to 1.5 meters long) dinosaurs had uniquely shaped cheek teeth that were chisel shaped with tiny cusps (denticles) restricted to the apex of the crown (figures 7.2 and 7.3). The name "heterodontosaur," which means "different-toothed lizard," is based

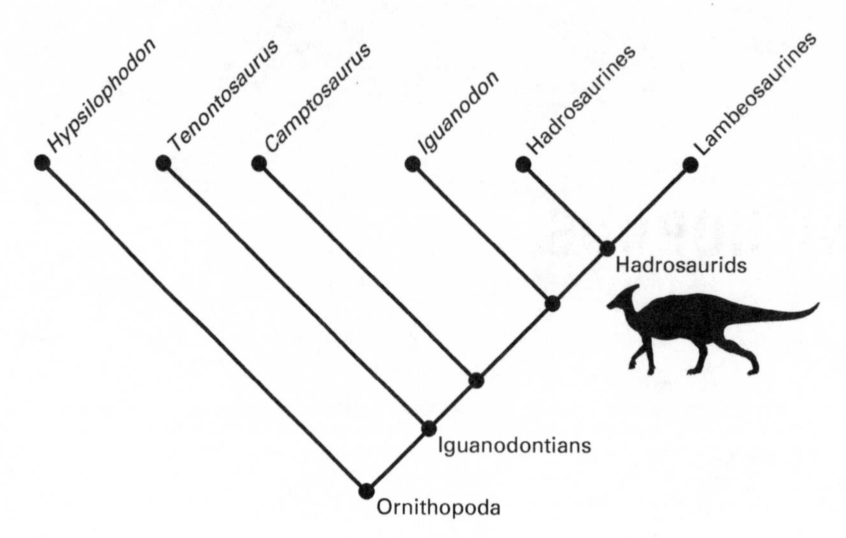


**FIGURE 7.1**
This cladogram of the ornithopods divides them into several clades. (Drawing by Network Graphics)

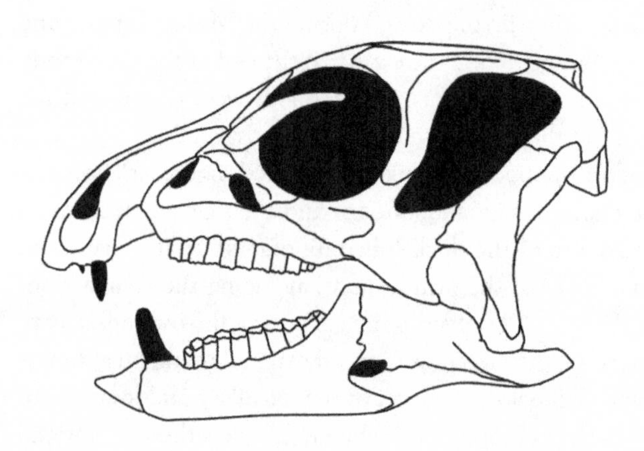

**FIGURE 7.2**
This skull of *Heterodontosaurus* well represents a primitive ornithischian. (© Scott Hartman)

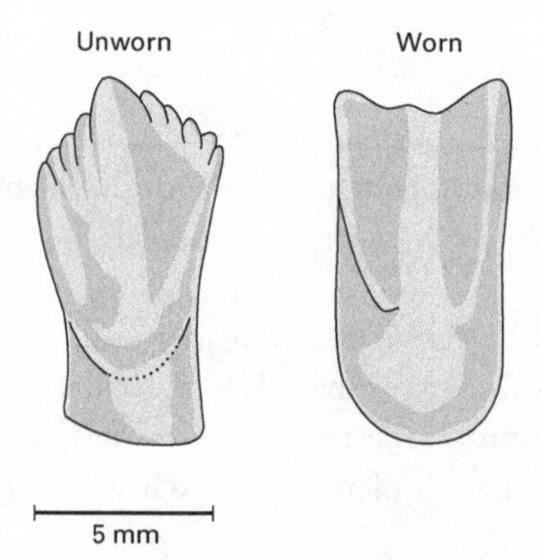


**FIGURE 7.3**
The chisel-shaped cheek teeth of heterodontosaurs are unique among dinosaurs. (Drawing by Network Graphics)

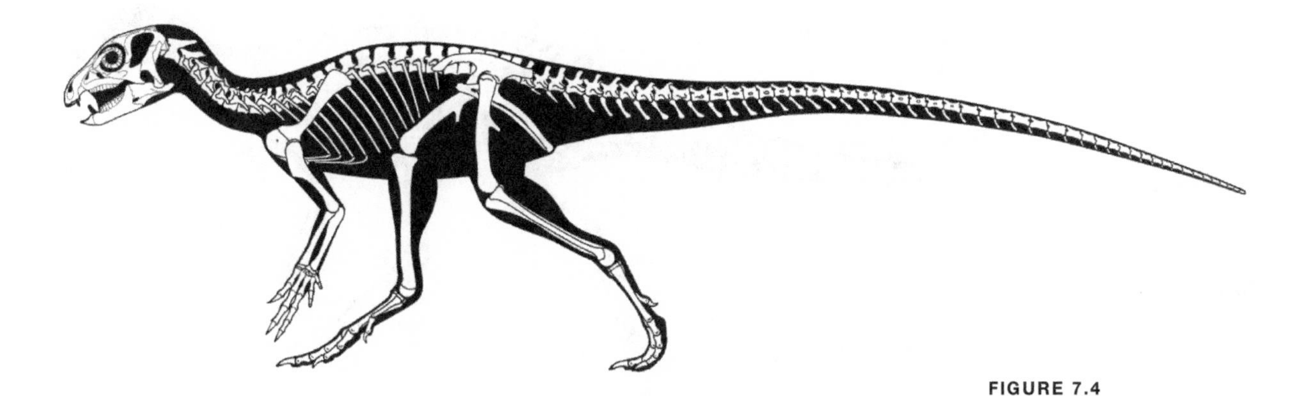

**FIGURE 7.4**
*Heterodontosaurus*, from the Lower Jurassic of southern Africa, was a typical heterodontosaur. (© Scott Hartman)

on the canine-like teeth in the front of the mouth (see figure 7.2), which had a markedly different shape from the chisel-shaped cheek teeth.

***Heterodontosaurus*** was a characteristic heterodontosaur (figure 7.4). This 1.2-meter-long dinosaur had short forelimbs and relatively long hind limbs, indicating it was a biped. The cheek teeth formed a **dental battery** set in massive lower jaws hinged to a solidly built skull. The teeth in this dental battery were usually heavily worn. The heavy wear and the chisel shape of the teeth indicate this dinosaur was a plant-eater. Paleontologists disagree as to how *Heterodontosaurus* moved its jaws when chewing; either the lower jaw was moved backward and forward, or it was rotated against the upper jaw when the mouth closed.

The long tail and short neck of *Heterodontosaurus* were features typical of primitive ornithischians. Ossified tendons were present only in the back region. A very bird-like feature of the hind-limb skeleton of *Heterodontosaurus* was the fusion of the tibia and fibula and their fusion to the tarsals. This fusion stabilized the lower leg and ankle and suggests that *Heterodontosaurus* was a fast runner. In contrast, the hands of *Heterodontosaurus* were stout and flexible; they may have been used to dig up or grasp vegetation.

Although *Heterodontosaurus* was a plant-eater, it had large tusks near the front of its jaws (see figure 7.2). These were probably used for defense and display, as are the tusks of some living, large plant-eating mammals, such as pigs. Some heterodontosaur fossils lack these tusks, suggesting that they may have been a sexually dimorphic feature. Males presumably had tusks, and females did not. The tusks might also have been used in feeding, helping the dinosaur to stab and tear vegetation.

Heterodontosaurs are known from the Upper Triassic of Argentina, the Lower Jurassic of southern Africa, the Jurassic of North America, the Cretaceous of Asia, and possibly the Cretaceous of Europe. Their spotty fossil record suggests that many heterodontosaurs remain to be discovered. Most intriguing is ***Tianyulong***, a heterodontosaur recently described from the Lower Cretaceous of northeastern China. The fossil of this dinosaur has filaments preserved in the skin below the neck and above the back and tail. Although these have been called "protofeathers," their actual relationship to feathers awaits further study.

**FIGURE 7.5**
Early Cretaceous *Hypsilophodon* was a typical primitive ornithopod. (© Scott Hartman)

## PRIMITIVE ORNITHOPODS

The most primitive ornithopods were long assigned to Hypsilophodontidae, a single family of small to medium-size (2 to 4 meters long) bipedal ornithischians. However, this formerly monophyletic family is now regarded as an array of primitive ornithopods, some more closely related to iguanodontians than others. Late Jurassic to Upper Cretaceous fossils of these primitive ornithopods have been discovered, and the fossils of primitive ornithopods are found on every continent. Best known and typical of the primitive ornithopods is ***Hypsilophodon*** (figure 7.5).

The limb proportions of *Hypsilophodon* were those of a bipedal runner. The hind limbs were much longer than the forelimbs, and the distal segments of the hind limb, especially the ankle, were elongate. The long tail, stiffened by calcified tendons, would have counterbalanced the dinosaur while it ran. Although some paleontologists previously believed that *Hypsilophodon* was a tree-climbing dinosaur, its skeleton is much more that of a fast ground runner.

The skull of *Hypsilophodon* resembled that of *Heterodontosaurus* but lacked the large tusks. The eyes were large, and a narrow, horny beak was present at the tip of both the upper and lower jaws. The massive jaws of *Hypsilophodon* supported a dental battery of interlocking cheek teeth that wore down to produce a continuous, inclined cutting edge. Unlike heterodontosaurs, there was a hinge in the skull of *Hypsilophodon* that ran from near the front of the tooth row to near the jaw joint on the posterior corner of the skull. This hinge allowed the jaw to move from side to side as the dinosaur chewed (box 7.1).

Primitive ornithopods are extremely well known from complete skeletons and eggs. The zenith of their diversity was during the Late Jurassic and Early Cretaceous when these small, bipedal herbivores lived on all continents. Only two Late Cretaceous primitive ornithopods are known, *Thescelosaurus* and *Parksosaurus*; their fossils have been found in western North America. The success of the primitive ornithopods was probably due in large part to their ability to grind vegetation.

## Box 7.1

### Grinding: The Key to Ornithopod Success

Many living mammals that eat plants grind their food by moving their lower jaws from side to side while chewing. This transverse movement of the lower jaw is achieved by arranging the jaw muscles and their points of attachment to pull the jaw laterally and medially. In tandem with teeth that provide flat surfaces for grinding and sharp edges for tearing, side-to-side movement of the lower jaw efficiently mills vegetation prior to swallowing.

Ornithopod dinosaurs efficiently milled the vegetation they ate with a mechanism analogous to, but different from, that of living mammals. *Hypsilophodon* is a good example. First, instead of moving its lower jaws transversely, *Hypsilophodon* had a joint within its skull that allowed the upper jaws to move sideways while chewing (box figure 7.1). This joint was essentially a diagonal hinge between the premaxillary, the upper jaw bone at the front of the mouth, and the maxillary, the upper jaw bone behind it. When the dinosaur closed its mouth, the teeth sheared past each other and the maxillary rotated slightly outward while the lower jaw rotated inward. Thus, while *Hypsilophodon* chomped on vegetation, the outwardly rotating upper jaw produced a grinding movement.

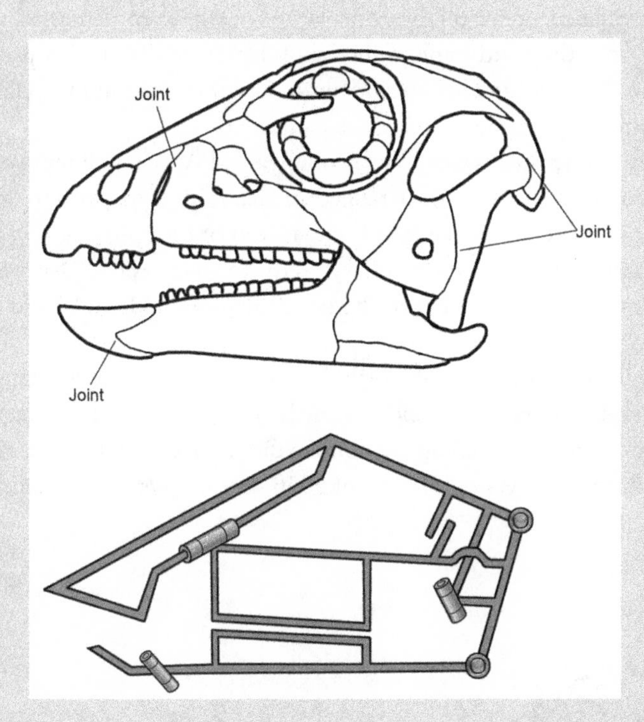


**BOX FIGURE 7.1**
The skull of *Hypsilophodon* had a joint that allowed the upper jaws to rotate outward when the mouth was closed. The bottom diagram shows the joints between the main elements of the skull of the dinosaur. (© Scott Hartman)

*Hypsilophodon* had broad, chisel-like teeth that locked together along their crowns to form a continuous cutting edge. A cheek pouch may have been present alongside the teeth to catch and hold food while chewing. This sophisticated mechanism for milling vegetation was very different from the generalized teeth and jaws of primitive ornithischians, such as the heterodontosaurs, dinosaurs that were otherwise rather similar to *Hypsilophodon*. The mechanism of grinding vegetation must have been one of the keys to ornithopod success.

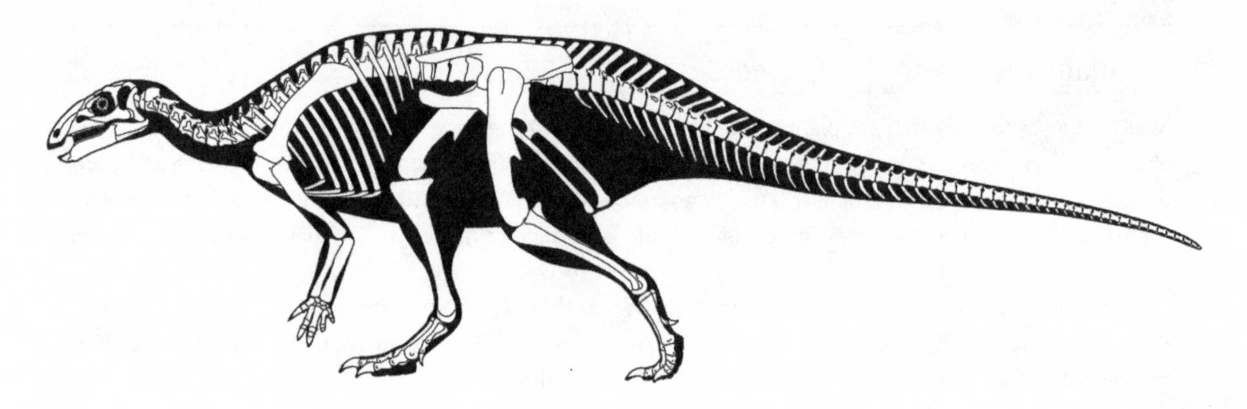

**FIGURE 7.6**
Late Jurassic *Dryosaurus* was a primitive iguanodontian. (© Scott Hartman)

## IGUANODONTIANS

All of the advanced ornithopods are classified in **Iguanodontia.** They are distinguished by their transversely widened premaxillaries that lacked teeth, deep dentaries, and the loss of a phalanx from the third digit of the hand. Here, we distinguish primitive iguanodontians, well represented by *Tenontosaurus, Camptosaurus,* and the classic *Iguanodon,* from the advanced iguanodontians, the hadrosaurids.

***Dryosaurus***, from the Upper Jurassic of North America and Africa, well represents the primitive iguanodontians (figure 7.6). It ranged in size from 2 to 3 meters long and had a short skull, long neck, long tail, and three instead of four phalanges in the third finger of the hand, and lacked premaxillary teeth. In many ways, *Dryosaurus* represents an ideal transitional form between the small "hypsilophodontids" and the large iguanodontids.

***Camptosaurus***, from the Upper Jurassic of North America and Europe, is well known from many complete skeletons and is the oldest heavily built ornithopod (figure 7.7). This relatively small (5 to 7 meters long) iguanodontian did not have a fully developed "thumb" spike and had small hooves on the tips of all its toes. *Camptosaurus* had four

**FIGURE 7.7**
Late Jurassic and Early Cretaceous *Camptosaurus* is the oldest iguanodontian known from a complete skeleton. (© Scott Hartman)

digits on the hind foot, massive hind limbs, and limb proportions that suggest it was a facultative quadruped. The fused wrist bones provide strong evidence of stable hand walking. The neck of *Camptosaurus* was relatively long, and its head was relatively small for an iguanodontian. But, it had the long, horse-like snout and broad, toothless beak characteristic of all iguanodontians.

Iguanodontians have a special place in the history of dinosaur studies because ***Iguanodon*** was one of the first dinosaurs to be described scientifically, by Gideon Mantell in 1825 (see chapter 11). *Iguanodon* was as much as 10 meters long and mostly a bipedal herbivore that lived principally during the Late Jurassic and Early Cretaceous in Europe and possibly Asia.

Best known from the Lower Cretaceous of Belgium, *Iguanodon* was a much larger and more specialized iguanodontian than *Camptosaurus* (figure 7.8; box 7.2). Like *Camptosaurus*, it had massive hind limbs. But, unlike *Camptosaurus*, *Iguanodon* had long forelimbs that were 70 to 80 percent as long as its hind limbs. The wrist bones of *Iguanodon* were fused, and the central three digits of the hand ended in hooves. The large conical "thumb" spike on the first digit must have been used as a defensive weapon, and the fifth digit was long and slender.

*Iguanodon* had a large head with a long snout, a long neck and long tail, and an extensive boxwork of ossified tendons that extended along the back from the shoulder region to the middle of the tail (see figure 7.8). The limb proportions and modifications of the hands of *Iguanodon* suggest that it did much more quadrupedal walking than did *Camptosaurus* or the smaller ornithopods. It seems likely, however, that *Iguanodon* could rear up on its hind limbs and swing its spike-shaped thumb like a knife fighter when attacked.

The skulls of iguanodontians varied greatly, but all had long snouts and toothless beaks. These long snouts included many more teeth (as many as 29 in the maxillary of *Iguanodon*) than did the snouts of heterodontosaurs and the primitive ornithopods. The iguanodontian beak was used to crop vegetation. The teeth were leaf shaped with long ridges on the sides and small cusps on the cutting edges and resembled the teeth of living iguanas: hence the name of the dinosaur taxon (see chapter 11). Paleontologists are certain that iguanodontians were plant eaters.

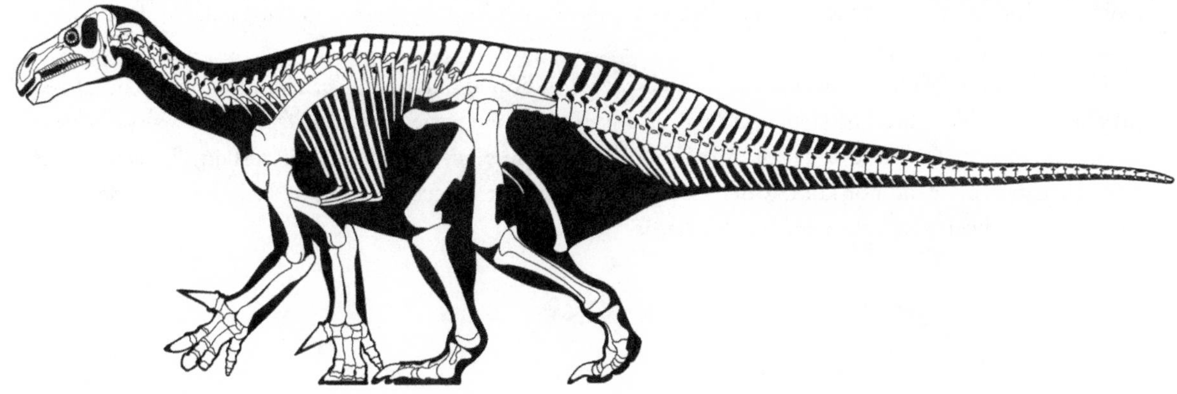

**FIGURE 7.8**
Early Cretaceous *Iguanodon* was a powerful quadrupedal walker. (© Scott Hartman)

## Box 7.2

### An *Iguanodon* Graveyard at Bernissart

Bernissart is a small mining town in southwestern Belgium. In 1878, miners there discovered a crevice full of clay that cut across a bed of coal and was full of dinosaur bones. This crevice, at a depth of 322 meters, was excavated from 1878 until 1881. The fossils of 39 individuals of *Iguanodon*, many of them complete skeletons (box figure 7.2), were recovered, as were the fossils of many plants, fishes, and other reptiles. Bernissart thus stands as one of the greatest dinosaur localities on Earth.

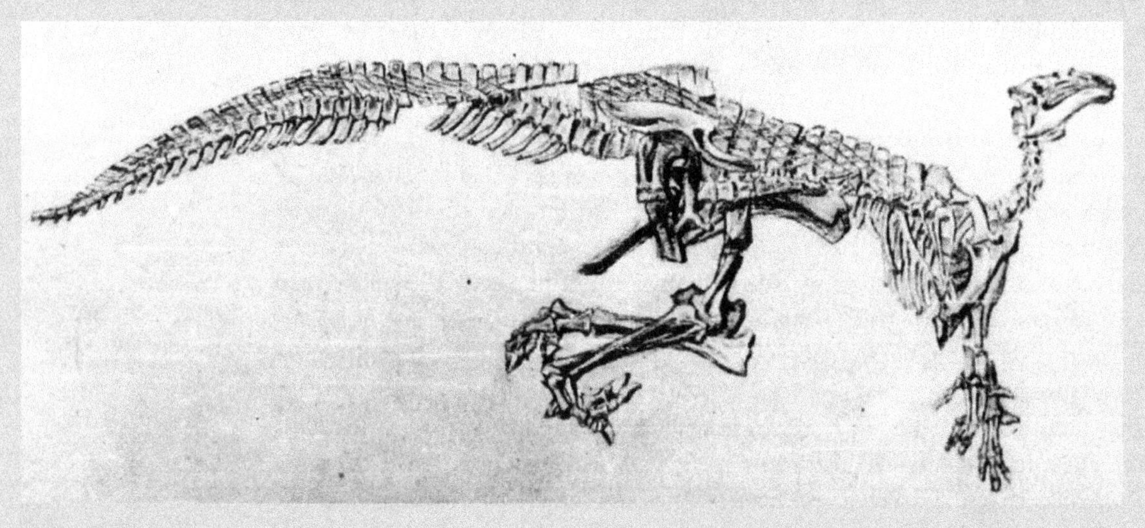

**BOX FIGURE 7.2**
Sketch of one of the many complete skeletons of *Iguanodon* from Bernissart as it was found in the rock. (From L. Dollo. 1882. Première note sur les dinosauriens de Bernissart. *Bulletin du Musée royal d'histoire naturelle de Belgique* 1:55–74)

For many years, paleontologists thought a single catastrophe led to the accumulated *Iguanodon* skeletons at Bernissart. One popular image shows the dinosaurs falling off a cliff into the crevice, which later filled with clay. Indeed, the "evidence" of a single catastrophe at Bernissart suggested to some paleontologists that *Iguanodon* lived in groups ("herds").

Careful restudy of the Bernissart *Iguanodon*, however, indicated that the *Iguanodon* skeletons accumulated in groups on at least three different occasions. The skeletons were buried and fossilized in a marshy environment, but the precise cause of death of the dinosaurs remains uncertain. What is certain, though, is that a single catastrophe did not kill the Bernissart *Iguanodon*, so the accumulation of skeletons here does not necessarily indicate that *Iguanodon* lived in herds.

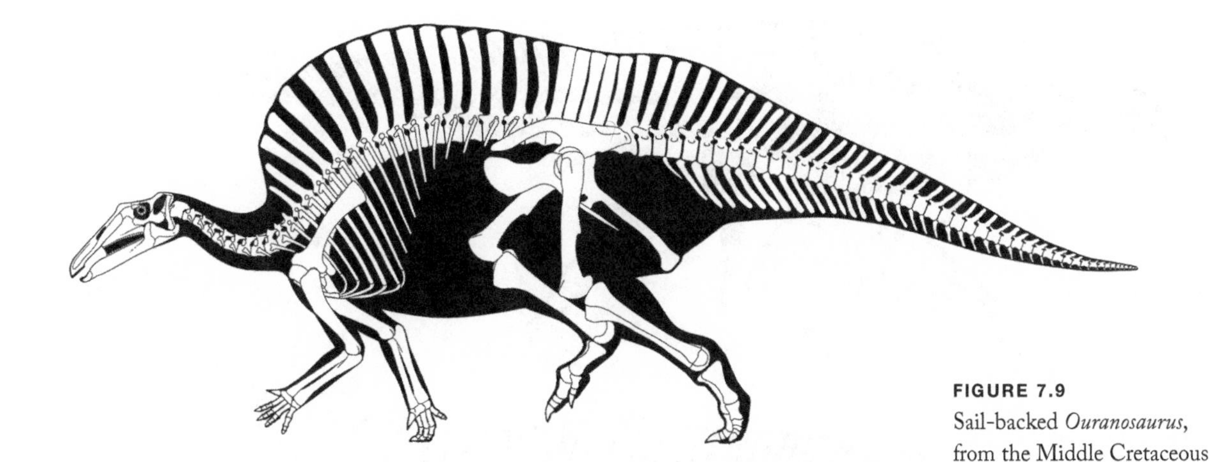

**FIGURE 7.9**
Sail-backed *Ouranosaurus*, from the Middle Cretaceous of Niger, was one of the most distinctive iguanodontians. (© Scott Hartman)

The zenith of primitive iguanodontian diversity was during the Early to Middle Cretaceous, and iguanodontian fossils are known from Europe, North America, Africa, Asia, and Australia. A particularly unusual iguanodontian from this time period is ***Ouranosaurus*** from Niger with its very distinctive head and long neural spines that formed a sail along its back (figure 7.9). Clearly, primitive iguanodontians were a diverse, widespread, and significant group of medium to large herbivorous dinosaurs of the Early Cretaceous world. There was a marked decline of primitive iguanodontians during the Late Cretaceous, probably because of the appearance of the hadrosaurids.

## HADROSAURIDS

Advanced iguanodontians belong to a single family: **Hadrosauridae**. Hadrosaurids, popularly known as the duck-billed dinosaurs, first appeared during the Late Cretaceous and thus were the last major group of ornithopods to evolve. Much more is probably known about hadrosaurids than about any other single group of dinosaurs. Their outstanding fossil record includes many complete skeletons, eggs, footprints, and coprolites and several mummified individuals with skin intact.

The hadrosaurids were large ornithopods, 7 to 12 meters long (figure 7.10), characterized by their broad beaks lacking teeth, intricate dental batteries in which three or more replacement teeth existed for each tooth position (figure 7.11), and the loss of the first digit (thumb) on the hands, among other features. The jaws were deep and long, and the jaw muscles were evidently powerful. Hadrosaurids chomped their food and achieved a side-to-side motion of the jaws by slightly rotating the upper tooth rows outward when the mouth was closed.

Hadrosaurid teeth were arranged in dental batteries consisting of literally hundreds of teeth cemented together in each jaw (see figure 7.11). The teeth thus formed washboard-like grinding surfaces that milled the vegetation cropped by the horny beak

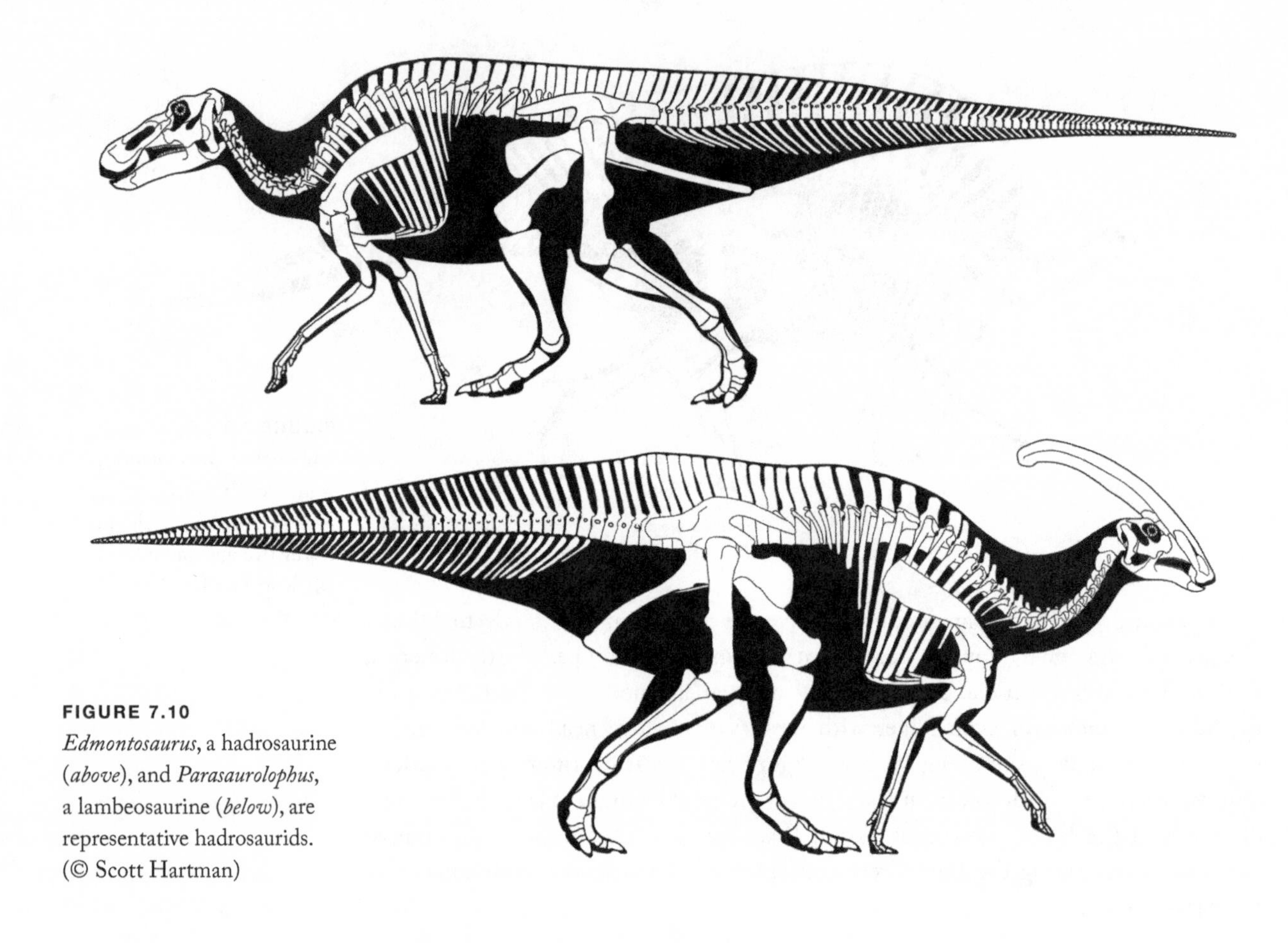

**FIGURE 7.10**
*Edmontosaurus*, a hadrosaurine (*above*), and *Parasaurolophus*, a lambeosaurine (*below*), are representative hadrosaurids. (© Scott Hartman)

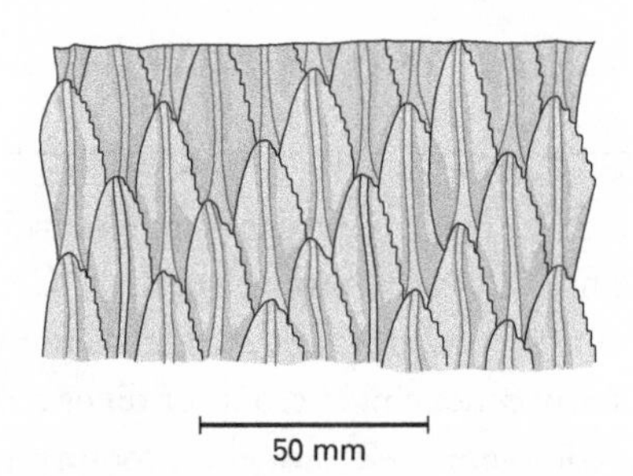


**FIGURE 7.11**
The complex dental batteries of hadrosaurids included three or more replacement teeth at each tooth position. (Drawing by Network Graphics)

at the front of the mouth. The hadrosaurid jaw mechanism could process resistant, fibrous vegetation and even twigs. Indeed, the two mummified hadrosaurids discovered in Alberta, Canada, have stomachs full of conifer needles and twigs, seeds, and other tough plant debris, and hadrosaur coprolites are packed with conifer debris.

The skeletons of hadrosaurids were very similar to those of primitive iguanodontians but lacked thumbs, differed in the shape of the pelvis, and had as many as 8 to 10 sacral vertebrae. Few differences are evident among the skeletons of different types of hadrosaurids, so identifying them is done almost exclusively by comparing their skulls.

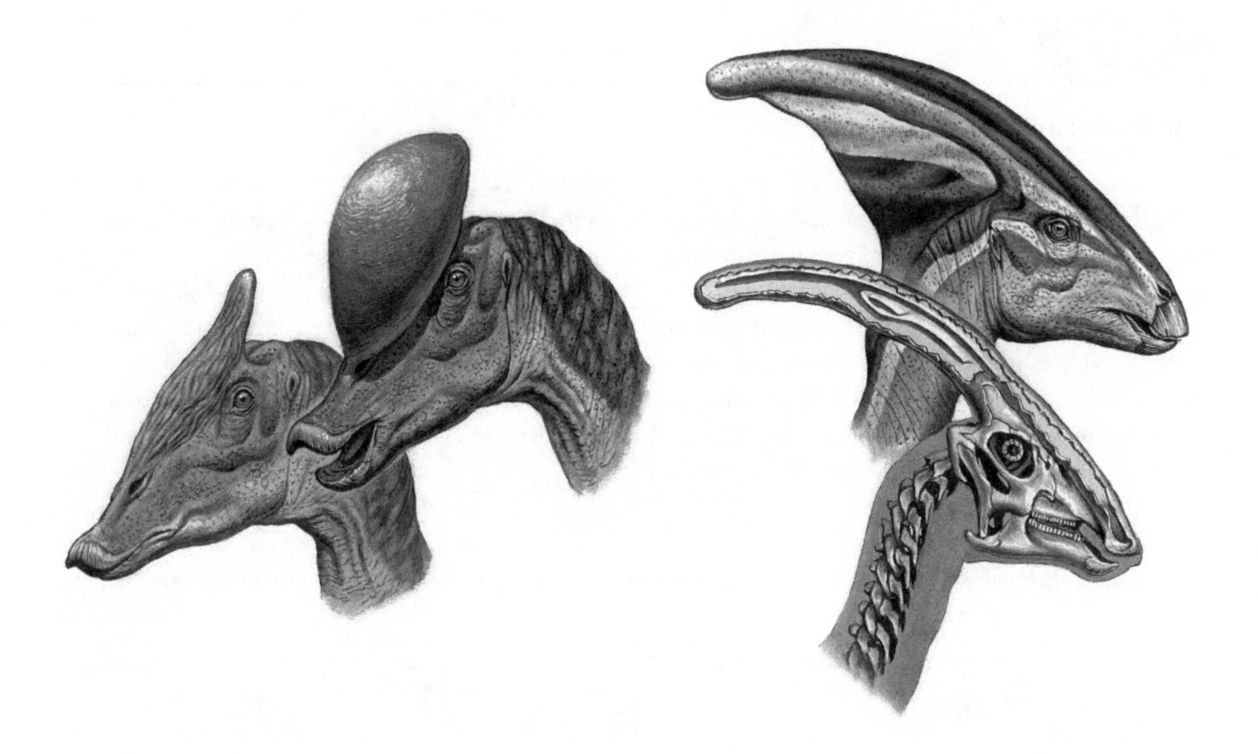

**FIGURE 7.12**
The flat skulls of hadrosaurines (*far left*) contrast with the crested skulls of lambeosaurines (*center* and *right*), among which there was great variety. (© Mark Hallett. Reproduced with permission of Mark Hallett Paleoart)

Although the variation in skull type among hadrosaurids is astounding, just two hadrosaurid subfamilies are generally recognized: **Hadrosaurinae** and **Lambeosaurinae** (figure 7.12). Hadrosaurines are the more primitive subfamily and had rather flat skull roofs. They are sometimes called the "Roman-nosed" hadrosaurs after their large and long nasals that often peaked near the posterior end of the nostrils. ***Edmontosaurus***, from the Upper Cretaceous of North America, is a characteristic hadrosaurine (see figure 7.10).

The lambeosaurines are the advanced subfamily, often referred to as the "crested" hadrosaurs. They are distinguished by the convoluted **tubes and crests** on the tops of their skulls, which contained the modified nasal passages and a nasal cavity relocated above the orbits. Other distinctive lambeosaurine features include elongated neural spines on the vertebrae and relatively robust limbs. The varied crests and tubes on lambeosaurine heads likely served as visual display and probably also acted as resonators to produce distinctive calls (box 7.3). Many of the differences in crest and tube size and shape reflect growth and sexual dimorphism (figure 7.13). ***Parasaurolophus***, from the Upper Cretaceous of North America, is a representative lambeosaurine (see figure 7.10).

Most of the features of the hadrosaurid skeleton suggest that these dinosaurs were powerful quadrupedal walkers also well suited to bipedal locomotion. The hadrosaur mummies from Alberta, however, have flaps of skin between their fingers that suggest a webbed hand. And, the tall, paddle-like tail of hadrosaurids looks suited (though its flexibility was limited) to propelling these dinosaurs in water. These features, and the fact that many hadrosaurid fossils, including complete, articulated skeletons, are found in rocks deposited in rivers, lakes, swamps, and even the sea, have led some

## Box 7.3

### A Lambeosaurine Symphony?

The crests and tubes on the heads of lambeosaurines have elicited speculation about their function for decades (box figure 7.3). An early idea was that the crests and tubes allowed for feeding underwater either by acting as snorkels (but no upward openings were present in the crests and tubes), as air-storage tanks (but the volume of air that could be stored in the crests was very small relative to lung volume), or as "air locks" to prevent water from entering the lungs (although how this "air lock" would have worked is unclear). Other, more plausible suggestions are that the hollow spaces inside the tubes and crests provided space for glands or, by increasing the surface area for tissue inside the nasal cavity, warmed and moistened inhaled air and improved the sense of smell. Another suggestion is that the tubes and crests were used to deflect foliage when lambeosaurines crashed through dense forests.

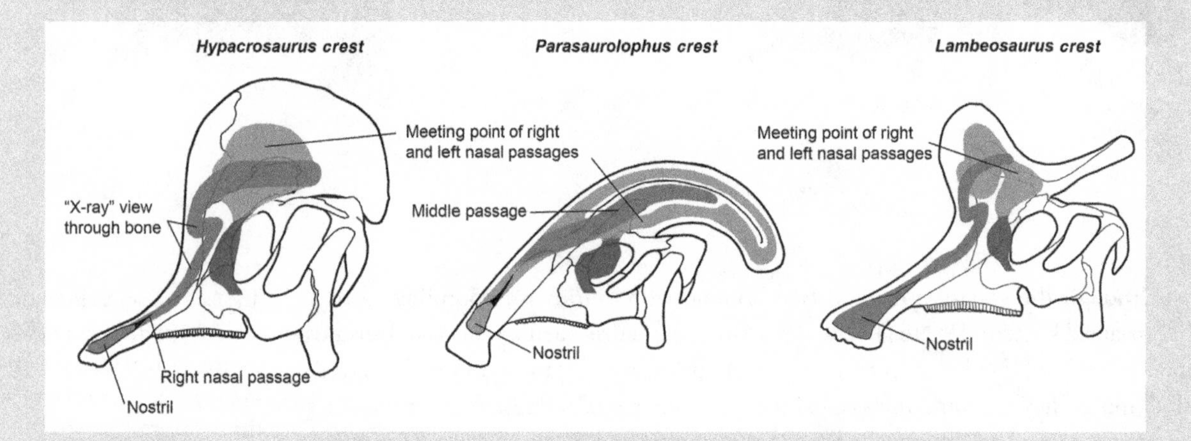


**BOX FIGURE 7.3**
The tubes and crests on lambeosaurine skulls were hollow and connected to the nasal passages.
(© Scott Hartman)

These plausible functions of the crests and tubes do not exclude what may have been their most important function—as signaling devices. This signaling would have been both visual and auditory. The fact that particular shapes of crests and tubes were specific to particular types of lambeosaurines, and that the crests of males and females of a specific type of lambeosaurine were of different sizes and shapes, strongly supports their identification as visual signaling devices. Distinctive-looking tubes and crests thus allowed lambeosaurines to recognize members of their own species and to distinguish males from females.

Another kind of display was also possible with the lambeosaurine crests and tubes because they were hollow and connected to the air passages. Such hollow structures could have acted as **resonating chambers**, with differently shaped tubes and crests producing distinctive sounds like the differently shaped wind instruments of a symphony orchestra. So, not only could lambeosaurines identify each other visually by their crests and tubes, they could signal to others with distinctive sounds. The Late Cretaceous landscapes they inhabited must have been noisy places!

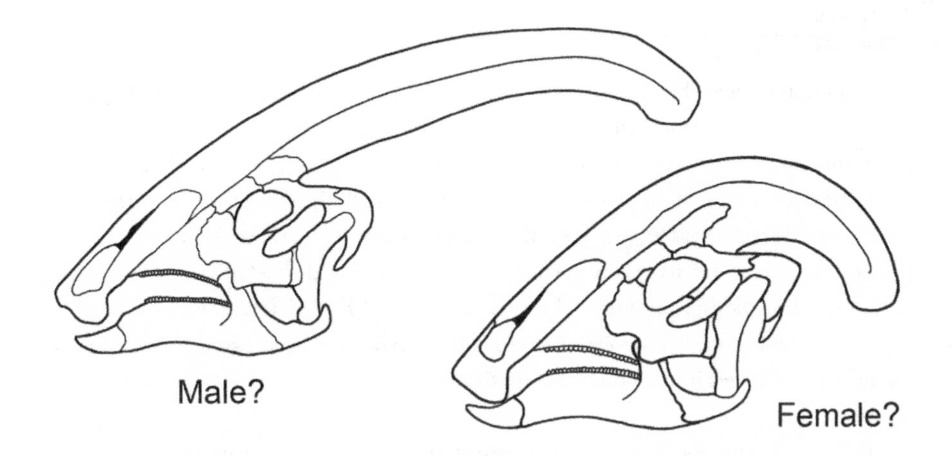


**FIGURE 7.13**
The tubes and crests on the skulls of lambeosaurines changed during growth and were sexually dimorphic. (© Scott Hartman)

paleontologists to view hadrosaurids as aquatic. It seems most certain, however, that hadrosaurids were very well suited to living on dry land and may have entered water only for defense or feeding. Much more has been suggested about hadrosaurid behavior, especially based on their nests of eggs, and is discussed in chapter 13.

Hadrosaurids first appeared in Asia during the Late Cretaceous. Later in the Late Cretaceous, they had spread to North America, Asia, Europe, South America, and Antarctica, though they were most diverse and abundant in Asia and North America. Hadrosaurids lived in a variety of habitats, from inland floodplains to coastal swamps and jungles. In the New World, their distribution extended from Alaska to Argentina. Hadrosaurids were among the dominant plant-eating dinosaurs of the Late Cretaceous and were among the last dinosaurs to become extinct.

## ORNITHOPOD EVOLUTION

As mentioned at the beginning of this chapter, the ornithopods were one of the most long-lived and diverse groups of dinosaurs. The dental battery was a hallmark of the ornithopods and one of the keys to their great success. They first appeared during the Middle Jurassic as small bipeds and were widespread by the Late Jurassic.

Indeed, during the Late Jurassic, it would have been virtually impossible to visit any place on Earth without encountering an ornithopod dinosaur. Small to medium-size primitive ornithopods and large iguanodontians were very diverse and abundant plant-eating dinosaurs of the Late Jurassic through to the Early Cretaceous and had an essentially worldwide distribution.

The hadrosaurids arose by Late Cretaceous time. Their first appearance coincided with a marked decline in the types and numbers of other ornithopods. It is thus tempting to believe that hadrosaurids contributed to the demise of their more primitive relatives. Hadrosaurids were extremely diverse and abundant during the Late Cretaceous and inhabited a variety of environments until their own demise 66 million years ago.

## Summary

1. Ornithopods were bipedal ornithischian dinosaurs that lived from the Middle Jurassic to the Late Cretaceous. Their fossils are found on all continents.
2. Heterodontosaurs were primitive ornithischians considered by some paleontologists to be closely related to ornithopods. They were small bipedal dinosaurs with unique, chisel-like teeth best known from the Lower Jurassic of southern Africa.
3. *Hypsilophodon* and its relatives were small bipedal ornithopods that were especially diverse and widespread during the Late Jurassic and Early Cretaceous.
4. The success of *Hypsilophodon* and its relatives may have been based in part on their ability, via hinges in the skull, to move their dental batteries sideways in order to grind vegetation.
5. Primitive iguanodontians are well represented by *Tenontosaurus*, *Camptosaurus*, and the famous *Iguanodon*.
6. The primitive iguanodontians were large, facultatively bipedal ornithopods that were particularly successful during the Late Jurassic and Early Cretaceous.
7. Hadrosaurids, the duck-billed dinosaurs, first appeared during the Late Cretaceous and were particularly diverse and abundant in Asia and North America.
8. The appearance of hadrosaurids nearly coincides with the decline of the other ornithopods.
9. Hadrosaurids encompass two subfamilies: the hadrosaurines with flat skulls and the lambeosaurines with skulls bearing dorsal tubes or crests.
10. The hollow tubes and crests on lambeosaurine skulls were most likely signaling devices used for both visual display and as resonating chambers to produce distinctive sounds.
11. Hadrosaurids were powerful quadrupedal walkers that may also have been amphibious.

## Key Terms

*Camptosaurus*
dental battery
*Dryosaurus*
*Edmontosaurus*
Hadrosauridae
Hadrosaurinae
*Heterodontosaurus*
*Hypsilophodon*
*Iguanodon*
Iguanodontia
Lambeosaurinae
ornithopod
*Ouranosaurus*
*Parasaurolophus*
resonating chamber
*Tianyulong*
*tubes and crests*
*Uteodon*

## Review Questions

1. What are the distinctive features of ornithopods?
2. To what feature(s) might you attribute the success of ornithopods?
3. How did primitive ornithopods differ from heterodontosaurs, and how might these differences explain the greater success of the primitive ornithopods?
4. How did hadrosaurids resemble and differ from primitive ornithopods?

5. What features of the skeletons of different groups of ornithopods identify them as either bipeds or facultative bipeds?
6. What defensive strategies did ornithopods employ?
7. How do the jaw, tooth, and skull structures of the various ornithopods differ from each other? Is a progression in the evolution of chewing mechanisms evident among the ornithopods?
8. Why do paleontologists think the crests and tubes of lambeosaurines functioned as signaling devices?

## Further Reading

Butler, R. J., and P. M. Barrett. 2012. Ornithopods, pp. 551–566, in M. K. Brett-Surman, T. R. Holtz, Jr., and J. O. Farlow, eds., *The Complete Dinosaur*. 2nd ed. Bloomington: Indiana University Press. (Concise review of the ornithopods)

Case, J. A., et al. 2000. The first duck-billed dinosaur (family Hadrosauridae) from Antarctica. *Journal of Vertebrate Paleontology* 20:612–614. (Describes an Upper Cretaceous hadrosaur tooth found in Antarctica)

Eberth, D. A., and D. C. Evans, eds. 2014. *Hadrosaurs*. Bloomington: Indiana University Press. (A collection of 36 technical articles presenting recent research on hadrosaurs)

Godefroit, P., ed. 2012. *Bernissart Dinosaurs and Early Cretaceous Terrestrial Ecosystems*. Bloomington: Indiana University Press. (A collection of 33 technical articles on Bernissart and much more about Early Cretaceous dinosaurs and ecosystems)

Prieto-Márquez, A. 2010. Global phylogeny of Hadrosauridae (Dinosauria; Ornithopoda). *Zoological Journal of the Linnean Society of London* 159:435–502. (Recent cladistic analysis of hadrosaur phylogeny)

Weishampel, D. B. 1981. Acoustical analysis of potential vocalization in lambeosaurine dinosaurs (Reptilia: Ornithischia). *Paleobiology* 7:252–261. (A technical analysis of the acoustical properties of lambeosaur cranial crests)

Weishampel, D. B. 1984. Evolution of jaw mechanisms in ornithopod dinosaurs. *Advances in Anatomy, Embryology and Cell Biology* 87:1–109. (A detailed examination of the skulls and teeth of ornithopods)

Zheng, X., H. You, X. Xu, and Z. Dong. 2009. An Early Cretaceous heterodontosaurid dinosaur with filamentous integumentary structures. *Nature* 458:333–336. (Describes *Tianyulong*)

## Find a Dinosaur!

Ornithopod dinosaurs can be found at many of the world's great natural history museums. In Brussels, Belgium, the Royal Belgian Institute of Natural Science displays 30 of the *Iguanodon* skeletons found during the nineteenth century at Bernissart, Belgium, by miners. Do take note of the large thumb spikes on those *Iguanodon* skeletons, which were originally thought to be nose horns. In North America, hadrosaurs (duck-billed dinosaurs) are mainstays of the dinosaur exhibits at the American Museum of Natural History (New York) and the Museum of the Rockies (Bozeman, Montana). The Bozeman display features the well-known Upper Cretaceous hadrosaur *Maiasaura* and its nests of eggs and baby hadrosaurs found in Montana. In Albuquerque, New Mexico, the bizarre tube-headed hadrosaur *Parasaurolophus* is featured as both skull and sculpture at the New Mexico Museum of Natural History and Science. Most notable, however, are the many superb hadrosaur skeletons and skulls on display in Canada at the Royal Ontario Museum (Toronto, Ontario) and the Royal Tyrrell Museum of Palaeontology (Drumheller, Alberta). And, if you are in Thailand, the Sirindhorn Museum and Phu Kum Khao Dinosaur Excavation Site near Sahatsakhan displays replicas of ornithopod skeletons and many other dinosaurs.

# 8

# STEGOSAURS AND ANKYLOSAURS

STEGOSAURUS is one of the most familiar dinosaurs, and the armadillo-like *Ankylosaurus* is also quite well known. Paleontologists consider the stegosaurs (plated dinosaurs) and ankylosaurs (armored dinosaurs) to be closely related and have united them in a group with the unwieldy name **Thyreophora** (shield bearers) (figure 8.1). Thyreophorans were a diverse group of armored, primarily quadrupedal ornithischians of Jurassic and Cretaceous age with a virtually worldwide distribution. The key evolutionary novelty of thyreophorans is the presence of one or more rows of **armor plates** in the skin above or alongside the vertebral column. In this chapter, I review the anatomy and evolution of thyreophoran dinosaurs.

## PRIMITIVE THYREOPHORANS

Not all thyreophorans can be assigned to the stegosaurs or ankylosaurs. The most primitive of the thyreophorans are also among the most primitive ornithischians. Two of these dinosaurs—*Scutellosaurus* and *Scelidosaurus*—provide us with a look at the early diversity of the Thyreophora.

***Scutellosaurus*** was a relatively small (about 1.3 meters long) generalized ornithischian (figure 8.2). It had a small skull and jaws, cheek teeth positioned on the jaw margins, a short neck, hind limbs only slightly longer than the forelimbs, and a long tail. These are primitive ornithischian characteristics and explain why some paleontologists consider *Scutellosaurus* to be closely related to primitive ornithischian dinosaurs like *Lesothosaurus*. However, *Scutellosaurus* had an extensive body covering of bony plates set in the skin (figure 8.3), which is an evolutionary novelty of thyreophorans that justifies the inclusion of *Scutellosaurus* in this group. Evolutionary novelties of the skull and lower jaw, including the sinuous lower cheek tooth row, distinguish stegosaurs and ankylosaurs from the most primitive thyreophorans.

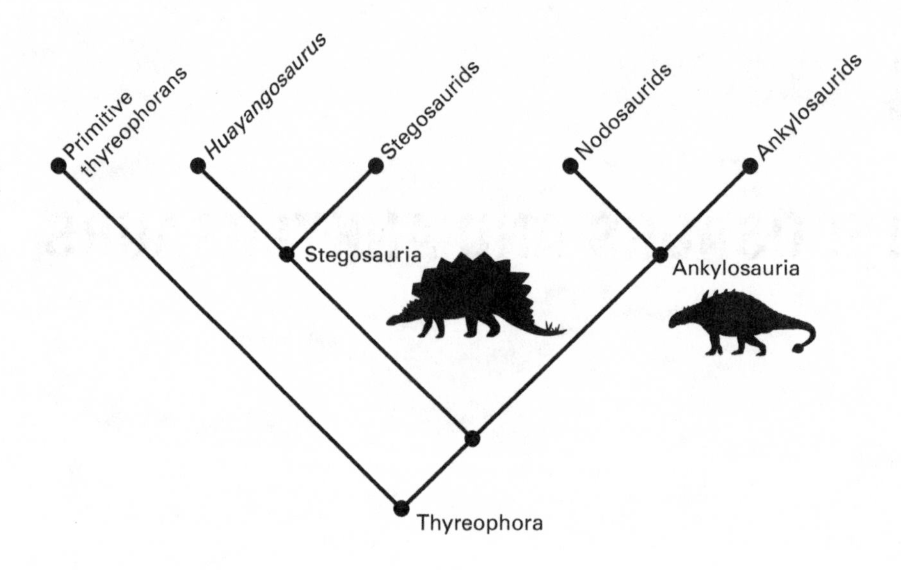


**FIGURE 8.1**
Thyreophoran dinosaurs include the stegosaurs, the ankylosaurs, and some primitive forms, as shown in this cladogram. (Drawing by Network Graphics)

Of course, it is possible that *Scutellosaurus* evolved its body armor independently of the evolution of body armor in thyreophorans. In this case, the similarity in body armor between *Scutellosaurus* and thyreophorans would reflect evolutionary convergence and not a close phylogenetic relationship. At present, though, the body armor of *Scutellosaurus* is best viewed as genuinely thyreophoran, especially because it shows

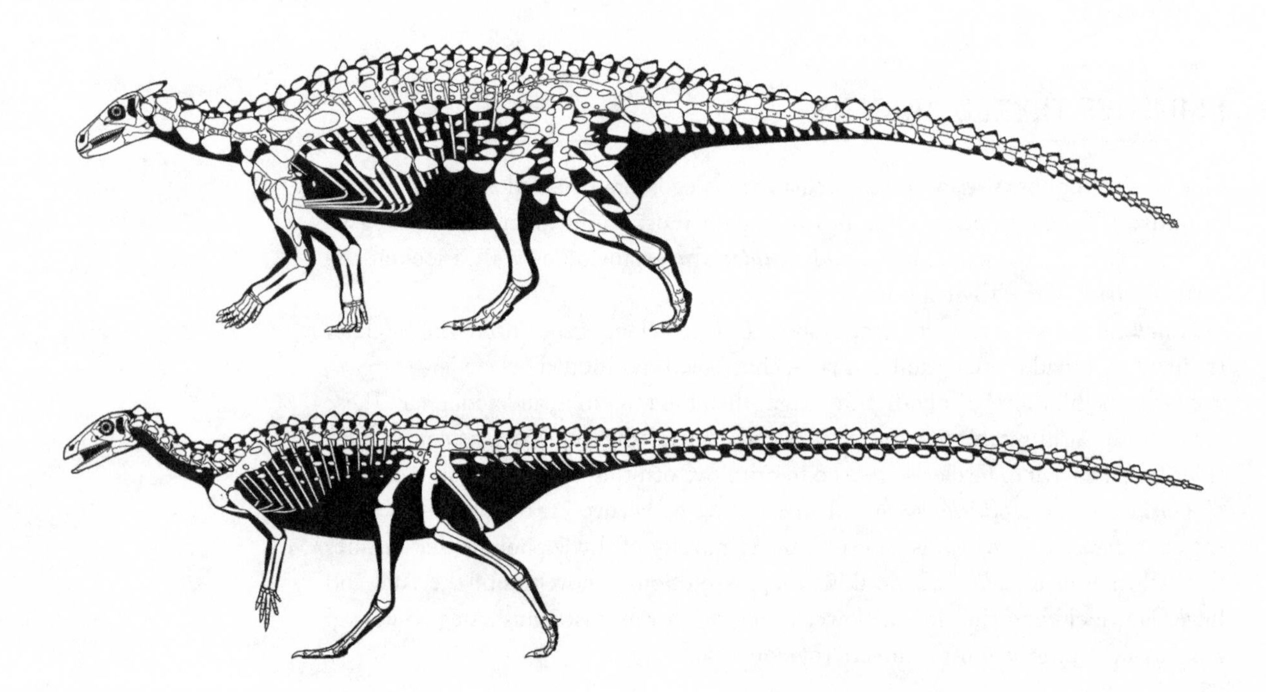

**FIGURE 8.2**
Early Jurassic *Scelidosaurus* (*above*) and *Scutellosaurus* (*below*) were primitive Jurassic thyreophorans that lacked the evolutionary novelties of either stegosaurs or ankylosaurs. (© Scott Hartman)

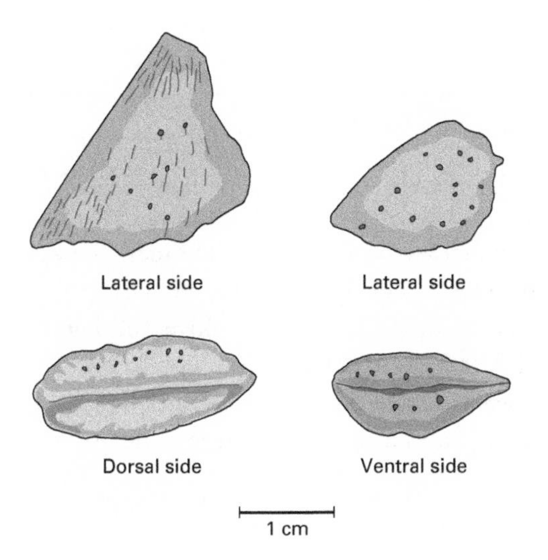


**FIGURE 8.3**
The armor plates of *Scutellosaurus* were characteristically thyreophoran: they were keeled dorsally and excavated ventrally. (Drawing by Network Graphics)

detailed and unique similarity—dorsal keels and ventral excavations—to the body armor of certain other thyreophorans.

***Scelidosaurus*** is another primitive thyreophoran (see figure 8.2), but it was very different from *Scutellosaurus*. Much larger (at least 4 meters long), *Scelidosaurus* had a small skull with simple, leaf-shaped teeth that ran to the tip of the snout. Its limbs were massive, and the fore- and hind limbs were of nearly equal length. The skull of *Scelidosaurus* lacked armor plating, but its back was covered by numerous bony plates embedded in the skin. These plates were remarkably similar to those of the ankylosaurs. The extremely broad sacrum of *Scelidosaurus* is another feature that unites it with thyreophorans. Because no complete articulated skeleton of *Scelidosaurus* has been discovered, the arrangement of the armor plates on its back remains somewhat uncertain. The most complete known skeleton, described by Richard Owen in 1863, suggests that the plates formed a broad covering across the back and sides of the dinosaur as in later ankylosaurs (see figure 8.2)

As in *Scutellosaurus*, much of the anatomy of *Scelidosaurus* is primitively ornithischian. But this quadruped (possibly a facultative biped), with its extensive body armor and broad sacrum, can be more readily allied with the thyreophorans. Indeed, some paleontologists have gone so far as to assign *Scelidosaurus* to the stegosaurs or the ankylosaurs. But, because the specific evolutionary novelties that distinguish stegosaurs and ankylosaurs are not present in *Scelidosaurus*, it is best regarded as neither stegosaur nor ankylosaur, but as a primitive thyreophoran.

*Scutellosaurus* fossils are known from the Lower Jurassic of Arizona, whereas those of *Scelidosaurus* are known from the Lower Jurassic of Great Britain. Another primitive thyreophoran is ***Emausaurus***, known from a skull and partial skeleton collected in the Lower Jurassic of Germany. Known primitive thyreophorans were so different from each other and so widely separated geographically that we can be sure much more remains to be discovered about the origin and early evolution of the shield-bearing dinosaurs.

## STEGOSAURS

The **Stegosauria**, meaning "plated lizards" because of the armor plates on these dinosaurs' backs, were medium-size to large (up to 9 meters long), quadrupedal, herbivorous ornithischians. They had small heads; short and massive forelimbs; long, columnar hind limbs; and short, stout feet that bore hooves on the ends of the toes. The key evolutionary novelty of stegosaurs was the vertical bony plates and spines arranged in single or double rows along the neck, back, and tail. The most primitive well-known stegosaur is *Huayangosaurus*, from the Middle Jurassic of China. A poorly known but even more primitive stegosaur may be ***Gigantspinosaurus***, from the Upper Jurassic of China. Other more advanced stegosaurs are placed in the family Stegosauridae.

### The Genus *Huayangosaurus*

***Huayangosaurus***, from the Middle Jurassic of Sichuan Province, China, is the most completely known primitive stegosaur (figure 8.4). This 4.3-meter-long stegosaur had spike-shaped armor along the midline of its body and additional rows of small armor plates along each side of the row of spikes. Unlike more advanced stegosaurids, *Huayangosaurus* had a rather deep (tall) skull with a short snout. The eye sockets were located relatively far forward, above the posterior cheek teeth. *Huayangosaurus* was smaller and less massive than stegosaurids and had fore- and hind limbs that were not as equal in length. Other differences between *Huayangosaurus* and stegosaurids may be found in details of the structure of the shoulder and hip girdles.

*Huayangosaurus* is known from several complete skeletons and is thus one of the best-known stegosaurs. However, to examine the origin of stegosaurs, we must still bridge the considerable anatomical gap between primitive Early Jurassic thyreophorans, such as *Scelidosaurus*, and primitive stegosaurs, such as *Huayangosaurus*.

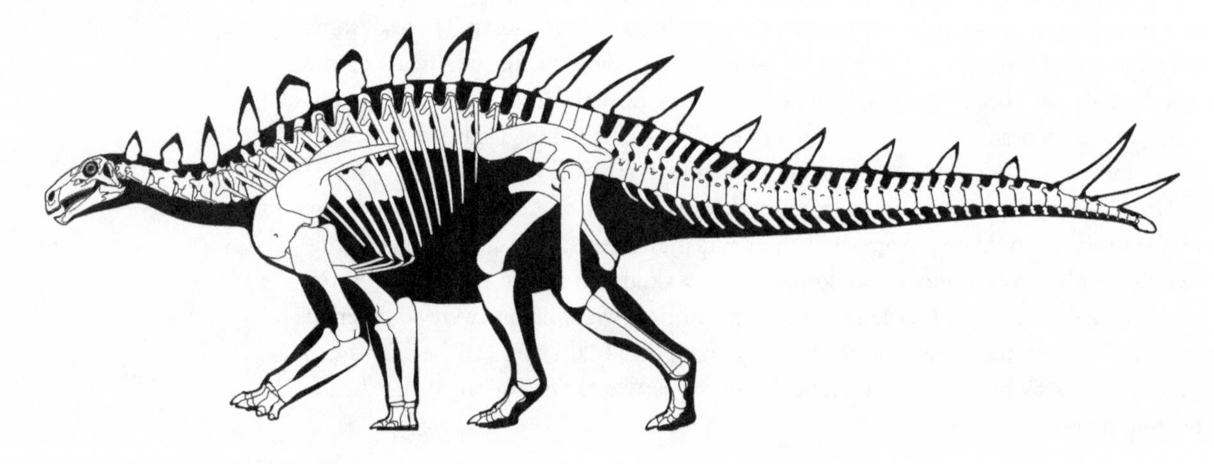

**FIGURE 8.4**
*Huayangosaurus*, from the Middle Jurassic of China, was one of the most primitive stegosaurs. (© Scott Hartman)

## Stegosaurids

All stegosaurs other than *Gigantspinosaurus* and *Huayangosaurus* are placed in the family **Stegosauridae**. Stegosaurids are easily distinguished from *Huayangosaurus* by their relatively low skulls (that is, not "tall" when measured top to bottom) with long snouts and posteriorly located eye sockets (behind the cheek-tooth row), larger size, more massive skeletons, and relatively long hind limbs. Best known and typical of the stegosaurids is *Stegosaurus*, from the Upper Jurassic of the western United States (figure 8.5). Other well-known stegosaurids are *Kentrosaurus*, from the Upper Jurassic of Tanzania, and *Tuojiangosaurus*, from the Upper Jurassic of China (figure 8.6). The remaining stegosaurids are known from much less complete fossils.

## The Genus *Stegosaurus*

Like all stegosaurids, *Stegosaurus* had only a midline row of armor plates. The exact arrangement of these plates has been debated (box 8.1), but the alternating, offset row of plates seen in most skeletal reconstructions is favored here (see figure 8.5).

*Stegosaurus* had a low, slender and, for an animal that weighed 1 to 2 tons, very small head. The tip of the snout formed a narrow, toothless beak. The cheek teeth were leaf-shaped with small denticles on the edges, as in many groups of plant-eating dinosaurs. These teeth did not, however, form the dental batteries seen in the ornithopods (see chapter 7). The brain in the skull of *Stegosaurus* was truly tiny and is estimated to have weighed about 70 to 80 grams (2.5 to 2.9 ounces). Indeed, the small skull and tiny brain of *Stegosaurus* have led to the popular notion that it was one of the "stupidest" dinosaurs (box 8.2).

**FIGURE 8.5**
Late Jurassic *Stegosaurus* was characteristic of the more advanced stegosaurs, the stegosaurids. (© Scott Hartman)

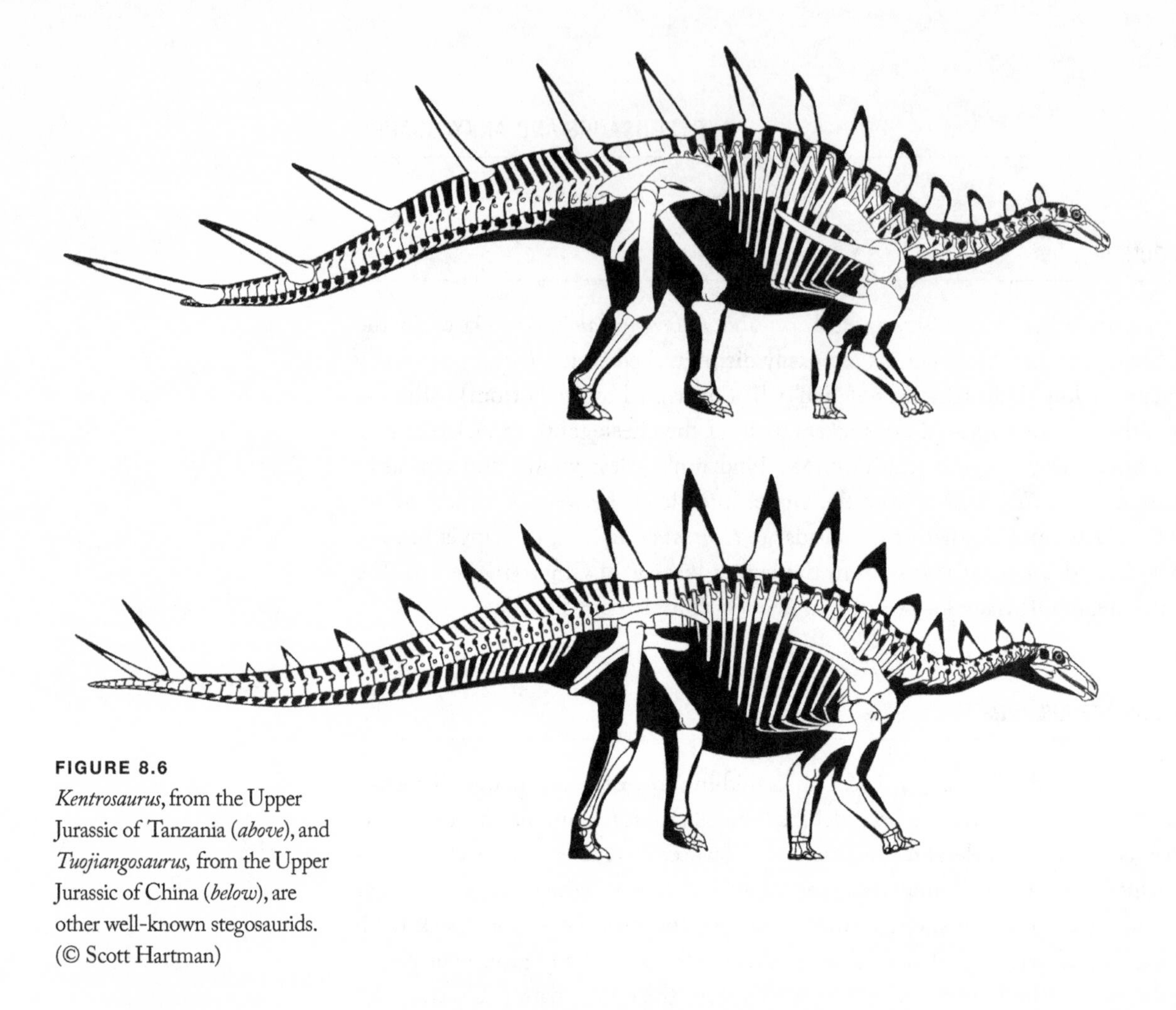

**FIGURE 8.6**
*Kentrosaurus*, from the Upper Jurassic of Tanzania (*above*), and *Tuojiangosaurus*, from the Upper Jurassic of China (*below*), are other well-known stegosaurids. (© Scott Hartman)

## Box 8.1

### *Stegosaurus* Plates: Two Rows or One?

As in other thyreophorans, the armor plates of *Stegosaurus* were anchored in the skin and not attached to other bones, so they usually fell out of place before fossilization occurred. One of the oldest clues to their arrangement remains the best clue, even today, and is seen in the remarkably well-preserved fossil skeleton of a *Stegosaurus* from Colorado described in 1901 (box figure 8.1A). This skeleton was found lying on its side and shows some overlap in the anterior and posterior ends of successive plates. But, rather than clearly demonstrating the arrangement of plates on the back of *Stegosaurus*, it has fostered two opposing views.

Prior to the discovery of the Colorado specimen, Yale University paleontologist Othniel Charles Marsh reconstructed the skeleton of *Stegosaurus* with its plates arranged in a single row (box figure 8.1B). The overlapping plates of the Colorado specimen, however, suggested to Smithsonian paleontologist Frederick Lucas that the plates were arranged in an alternating pattern in two rows (see figure 8.5). But, those who defended Marsh's original reconstruction suggested that the overlap of the plates of the Colorado specimen was a result of distortion of the skeleton after the dinosaur died.

In 1987, paleontologist and artist Stephen Czerkas reviewed ideas about the arrangement of *Stegosaurus* plates and argued they were arranged in a single row, as Marsh originally believed. Czerkas pointed out that on the Colorado skeleton, only the anterior and posterior tips of the plates overlap. If there had been two rows of plates, he reasoned, we would see more overlap. Czerkas also noted, as had some earlier paleontologists, that the neck, back, and tail of *Stegosaurus* were long enough to allow the plates to be arranged in a

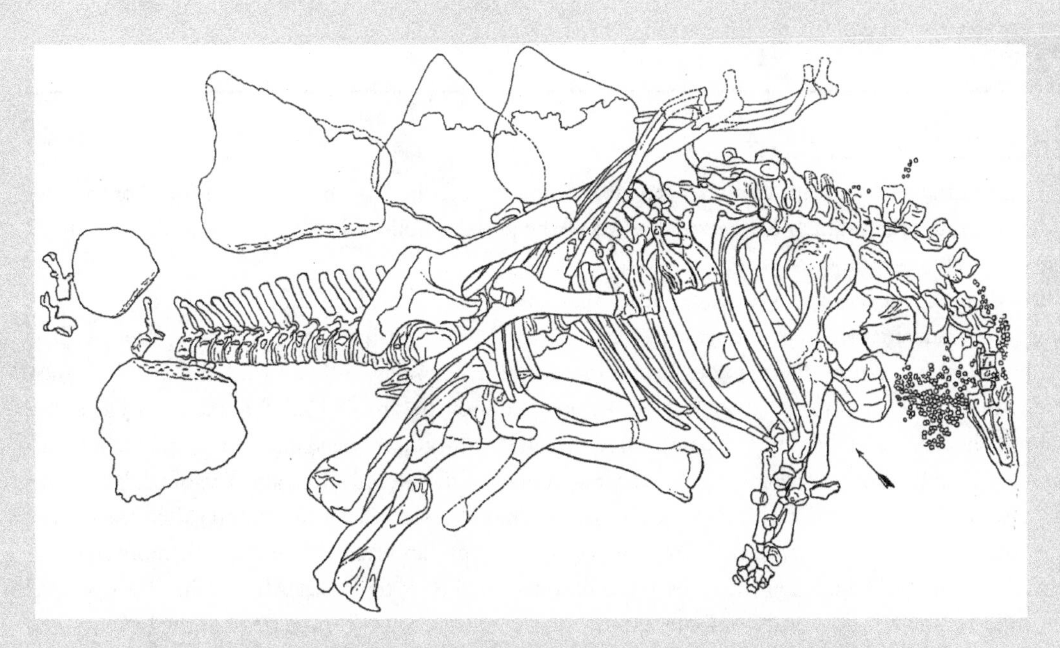

**BOX FIGURE 8.1A** This skeleton of *Stegosaurus* from Colorado, shown here as it was found in the rock, is the best direct evidence of the arrangement of the plates of *Stegosaurus*. (From C. W. Gilmore, and the United States National Museum. 1914. *Osteology of the Armored Dinosauria in the United States National Museum with Special Reference to the Genus Stegosaurus*. Washington, D.C.: Government Printing Office)

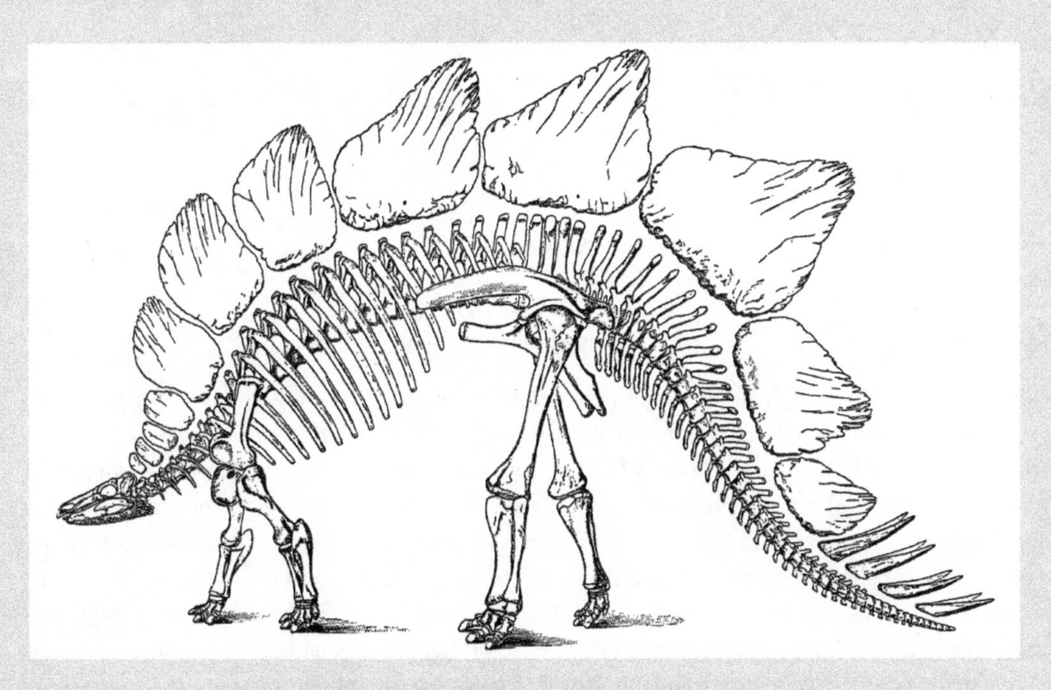

**BOX FIGURE 8.1B** The first reconstruction of *Stegosaurus* by Othniel Charles Marsh, in 1891, showed a single row of plates that do not overlap. (From E. H. Colbert. 1984. *The Great Dinosaur Hunters and Their Discoveries*. Mineola, N.Y.: Dover. Reprinted by permission of Dover Publications, Inc.)

single row with just a slight overlap of plates in the neck, shoulder, and back regions. Yet, despite the force of Czerkas's arguments, most paleontologists still find impressive the argument that the plates of *Stegosaurus* functioned best as radiators when arranged in two alternating rows, as shown in figure 8.5. Indeed, recently discovered skeletons of *Stegosaurus* from Colorado and Utah confirm the overlap of the plates.

## Box 8.2

### The Brain of *Stegosaurus*

The small size of the brain of *Stegosaurus* is legendary. Variously described as "the size of a walnut" or "smaller than a kitten's brain," the brain of *Stegosaurus* has led to the popular notion that it was one of the "dumbest" dinosaurs.

Our knowledge of the brain of *Stegosaurus*, or for that matter the brain of any dinosaur, rests on **endocasts** (short for "endocranial casts"). An endocast is a cast (replica) of the inside of a dinosaur's braincase. It is prepared by carefully cleaning the rock out of the interior of a dinosaur skull and then filling the cleaned braincase with soft rubber (usually latex) that can be pulled out of the opening for the spinal cord after the rubber cures. The flexible rubber must be compressed to pull it out through the spinal cord opening, but then it regains its shape. The resulting endocast is an accurate replica of the overall shape and size of the brain cavity, and it also records the positions of the major nerves and blood vessels that entered and exited the brain.

The endocast of the brain of *Stegosaurus* reveals a typically reptilian brain that had much more in common with the brain of a lizard than with the brain of a mammal (box figure 8.2). The brain of *Stegosaurus* was long and low, and it lacked much of a curve in the region of the cerebrum. The front of the brain, which included the olfactory bulbs associated with the sense of smell, was extraordinarily large, but the cerebrum and cerebellum were small, while the medulla was long and large.

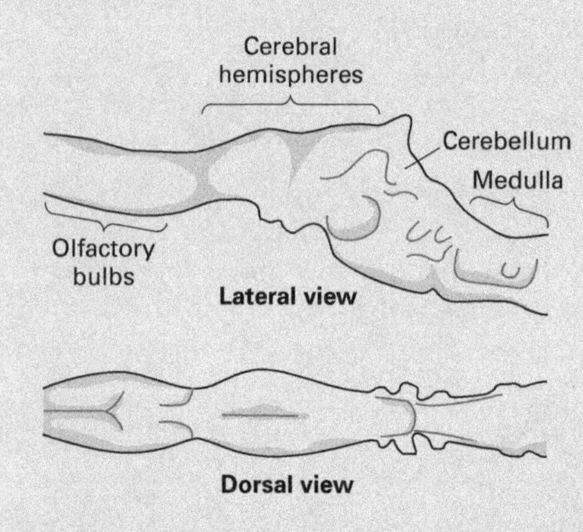


**BOX FIGURE 8.2**
The endocast of *Stegosaurus* resembles that of a living lizard. (Drawing by Network Graphics)

The endocast of *Stegosaurus* only displaces about 56 milliliters of water. Computer analysis of a different *Stegosaurus* skull estimated an endocast volume of about 64 milliliters. A living lizard enlarged to the size of *Stegosaurus*, however, would have an endocast volume of about 110 milliliters. The endocast of a living, fully grown house cat displaces about 30 milliliters of water. The average walnut from a grocery store displaces only 15 to 20 milliliters of water. And, when we look at the endocasts of a variety of dinosaurs, *Stegosaurus* does have one of the smallest brain volumes relative to its body size (see chapter 14).

*Stegosaurus* thus does not stand out as particularly "brainy" by any standard. Nevertheless, its brain was bigger than a walnut or a kitten's brain, so the intelligence of *Stegosaurus* has been somewhat maligned. Furthermore, stegosaurs were very successful land animals for at least 100 million years, more than 30 times the duration of the human species. Perhaps the best conclusion we can reach about stegosaur intelligence is that these dinosaurs were as intelligent as they had to be!

Some who have wondered how a 1- to 2-ton *Stegosaurus* functioned with a 2- to 3-ounce brain have pointed to the enlarged spinal cavity in its hip region as the possible location of a "second brain." However, no living reptile has a "second brain" located in the hip region. This enlargement, which is a cavity at least 20 times the size of a stegosaur brain, may have been an area where extra nerves met from the tail and hind limbs. But, a more likely explanation is that this cavity of *Stegosaurus* was for the storage of fat and sugar, as is seen today in some birds, such as ostriches.

The long neck and tail of *Stegosaurus* are typical features of stegosaurids. The powerfully built forelimbs of this dinosaur featured a shoulder blade and humerus with thick bony crests for the attachment of massive shoulder and upper arm muscles. This structure suggests a semi-sprawling posture for the forelimb when walking. The forefoot had five short, broad toes with hoof-like tips. In contrast, the hind limbs of *Stegosaurus* were extremely long and pillar-like. The long, columnar femur lacking broad flanges of bone suggests an upright hind limb posture when walking. Unlike the forefeet, the hind feet of *Stegosaurus* bore three short, stout toes with hooves.

The armor of *Stegosaurus* included numerous, small, knob-like plates that were distributed in the skin over most of the body. Most prominent, though, were the plates along the backbone. The plates immediately behind the head were small, flat and had irregularly shaped edges. Large, diamond-shaped plates followed and extended down the back onto the tail. The plates were largest over the hips. The plates were not solid sheets of bone; instead, their surfaces were covered with grooves and channels, and their interiors contained large canals (figure 8.7). Clearly, blood could flow into and around the plates of *Stegosaurus*. The row of plates was followed by two pairs of long spikes located at the end of the tail.

## Plate Function

For nearly a century, paleontologists believed that *Stegosaurus* used its plates as defensive armor. Various illustrations from earlier times showed *Stegosaurus* protecting itself from attacking predators with its plates. Some drawings even showed the plates lying nearly flat against the sides of the dinosaur's body, although even, in this situation, relatively little of the back and flanks of the dinosaur would have been shielded (figure 8.8).

**FIGURE 8.7**
The plates of *Stegosaurus* were not solid sheets of bone, but were instead covered with grooves and canals for blood vessels. (Drawing by Network Graphics)

**FIGURE 8.8**
This restoration of *Stegosaurus* with its plates lying flat shows just how little of the body and flanks of the dinosaur would have been covered by the plates. (Drawing by Network Graphics)

There are two reasons to question the idea that the plates of *Stegosaurus* were defensive armor. The first, as already stated, is that these plates, however they were arranged, only covered a small portion of the back and flanks of the dinosaur, leaving the belly and legs entirely unprotected. The second is the vascularization of the plates, which suggests that they were regularly and extensively filled with blood. It simply doesn't make sense to expose a blood-filled structure to an attacker.

A much more reasonable interpretation of the function of the plates of *Stegosaurus* stems from an elegant analysis undertaken by James Farlow and colleagues at Yale University during the late 1970s. These scientists argued, by analogy with living reptiles, that the plates of *Stegosaurus* functioned as radiators and solar panels that helped to regulate the dinosaur's body temperature. Pumping blood through the plates would have increased the surface area over which the blood was exposed so it could rapidly be cooled (in the shade) or heated (in the sun). A similar process cools the water in an automobile engine when the water is pumped into the radiator where numerous flat chambers increase its surface area. Living lizards also achieve rapid heating of their blood by flattening their bodies on warm rocks in the sun to increase the surface area over which their blood is exposed.

Farlow and colleagues tested the idea of *Stegosaurus* plates functioning as radiators and solar panels by modeling a *Stegosaurus* as a metal cylinder with slots on top into which they could place small metal plates. When this cylinder was heated, it was found to cool most quickly when the plates were diamond-shaped and arranged in two alternating rows. This strongly suggests that the plates of *Stegosaurus* functioned most effectively as radiators and solar panels when arranged in two alternating rows as shown in most reconstructions of *Stegosaurus* (see figure 8.5).

This analysis of the function of the plates of *Stegosaurus* seems convincing, but does it apply to the other stegosaurs? *Huayangosaurus* had spikes, not plates, and the arrangement and shape of the plates of other stegosaurids, such as *Kentrosaurus* and *Tuojiangosaurus*, are not as functionally optimal as *Stegosaurus*. Furthermore, the paired tail spikes of *Stegosaurus* do not fit into the model of thermoregulation and probably did function as defensive structures. Perhaps the best explanation is that stegosaur midline armor originally evolved as defensive spikes like those of *Huayangosaurus*. In some later stegosaurs, such as *Stegosaurus*, the shape and arrangement of the midline armor became modified for optimum efficiency as radiators and solar panels. The earlier stegosaurids may have had less efficient radiators and solar panels and may still have employed the plates, to some extent, in defense. Whatever else stegosaurs did with their midline armor, it likely also functioned in display as a device for recognizing members of the same species.

## Stegosaur Lifestyles and Evolution

Stegosaur skulls and teeth indicate they were plant-eaters that cropped vegetation with their horny beaks, then sliced it with their teeth before swallowing. Their limb structures suggest they were habitual quadrupeds that weighed as much as 2 tons. Stegosaurs probably could have reared up on their pillar-like hind limbs, but they browsed primarily on vegetation at a height of 1 meter or less above the ground.

Middle Jurassic stegosaur footprints are known primarily from the United States. The oldest stegosaur skeletons are from the Middle Jurassic of China and Europe.

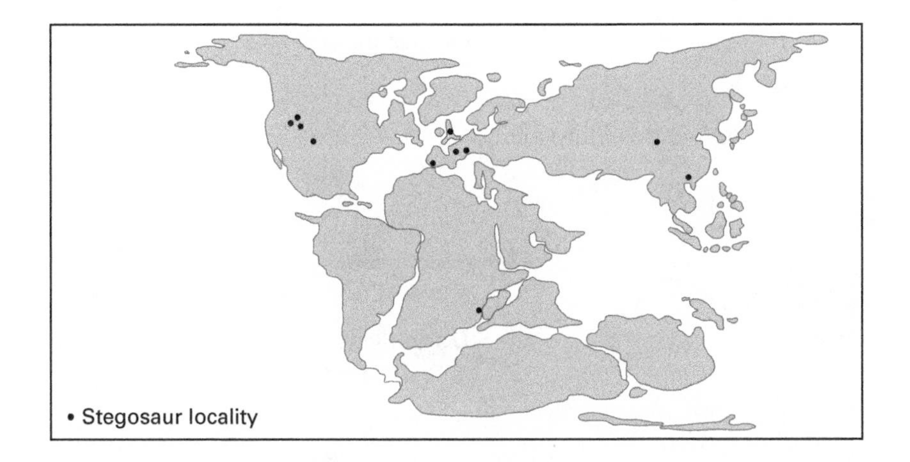


**FIGURE 8.9**
This map of stegosaur distribution shows that the stegosaurs lived nearly worldwide during the Late Jurassic. (Drawing by Network Graphics)

By the Late Jurassic, stegosaurs had achieved a nearly worldwide distribution and reached the zenith of their diversity, at least eight genera (figure 8.9). Stegosaurs were among the dominant large plant-eaters of the Late Jurassic, and their fossils are common in Upper Jurassic nonmarine rocks, especially in southern China, the western United States, and eastern Africa.

The Cretaceous record of stegosaurs is much sparser, and by the Cretaceous, the stegosaurs were clearly past their peak. The Cretaceous decline of stegosaurs is usually linked to the appearance of new types of plants (the flowering plants), new types of plant-eating dinosaurs (especially the large ornithopods), or both. Stegosaurs apparently were extinct by the beginning of the Late Cretaceous.

## ANKYLOSAURS

The ankylosaurs (meaning "fused lizards" because of the rod of fused vertebrae in their backs) were medium-size to large (up to 9 meters long), quadrupedal, plant-eating ornithischians. They had small heads with leaf-shaped, non-interlocking teeth, similar to the teeth of stegosaurs. The broadly arched ribs of ankylosaurs formed a very wide body, which was covered with small round or square armor plates that produced a fairly continuous shield over its dorsal surface (figure 8.10). Some ankylosaurs had spikes or spines as part of their body armor, and others had a club at the end of the tail. Their limbs were robust, the forelimbs were about two-thirds to three-fourths the length of the hind limbs, and their short, stout feet bore hooves.

The low skulls of ankylosaurs (in which the width between the eyes exceeds the height of the skull), the closure of the fenestrae in front of the eyes and on top of the skull, the added dermal bones on the skull, and the extensive body armor and modifications of the rib cage and hip girdle to form a base for this armor, are among the many distinctive evolutionary novelties of the **Ankylosauria**. Almost all of these primarily Cretaceous dinosaurs comprise two families: the **Nodosauridae** and the **Ankylosauridae**. However, some students of ankylosaurs assign the primitive ankylosaurs to a third family: the Polacanthidae (box 8.3).

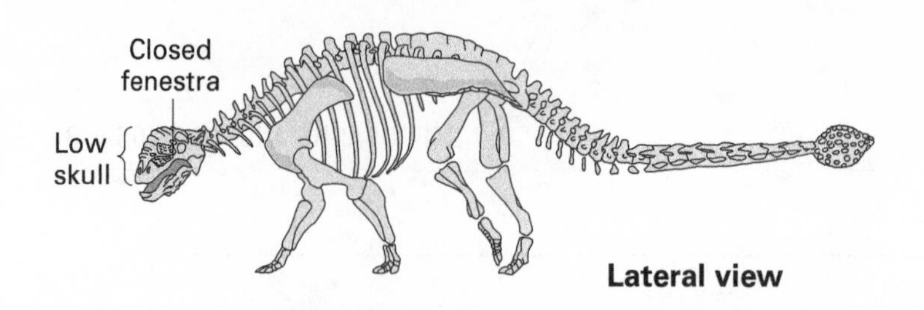


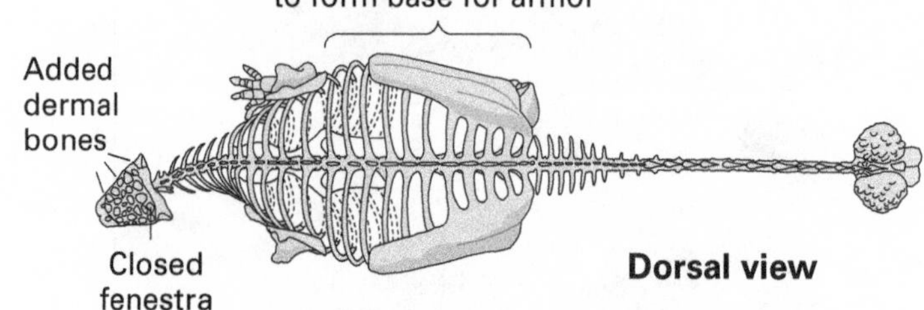


**FIGURE 8.10**
This ankylosaur skeleton (the ankylosaurid *Euoplocephalus*) displays the key evolutionary novelties of ankylosaurs. (Drawing by Network Graphics)

## Box 8.3

### Polacanthids

We know relatively little about what really happened during the early evolution of the ankylosaurs. Very recent discoveries, however, are rapidly filling the gap. One of the most interesting of these discoveries is the ankylosaur genus *Gastonia* (box figure 8.3). Named after its discoverer, Robert Gaston, *Gastonia* is known from more than 1,000 bones and pieces of armor collected in the Lower Cretaceous Cedar Mountain Formation of Utah. It was a 6-meter-long ankylosaur most similar to *Polacanthus,* particularly in its body armor. But, although the body of *Gastonia* shows features characteristic of nodosaurids (spiny armor and no tail club), its head shows characteristic ankylosaurid features, especially the triangular horns. *Gastonia* thus is not readily assigned to either the Nodosauridae or Ankylosauridae. Does it belong to another group of ankylosaurs?

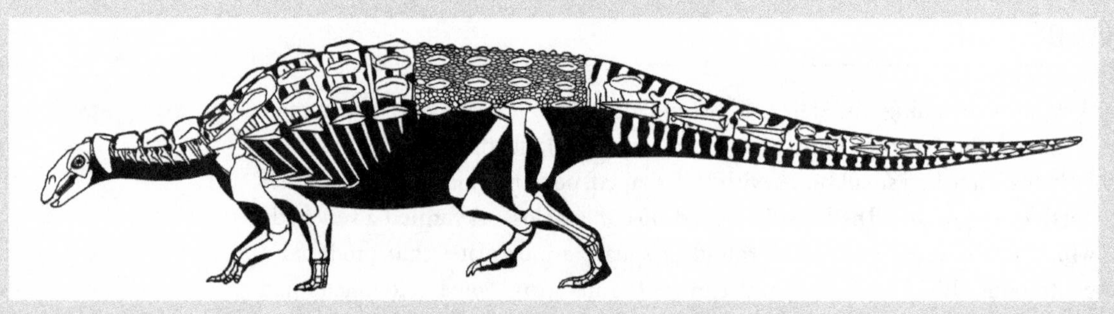

**BOX FIGURE 8.3**
*Gastonia* was a primitive ankylosaur from the Lower Cretaceous of North America. (© Scott Hartman)

Some paleontologists unite *Gastonia* with *Polacanthus* and some other ankylosaurs in the family Polacanthidae. In this analysis, ankylosaur evolution began with an early split into three groups: the nodosaurids, whose fossils first appear in the Lower Cretaceous; the Late Jurassic–Early Cretaceous polacanthids; and the Late Cretaceous ankylosaurids, which are more closely related to polacanthids than to nodosaurids. If this recent analysis is upheld, then *Gastonia* and other recent finds will have shaken the ankylosaur tree so hard that it now has three branches instead of two.

## Nodosaurids

Nodosaurid ankylosaurs are distinguished from ankylosaurids by their narrow skulls that lacked armor and horns at the posterior corners, the presence of spines in the armor, and the lack of ossified tail tendons or a tail club. *Edmontonia*, from the Upper Cretaceous of Montana, is typical of the family (figure 8.11). Other well-known nodosaurids are ***Nodosaurus***, from the Upper Cretaceous of Wyoming and Kansas; ***Panoplosaurus***, from the Upper Cretaceous of Alberta, Canada; and *Sauropelta*, from the Lower Cretaceous of the United States.

The skull of *Sauropelta* was small and narrow, had a pointed snout, and lacked horns at the posterior corners and extra armor that were characteristic of ankylosaurids (see figure 8.11). The lateral temporal fenestra was open, the cheek teeth were leaf-shaped, and the beak was toothless. As in stegosaurs and other ankylosaurs, the teeth of *Sauropelta* were not arranged into a dental battery.

The skeleton of *Sauropelta* represents a 6-meter-long nodosaurid. The neck was short, and the long tail had no club at its tip. The entire body was covered with regular bands of bony plates that formed a thick and heavy covering. Bony spikes were part of the body armor of *Sauropelta*, as in most other nodosaurids (see figure 8.11). The hind limbs of *Sauropelta* were longer than the forelimbs, and its feet were short, stout, and bore hooves.

**FIGURE 8.11**
The skeleton of *Edmontonia* is typical of nodosaurids. (© Scott Hartman)

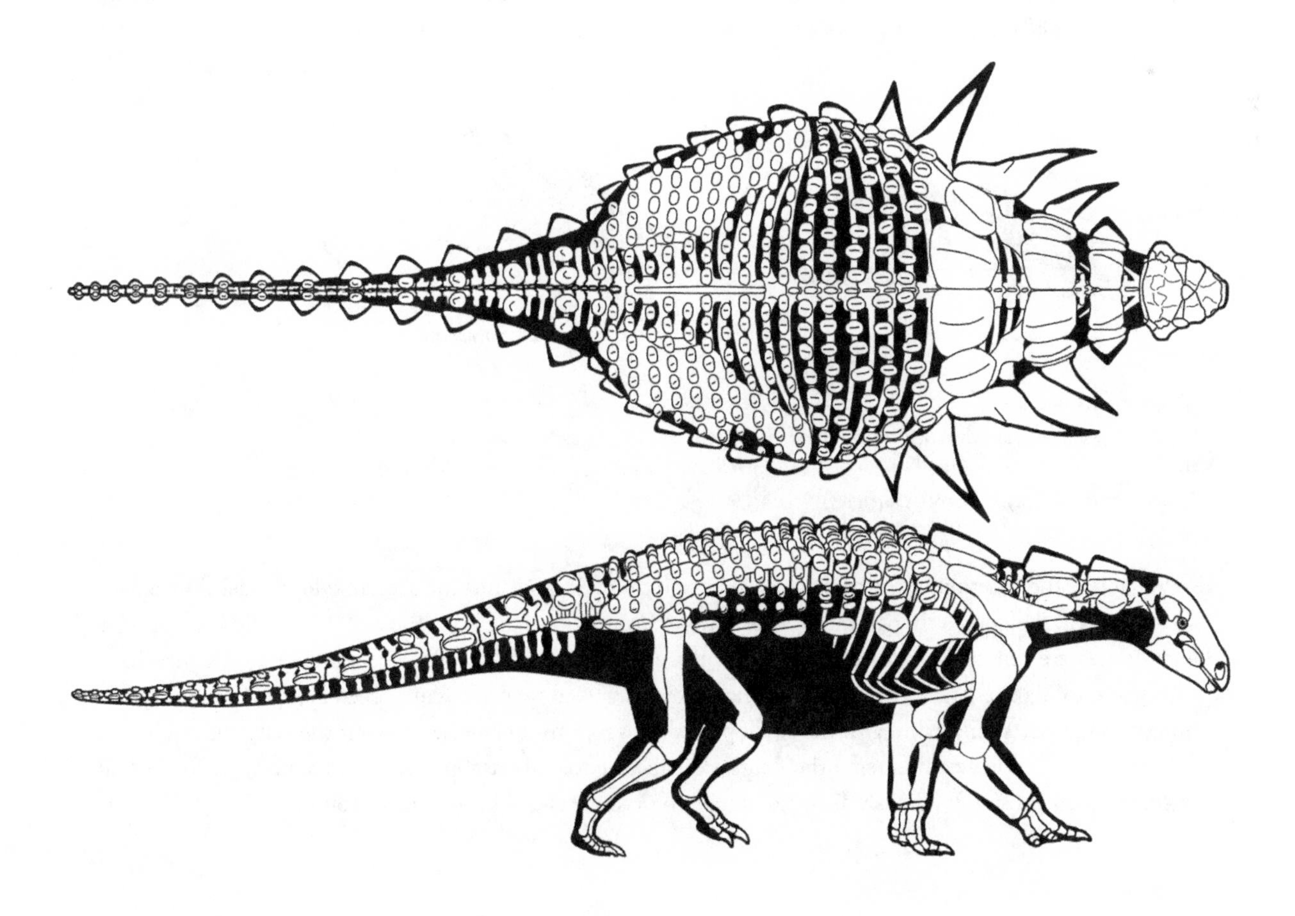

## Ankylosaurids

Ankylosaurids present a striking contrast to nodosaurids because of their wide, armored heads (as wide as they were long), which had long, triangular horns at the posterior corners and lateral temporal fenestrae completely covered with bony armor. The complex nasal passages of ankylosaurids also distinguish them from nodosaurids (box 8.4). Few or no spines were present in the body armor of ankylosaurids, and their

### Box 8.4

### Ankylosaurid Nasal Passages

Ankylosaurids are easily distinguished from nodosaurids by a variety of anatomical features, in particular their uniquely shaped nasal passages. As in most dinosaurs, the nasal passages of nodosaurids were paired tubes that ran from the nostrils directly back to the throat (box figure 8.4). By contrast, ankylosaurid nasal passages followed a folded, S-shaped path, and, on either side of the nasal passages, ankylosaur skulls were honeycombed with complex sinuses. Furthermore, in the Mongolian Cretaceous ankylosaurids *Pinacosaurus* and *Saichania*, thin, blade-like bones have been found in the nasal passages. These bones resemble the bones called "turbinals" present in the nasal passages of many mammals, including humans.

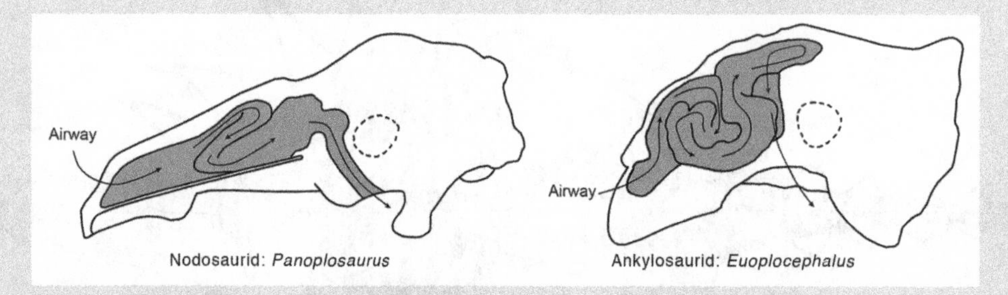


**BOX FIGURE 8.4**
Unlike nodosaurids, ankylosaurids had folded, S-shaped nasal passages, as shown in these cross-sections of the skull of the nodosaurid *Panoplosaurus* (*left*) and the ankylosaurid *Euoplocephalus* (*right*). (© Scott Hartman)

What was the function of the folded nasal passages, sinuses, and "turbinals" of ankylosaurids? As we have so often seen, they probably had multiple functions, not just a single one. The ankylosaurid nasal passages and sinuses are reminiscent of those in the skulls of hadrosaurian dinosaurs, so ankylosaurids may have used them to make characteristic noises, as has been suggested for hadrosaurs (see box 7.3). Other possible functions include strengthening of the skull, providing space for glands, improving the sense of smell, and moistening the air before it entered the lungs. Perhaps the folded nasal passages and sinuses performed all of these functions simultaneously and also helped to produce characteristic, nasal sounds.

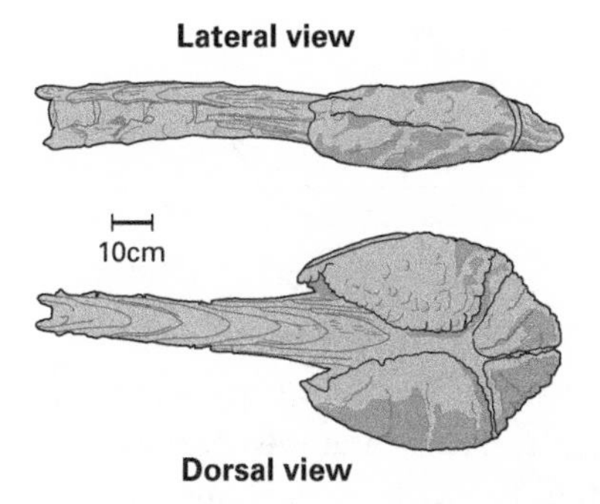


**FIGURE 8.12**
The tail club of ankylosaurids was made of two large and two small dermal plates. (Drawing by Network Graphics)

last few tail vertebrae were braced by ossified tendons. The tip of the ankylosaurid tail bore a club made up of two large and two small plates of bone (figure 8.12).

*Euoplocephalus*, from the Upper Cretaceous of Canada and the United States, is a characteristic ankylosaurid that displayed all the features typical of the family (see figure 8.10). This medium-size to large ankylosaurid (6 to 7 meters long) had a limb skeleton similar to that of *Sauropelta* but more massive. Large slabs of bone covered the skull of *Euoplocephalus* to armor it completely.

The body armor of *Euoplocephalus* consisted of bands of bony plates that ran across the body and tail, as well as smaller bony studs planted throughout the skin. Probably weighing as much as 6 tons, *Euoplocephalus* is one of the best-known ankylosaurids. Other well-known ankylosaurids include *Ankylosaurus*, from the Upper Cretaceous of the United States and Canada, and *Pinacosaurus* and *Saichania*, from the Upper Cretaceous of China and Mongolia. Our knowledge of ankylosaur diversity is growing rapidly, with new discoveries every year (box 8.5).

## Ankylosaurs: Mesozoic Tanks

Modern military tank design is based on balancing three factors: speed, armor, and firepower. An increase in tank speed results from a decrease in armor and firepower (weight) and vice versa. The ankylosaur design of a dinosaurian "tank" clearly sacrificed speed for the sake of heavy armor. Increased firepower, in the form of powerful jaws and large teeth, was also obviously avoided by ankylosaurs.

The robust and heavy limbs of ankylosaurs were not designed for speed, but for slow and powerful walking. The chassis upon which ankylosaur armor was hung consisted of a wide rib cage with broad ribs and a remarkably broad sacrum (see figure 8.10). Huge ilia hung down from the sacrum, draped over the upper thighs to provide additional protection. A heavy armor coating of round or rectangular plates was laid over this chassis. Plates arranged in rows, and not sutured to each other, helped the ankylosaur maintain some flexibility within this body covering.

Nodosaurids stopped with this design, so they probably maintained a modicum of speed. But, ankylosaurids went a step further in sacrificing speed for increased armor. Thus, the head of ankylosaurids acquired bony plating, and a tail club evolved to add some firepower to the slower ankylosaurid "tank."

## Box 8.5

### A New Ankylosaur

In July 2011, a joint field expedition of the New Mexico Museum of Natural History and Science (NMMNHS) and the State Museum of Pennsylvania (SMP) scoured the Upper Cretaceous badlands of northwestern New Mexico looking for dinosaur fossils. On a sandstone flat, SMP curator Robert M. Sullivan discovered an ankylosaur skull and part of its neck armor eroding from the 73-million-year-old rocks. The team excavated the fossil, encasing it in a plaster jacket, and transported it back to the NMMNHS in Albuquerque. There preparation began—cleaning, stabilization, and repair of the skull and armor plates. Uncovering the skull revealed a *Euoplocephalus*-like ankylosaurid, but one that differs in the configuration and shape of both the armor plates and the horns on the skull (box figure 8.5).

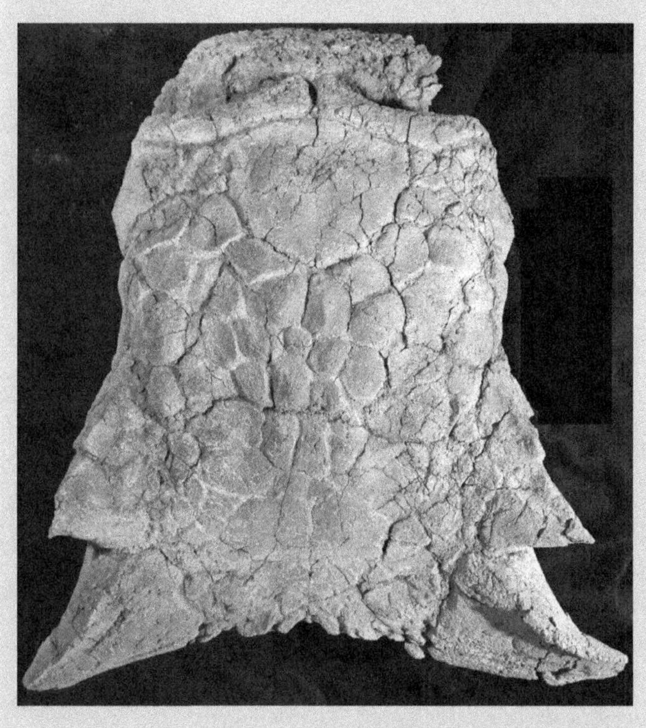

**BOX FIGURE 8.5**
The skull of *Ziapelta* has uniquely configured and shaped armor plates and horns.

Adding ankylosaur experts from the University of Alberta to the research team, it was decided to name a new kind of ankylosaurid. The researchers coined the name **Ziapelta** for the new dinosaur. *Zia* is a Native American sun symbol that is the logo on the New Mexico state flag, and *pelta* is Latin for "shield," a common suffix on ankylosaur generic names. A scientific article was written describing the new ankylosaurid and interpreting its significance for ankylosaur evolution. The manuscript was submitted to a scientific journal and after peer review, some revision, and editing, it was published in September 2014. Thus, bringing *Ziapelta* to the world, from discovery to publication, took a little more than three years!

There seems little doubt that, when attacked, an ankylosaur would not have tried to flee or counterattack. Instead, it would have squatted down, presenting a nearly impervious back to its attacker. Because of the ankylosaur's low-slung body and great weight, an attacker would have found it difficult to flip the "tank" over to get at its belly. When attacked, ankylosaurids would also have protected their blind backsides by swinging their massive tail clubs back and forth.

### Ankylosaur Evolution

The oldest known ankylosaur body fossil is that of *Sarcolestes*, from the Middle Jurassic of Europe. Although known only from part of a lower jaw, the teeth and an armor plate attached to the jaw identify *Sarcolestes* as an ankylosaur. Late Jurassic ankylosaurs are known from Europe and North America, notably ***Gargoyleosaurus***, described from a skull found in the Upper Jurassic of Wyoming.

Two types of Mesozoic "tanks" evolved among the ankylosaurs: the somewhat lighter nodosaurids and the very heavy ankylosaurids. But, the hallmark of ankylosaur evolution was conservatism. Once the basic heavily armored, tank-like body plan of these dinosaurs appeared, it was modified very little throughout their evolution.

It is clear that by the Cretaceous, the evolutionary split between nodosaurids and ankylosaurids had taken place. At least 18 genera of Cretaceous nodosaurids are known from Europe, North America, and Australia. Fragmentary ankylosaur fossils from South America and Africa may also be of nodosaurids, but these need to be reassessed. Nodosaurids reached their greatest diversity during the Early to Middle Cretaceous, but they did survive to the Late Cretaceous in the form of *Struthiosaurus* and *Edmontonia* in North America.

Ankylosaurids are well known from the Cretaceous of North America, Europe, and Asia. They encompass at least 18 genera, one of which, *Ankylosaurus*, at 6 to 7 meters long, was the largest ankylosaur. Ankylosaurids reached their greatest diversity during the Late Cretaceous but are not particularly abundant in the fossil record.

## Summary

1. Thyreophoran dinosaurs include two closely related groups, Stegosauria and Ankylosauria, as well as the primitive thyreophorans, *Scutellosaurus* and *Scelidosaurus*.
2. The presence above or alongside the vertebral column of one or more rows of dermal armor plates distinguishes thyreophorans from other dinosaurs.
3. The key evolutionary novelty of stegosaurs is the vertical bony plates and spines arranged in single or double rows along the neck, back, and tail.
4. *Huayangosaurus*, from the Middle Jurassic of China, is the oldest and most primitive stegosaur known from a complete skeleton.
5. More advanced stegosaurs are the stegosaurids, best represented by *Stegosaurus*, from the Upper Jurassic of the United States.
6. The plates of *Stegosaurus* were highly vascularized and probably had a temperature-regulation function.

7. Stegosaurs reached the zenith of their diversity, at least eight genera, and achieved a nearly worldwide distribution during the Late Jurassic. The Cretaceous decline of the stegosaurs and their extinction by Late Cretaceous time may have been due to the appearance of new vegetation (flowering plants) and/or the evolution of new types of plant-eating dinosaurs.
8. One of the key evolutionary novelties of ankylosaurs is their extensive body armor.
9. Two types of ankylosaurs, nodosaurids and ankylosaurids, had evolved by the Early Cretaceous and were distinguished from each other by differences in skull structure and body armor.
10. Ankylosaurs were the "tanks" of the Mesozoic and adopted a defensive strategy based on impervious armor.
11. Although the oldest ankylosaur is of Middle Jurassic age, the conservative evolution of this group of dinosaurs took place primarily during the Cretaceous.

## Key Terms

Ankylosauria
Ankylosauridae
armor plates
*Emausaurus*
endocast
*Euoplocephalus*
*Gargoyleosaurus*
*Gastonia*
*Gigantspinosaurus*
*Huayangosaurus*
Nodosauridae
*Nodosaurus*
*Panoplosaurus*
*Sauropelta*
*Scelidosaurus*
*Scutellosaurus*
Stegosauria
Stegosauridae
*Stegosaurus*
Thyreophora
*Ziapelta*

## Review Questions

1. What features identify a dinosaur as a thyreophoran?
2. Why is *Scutellosaurus* identified as a thyreophoran, and why is it easier to assign *Scelidosaurus* to the thyreophorans?
3. What are the evolutionary novelties of stegosaurs, and how does *Huayangosaurus* exemplify them?
4. How are stegosaurids distinguished from *Huayangosaurus*?
5. What do paleontologists believe were the arrangement and function of the plates of *Stegosaurus*? Why?
6. How does the evolution of stegosaurs differ from that of ankylosaurs in terms of timing, distribution, and amount of morphological change?
7. How are ankylosaurs distinguished from other dinosaurs? How are nodosaurid ankylosaurs distinguished from ankylosaurids?
8. What defensive strategy did ankylosaurs employ?

## Further Reading

Arbour, V. M., et al. 2014. A new ankylosaurid dinosaur from the Upper Cretaceous (Kirtlandian) of New Mexico with implications for ankylosaurid diversity in the Upper Cretaceous of western North America. *PLoS ONE* 9:e108804. (Names a brand new ankylosaurid, *Ziapelta*)

Carpenter, K., ed. 2001. *The Armored Dinosaurs*. Bloomington: Indiana University Press. (A collection of 21 technical articles presenting research on ankylosaurs)

Carpenter, K., 2012. Ankylosaurs, pp. 505–525, in M. K. Brett-Surman, T. R. Holtz, Jr., and J. O. Farlow, eds., *The Complete Dinosaur*. 2nd ed. Bloomington: Indiana University. (Concise review of the ankylosaurs)

Colbert, E. H. 1981. *A Primitive Ornithischian Dinosaur from the Kayenta Formation of Arizona*. Flagstaff: Museum of Northern Arizona Press. (Complete description of the primitive thyreophoran *Scutellosaurus*)

Czerkas, S. A. 1987. A reevaluation of the plate arrangement of *Stegosaurus stenops*, pp. 83–99, in S. J. Czerkas and E. C. Olson, eds., *Dinosaurs Past and Present*. Vol. 2. Los Angeles: Natural History Museum of Los Angeles County and University of Washington Press. (Reviews ideas about the plate arrangement of *Stegosaurus* and argues for a single row of plates)

Farlow, J. O., S. Hayashi, and G. J. Tattersall. 2010. Internal vascularity of the dermal plates of *Stegosaurus* (Ornithischia, Thyreophora). *Swiss Journal of Geosciences* 103:173–185. (The latest analysis of thermoregulation by *Stegosaurus* using its plates)

Farlow, J. O., C. V. Thompson, and D. E. Rosner. 1976. Plates of the dinosaur *Stegosaurus*: Forced convection heat loss fins? *Science* 192:1123–1125. (Proposes that the plates of *Stegosaurus* functioned like the cooling fins of radiators)

Galton, P. M. 2012. Stegosaurs, pp. 483–504, in M. K. Brett-Surman, T. R. Holtz, Jr., and J. O. Farlow, eds., *The Complete Dinosaur*. 2nd ed. Bloomington: Indiana University Press. (Concise review of the stegosaurs)

Maidment, S. C. R. 2010. Stegosauria: A historical review of the body fossil record and phylogenetic relationships. *Swiss Journal of Geosciences* 103:199–210. (Reviews stegosaur distribution, taxonomy, and phylogeny)

Maidment, S. C. R., D. B. Norman, P. M. Barrett, and P. Upchurch. 2008. Systematics and phylogeny of Stegosauria (Dinosauria: Ornithischia). *Journal of Systematic Palaeontology* 6:367–407. (Cladistic analysis of stegosaur phylogeny)

Thompson, R. S., J. C. Parrish, S. C. R. Maidment, and P. M. Barrett. 2012. Phylogeny of the ankylosaurian dinosaurs (Ornithischia: Thyreophora). *Journal of Systematic Palaeontology* 10:301–312. (Cladistic analysis of ankylosaur phylogeny)

## Find a Dinosaur!

Tiny heads, bulky bodies, weird plates in rows on their backs, and spiky tails identify stegosaurs as some of the most bizarre-looking dinosaurs. Thus, they are popular display fossils in many natural history museums. If you make it to southern China, the Zigong Dinosaur Museum (Sichuan) exhibits a skeleton of the most primitive known stegosaur, *Huayangosaurus*. Down the road, check out the mounted skeleton of the Late Jurassic stegosaur *Tuojiangosaurus* in the Chongqing Municipal Museum. But, closer to home, at the Natural History Museum in London, England, you will find a mounted replica of that skeleton of *Tuojiangosaurus*. Not far away, in Berlin, Germany, you can also see a similar Late Jurassic stegosaur from East Africa, *Kentrosaurus*, in the Museum für Naturkunde. Famous Yale paleontologist Othniel Charles Marsh coined the name *Stegosaurus* in 1877, so do make a pilgrimage to the mounted *Stegosaurus* skeleton at the Yale Peabody Museum of Natural History (New Haven, Connecticut). *Stegosaurus* is the official state fossil of Colorado, so you can also see one at the Denver Museum of Nature and Science. And, if you go to The Dinosaur Museum (Blanding, Utah), you can see a reconstruction of *Stegosaurus* with a single row of plates (Marsh's original idea) by famed dinosaur artist Stephen Czerkas. Between Blanding and New York and beyond, almost every natural history museum has at least a replica (cast) of a skeleton of *Stegosaurus*.

# 9

# CERATOPSIANS AND PACHYCEPHALOSAURS

THE horned dinosaurs, **Ceratopsia**, are among the most familiar dinosaurs to us. *Triceratops*, usually in combat with a *Tyrannosaurus rex*, has been a staple of dinosaur movies for decades. Less familiar are the dome-headed dinosaurs, **Pachycephalosauria**, close relatives of the ceratopsians. Here, I examine the anatomy, classification, evolution, and probable habits of these two groups of dinosaurs.

Most paleontologists have concluded that the ceratopsians and pachycephalosaurs share an evolutionary ancestry that justifies linking them into a single group called **Marginocephalia** (figure 9.1). This is despite the fact that, on first appearance, horned and dome-headed dinosaurs appear to be strikingly different. But, a few key evolutionary novelties—especially the development of at least a small **frill** (a shelf of bone projecting from the back of the skull)—suggest a close relationship between the two groups. Marginocephalian dinosaurs were primarily of Late Cretaceous age. They were plant-eating ornithischians closely related to ornithopods.

## CERATOPSIANS

Ceratopsian dinosaurs were one of the most diverse groups of plant-eating dinosaurs of the Late Cretaceous. They comprise at least two groups: the primitive ceratopsians, well represented by the genus *Psittacosaurus*, from the Early Cretaceous of Asia, and Neoceratopsia, from the Jurassic–Cretaceous of Asia and Cretaceous of North America. Evolutionary novelties that unite ceratopsians are the presence of a rostral (in the snout) bone in the skull, a skull with a narrow beak and flaring jugals (cheeks), a deep jugal with a distinct ridge, a frill, and a highly vaulted palate in the front of the mouth (figure 9.2). The morphology and distribution of ceratopsian fossils suggest the ceratopsians originated in Asia and spread to North America, where they were highly successful until their extinction at the end of the Cretaceous. They were among the last dinosaurs.

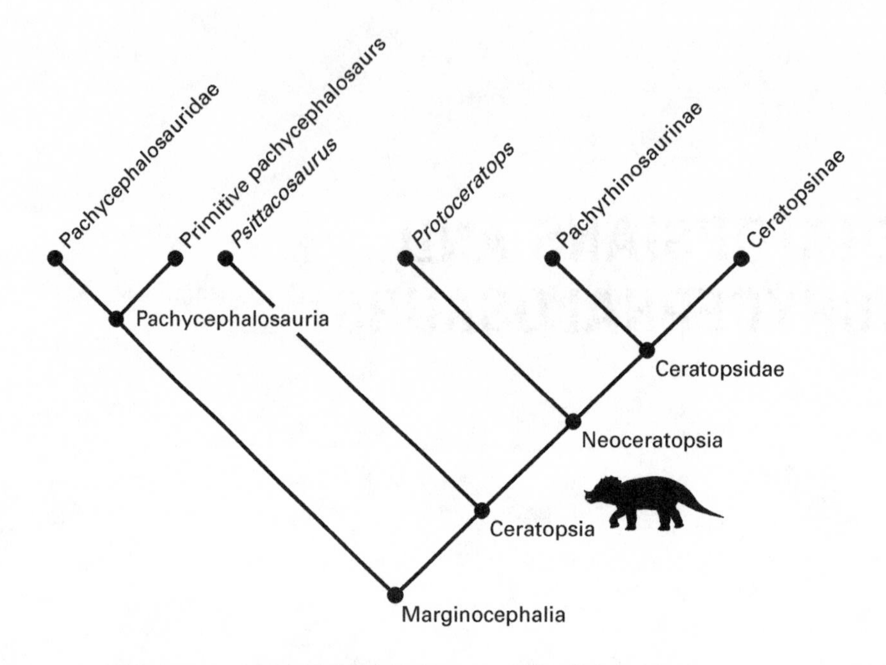


**FIGURE 9.1**
Ceratopsians and pachycephalosaurs share an ancestry and thus represent one group of dinosaurs, the Marginocephalia. (Drawing by Network Graphics)

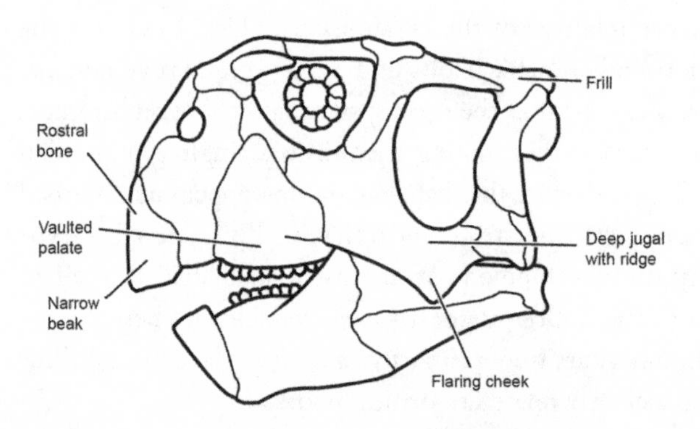


**FIGURE 9.2**
This skull of *Psittacosaurus* shows the key evolutionary novelties of the ceratopsians. (© Scott Hartman)

## The Genus *Psittacosaurus*

The "parrot dinosaur," ***Psittacosaurus***, well represents a primitive ceratopsian (figure 9.3). As many as seven species of *Psittacosaurus* are known from the Lower Cretaceous of eastern Asia, and their remains include many complete skulls and skeletons and even skin impressions (see box 13.1). Recently discovered early and primitive ceratopsians include the Late Jurassic *Yinlong* and *Chaoyangsaurus* from China. For many years, psittacosaurs had been allied with the ornithopods, but analysis of their excellent fossil record supports the identification of *Psittacosaurus* as an early ceratopsian.

*Psittacosaurus* possessed all the key evolutionary novelties of the Ceratopsia, even though it had only the most rudimentary of frills. Indeed, the posterior end of the skull roof just barely overhung the back end of the skull. The short snout, the high

**FIGURE 9.3**
Early Cretaceous *Psittacosaurus* was a characteristic primitive ceratopsian. (© Scott Hartman)

position of the nostrils, the tall rostrum ("snout") that superficially resembled a parrot's beak, and the reduction of the functional toes of the hand to three are diagnostic features of *Psittacosaurus* among ceratopsians.

The cheek teeth of *Psittacosaurus* had broad, flat wear surfaces with self-sharpening edges but did not occlude precisely. Their placement in the jaws was inset from the sides of the skull, suggesting the presence of cheek pouches. *Psittacosaurus* did not exceed 2 meters in length, and its skeletal structure was much more like that of a primitive ornithischian than other ceratopsians. In particular, the hind limb was longer than the forelimb, and the forefoot had only three functional toes, whereas the hind foot had four slender toes. The neck was short, and the tail was moderately long. Ossified tendons were present in some species of *Psittacosaurus* along the spine in the back and hip region.

The teeth of psittacosaurs were characteristic of plant eaters. They were usually well worn, and polished stones (gastroliths) associated with some psittacosaur skeletons suggest that significant amounts of vegetation were milled in the stomach. The forelimbs of *Psittacosaurus* were about 58 percent of the hind limb length, indicating this dinosaur was a facultative biped. Psittacosaurs were probably able to grasp with their hands; their first finger diverged from the other two (see figure 9.3). If this was the case, psittacosaurs were primarily bipeds, using the hands to grasp vegetation while eating.

The psittacosaurs were widespread and reasonably common dinosaurs in Asia during the Early Cretaceous. They represent well an early stage in ceratopsian evolution that preceded a much more diverse and impressive group of dinosaurs, the **Neoceratopsia**.

## Neoceratopsians

If you had visited a Late Cretaceous landscape in western North America, you would surely have seen a neoceratopsian. These dinosaurs were among the most diverse and abundant dinosaurs at that time and place. Neoceratopsians include an array of

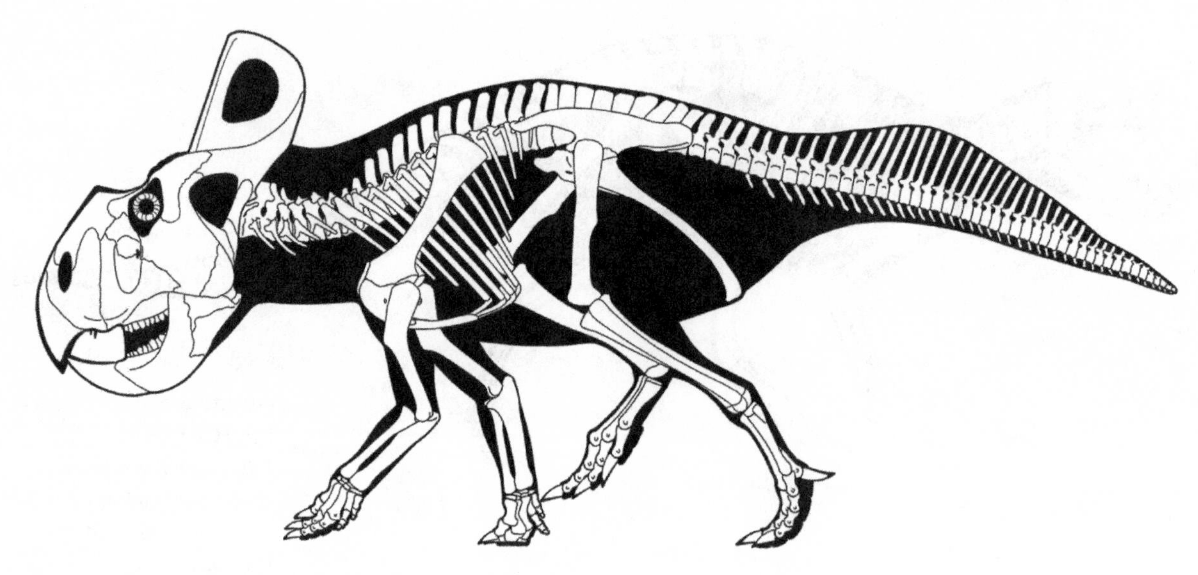

**FIGURE 9.4**
Late Cretaceous *Protoceratops* was a characteristic primitive neoceratopsian. (© Scott Hartman)

primitive genera, mostly Asian, as well as the Asian and western North American ceratopsids. Evolutionary novelties that distinguish the neoceratopsians from the psittacosaurs include an extremely large head, a broad and prominent frill, a pointed and sharply keeled rostrum, and limb structures associated with obligate quadrupedalism (figure 9.4).

The fossil record of neoceratopsians is one of the best of any group of dinosaurs. Many complete skeletons have been discovered, and well-preserved skulls are common.

## The Genus *Protoceratops*

Most primitive neoceratopsians were of Late Cretaceous age, but they truly represent a stage of ceratopsian evolution intermediate between the psittacosaurs and the ceratopsids. These small (1 to 2.5 meters long), primitive Late Cretaceous neoceratopsians include the well-known genera ***Protoceratops*** (see figure 9.4) and *Bagaceratops* from Mongolia and *Montanoceratops* and *Leptoceratops* from North America. Early Cretaceous neoceratopsians include *Liaoceratops* from China and *Aquilops* from North America.

*Protoceratops*, here considered a typical primitive neoceratopsian, had a relatively larger skull and a much longer frill than *Psittacosaurus*. The fore- and hind limbs of *Protoceratops* were of more nearly equal lengths than those of *Psittacosaurus*, and the limbs were more massive with broader feet. In these features, *Protoceratops* was much more like a ceratopsid than was *Psittacosaurus*.

Yet, unlike ceratopsids, the frill of *Protoceratops* was still rather short, *Protoceratops* had no horns, and its nostrils were small. These features identify *Protoceratops* as a neoceratopsian more primitive than the ceratopsids. Indeed, the ancestors of the ceratopsids must have looked similar to *Protoceratops*.

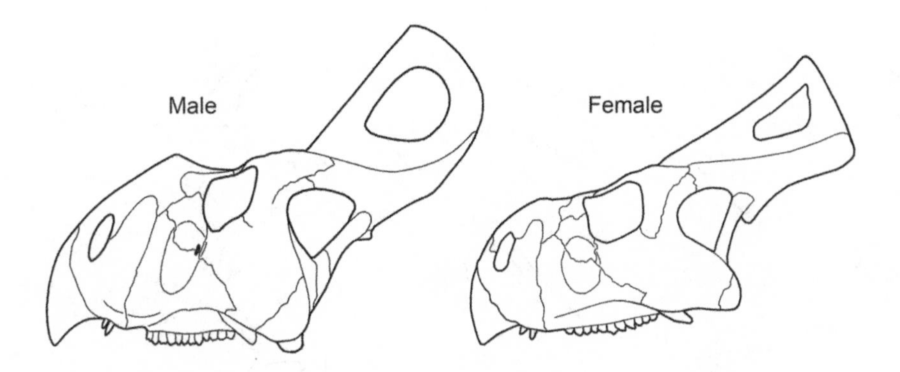


**FIGURE 9.5**
Differences in skull size and shape distinguished male and female *Protoceratops*. (© Scott Hartman)

## The Lifestyle of *Protoceratops*

Fossils of *Protoceratops* come from areas that were deserts during the Late Cretaceous. In those deserts, extensive dune fields were dotted by intermittent lakes and streams. The dozens of skeletons and skulls of *Protoceratops* found suggest that these dinosaurs nested communally and were gregarious.

Measurements of *Protoceratops* indicate that as adults, these dinosaurs had skulls of two different shapes (figure 9.5). One type of skull had a larger and more erect frill than the other, as well as a prominent bump on the snout. It seems likely, by analogy with living animals, that these two skull types represent males and females of a single species of *Protoceratops*.

We can thus envision groups of *Protoceratops* living in the deserts of Central Asia during the Late Cretaceous and foraging for the vegetation that grew in wetter places in the desert.

## Ceratopsids

The **Ceratopsidae** were large (4 to 8 meters long) habitual quadrupeds. Their distinctive evolutionary novelties include very large skulls (1 to 2.8 meters long), large nostrils, prominent frills, and a variety of horns (figure 9.6). Ceratopsids are known primarily from the Late Cretaceous of North America. One ceratopsid, *Turanoceratops*, is known from the Late Cretaceous of Asia.

Paleontologists recognize two different types of ceratopsids: **Pachyrhinosaurinae** and **Ceratopsinae** (figure 9.7). The pachyrhinosaurines are considered the more primitive of the two because they had a relatively short, high face and a short frill. They also had horns near the nostrils (nasal horns) that were much larger than the paired horns behind the eyes (postorbital horns). In contrast, the more advanced ceratopsines had long, low faces, long frills, and postorbital horns that were usually larger than the nasal horns. Ceratopsines include the largest of all ceratopsids, *Torosaurus*, *Triceratops*, and *Pentaceratops*, which had skulls as much as 2.8 meters long and bodies up to 8 meters long.

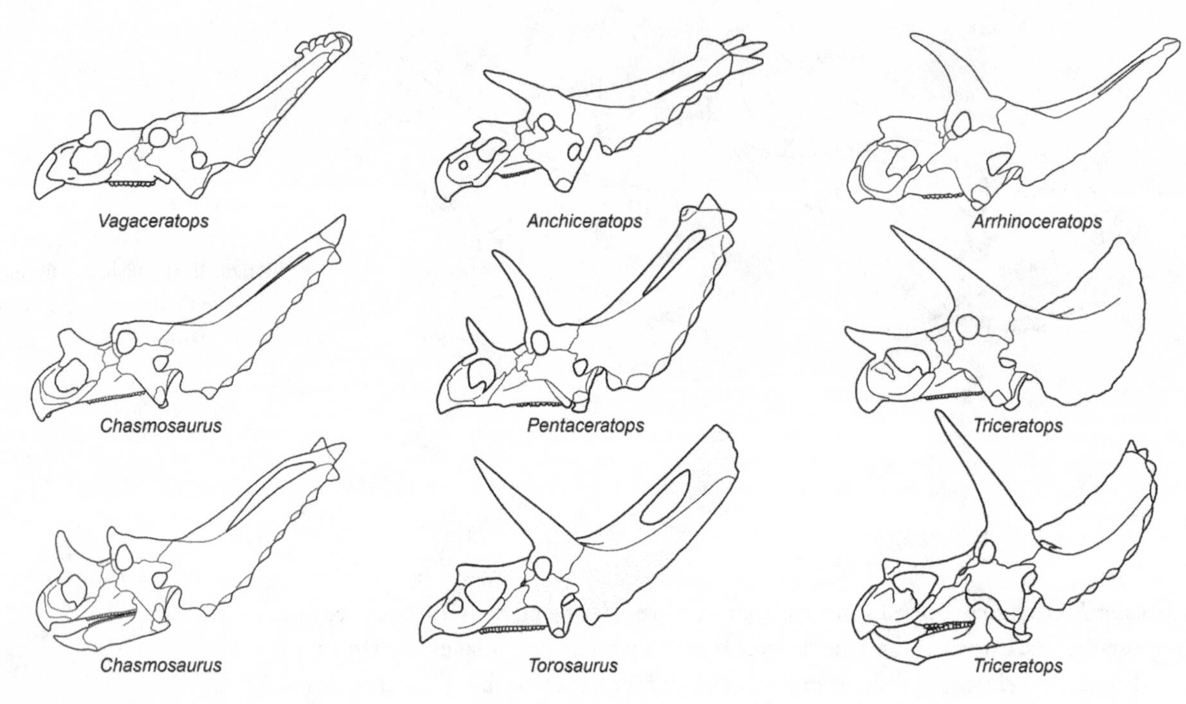


**FIGURE 9.6**
The skulls of ceratopsids encompassed a variety of horn types and frill shapes. (© Scott Hartman)

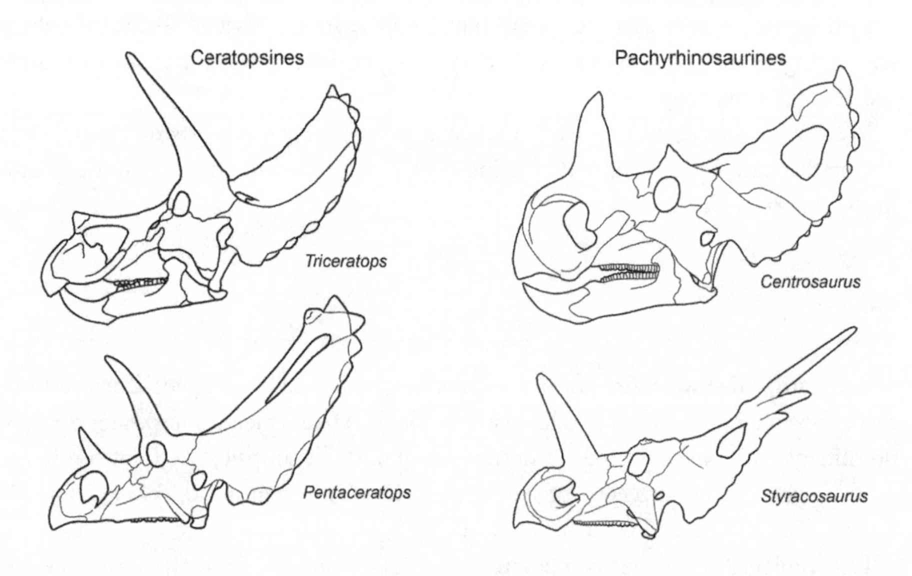


**FIGURE 9.7**
Paleontologists distinguish two types of ceratopsids, pachyrhinosaurines and ceratopsines, based on skull features. Pachyrhinosaurines had relatively short, high faces, short frills, and nasal horns larger than the postorbital horns. Ceratopsines had long, low faces, long frills, and postorbital horns larger than the nasal horns. (© Scott Hartman)

Pachyrhinosaurine fossils are known primarily from Montana, Alberta, and Alaska. Ceratopsines, however, were more widespread. Their fossils extend from Alberta to Texas, and fragmentary ceratopsid fossils from Alaska and Mexico may also be of ceratopsines. *Triceratops* is the most famous ceratopsine, and I examine it here as a typical ceratopsid.

## The Genus *Triceratops*

In many ways, ***Triceratops*** is a typical ceratopsine (figure 9.8). It had a long, low snout and large postorbital horns. But, unlike other ceratopsines (see figure 9.7), *Triceratops* had a short frill that lacked openings (fenestrae). The edge of the frill of *Triceratops* was lined with small, conical bones called epoccipitals, a feature seen in some other ceratopsids, including *Centrosaurus* and *Pentaceratops*.

As in all ceratopsids, the teeth of *Triceratops* formed a complex dental battery, in which adjacent teeth were locked together in longitudinal rows and vertical columns (figure 9.9). As the teeth along the cutting edge wore out, they were lost and replaced from below by new teeth. Ceratopsid teeth and dental batteries were thus very similar (convergent) to those of some hadrosaurids (see chapter 7).

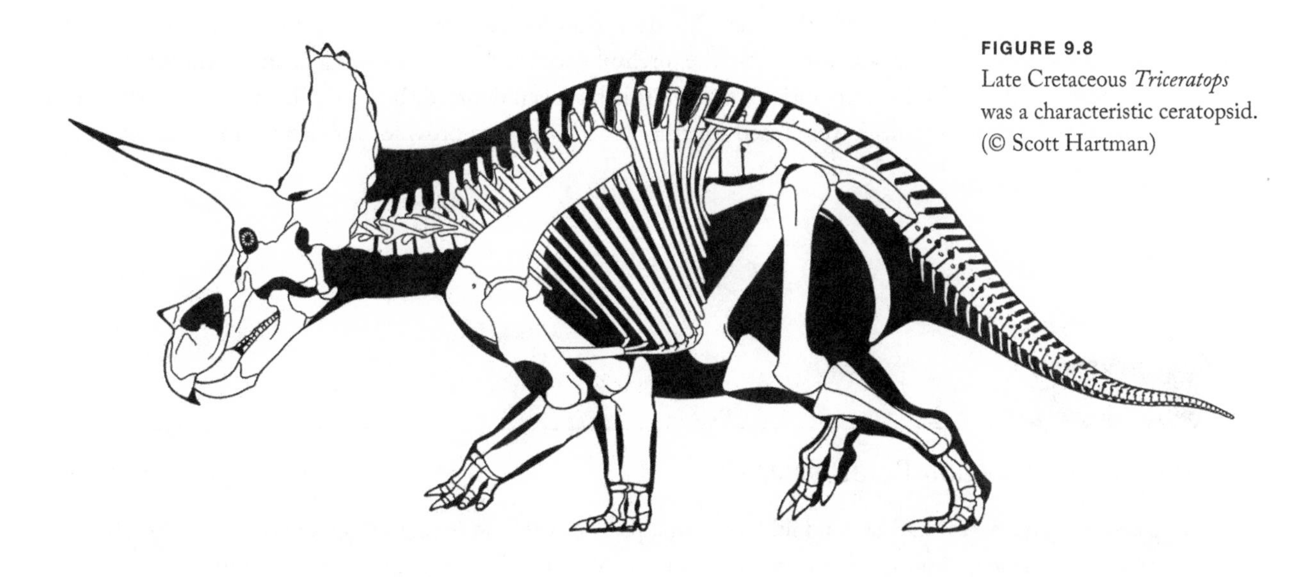

**FIGURE 9.8**
Late Cretaceous *Triceratops* was a characteristic ceratopsid. (© Scott Hartman)

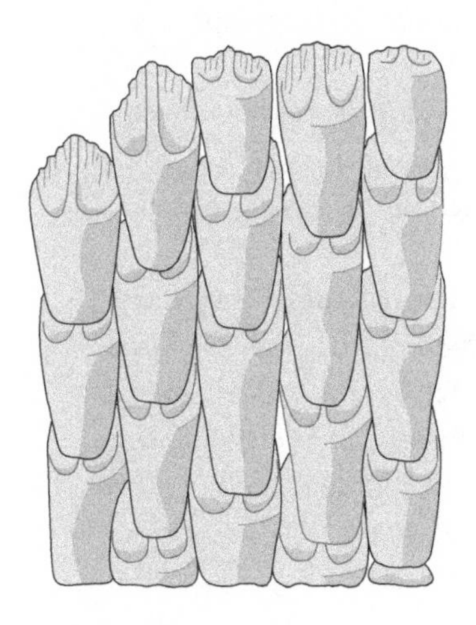

**FIGURE 9.9**
Ceratopsid teeth formed a complex dental battery. (Drawing by Network Graphics)

To move the jaws and operate its dental battery, *Triceratops* had large jaw-moving muscles anchored in the frill. Three forward-pointing horns projected from the skull: two large ones above each orbit and a much smaller one above the nostril. The bone of these horns, the frill, and much of the skull bore numerous grooves and channels for blood vessels. It seems likely that the horns of *Triceratops* had keratinous sheaths, as do the horns of living cows.

As in other ceratopsids, the forelimbs and hind limbs of *Triceratops* were of nearly equal length, thus identifying these dinosaurs as obligate quadrupeds. This interpretation is supported by the short, slender tail of *Triceratops*, which clearly was not used as a counterbalance for walking. Ossified tendons were present only in the hip region of *Triceratops*, and the pelvis was fused to the backbone along 10 sacral vertebrae, four or five more than was usual among dinosaurs.

The solid fusion of the backbone and hip of *Triceratops* suggests great stability and power when walking. The massive hind limbs with four broad, hooved toes are consistent with this idea. The deep rib cage of *Triceratops* supported a massive shoulder girdle and forelimbs. The forelimbs were held in a sprawling posture, whereas the hind limbs were upright, though this has been debated (box 9.1). The first four vertebrae of the neck were fused together (co-ossified) to provide extra support for the extremely heavy head.

## Box 9.1

### Upright or Sprawling Ceratopsians?

All paleontologists agree that the hind limbs of ceratopsians were held in an upright stance with the limbs essentially vertical and almost directly under the pelvis. But, sharp differences of opinion exist as to how the forelimbs were held by ceratopsians. Some paleontologists favor a semi-sprawling or sprawling forelimb in which the ceratopsian humerus was held obliquely or even parallel to the ground. Others favor a fully upright forelimb posture in which the humerus was held vertical to the ground (box figure 9.1A). Indeed, a ceratopsian trackway from Colorado indicates the dinosaur walked with nearly upright (semi-sprawling) forelimbs (see figure 12.6).

The most rigorous analyses of ceratopsian forelimb posture based on bones favor the sprawling or semi-sprawling stance. These studies argue that the shape of the ceratopsian humerus, the size and shape of the shoulder and elbow joints, and the inferred alignment of forelimb muscles based on these shapes, make a habitually upright forelimb in ceratopsians impossible.

We can easily gain some insight into these analyses by comparing the humerus of a living lizard (a sprawler), a living rhinoceros (an upright walker) and a *Triceratops* (box figure 9.1B). The *Triceratops* humerus most closely resembles that of the lizard in having broad flanges of bone designed for the attachment of large shoulder (deltoid) and chest (pectoral) muscles. In the lizard, these relatively large muscles are needed to hold and move the humerus horizontally while walking. In contrast, the rhinoceros

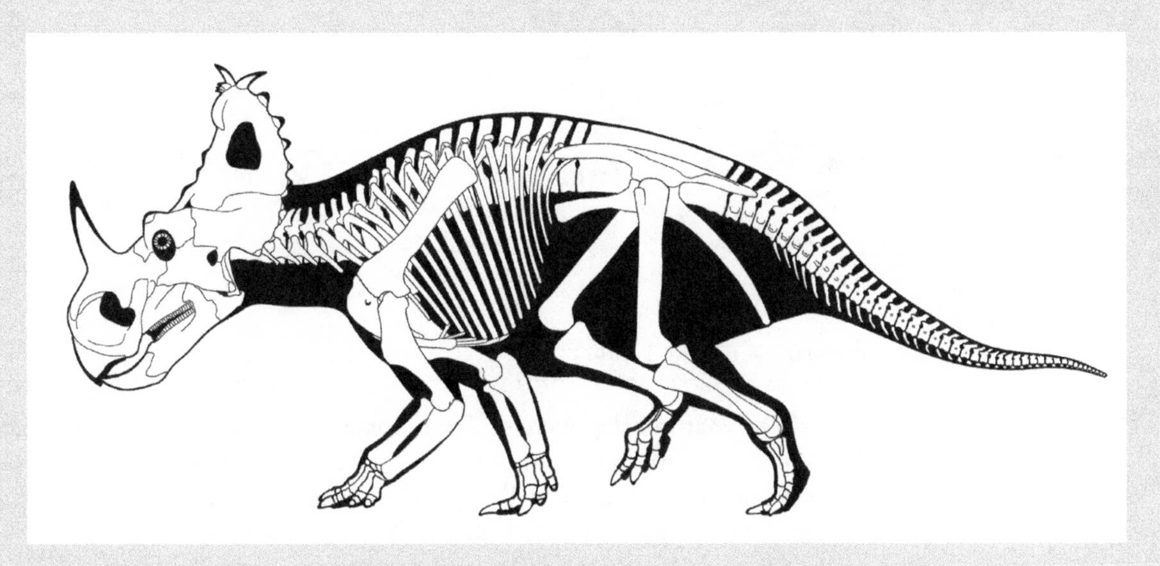

**BOX FIGURE 9.1A**
Some paleontologists have argued that ceratopsians held the forelimb vertically. (© Scott Hartman)

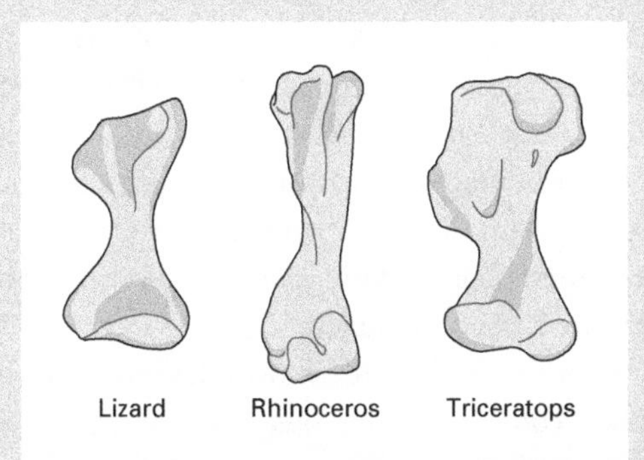


**BOX FIGURE 9.1B**
When the humeri of a living Komodo dragon, a living rhinoceros, and a *Triceratops* are compared, it is easy to see that the ceratopsian most resembles the lizard. (Drawing by Network Graphics)

does not have such relatively large shoulder and chest muscles (or bony attachments for them on its humerus), because it walks with its legs vertical. Given the shape of the humerus of *Triceratops* and other ceratopsians, it is most reasonable to conclude that these dinosaurs walked with the humerus in a semi-sprawling or sprawling posture.

Nevertheless, the only known ceratopsian footprints seem to indicate an upright forelimb posture. The bone and footprint evidence thus present a contradiction not easy to resolve. Perhaps the answer lies in suggesting that *Triceratops* held the forelimb with at least a bit of a sprawl and twisted the backbone and front part of the body, so that the forefeet were planted much closer together when walking than normally occurs when walking on a sprawling limb.

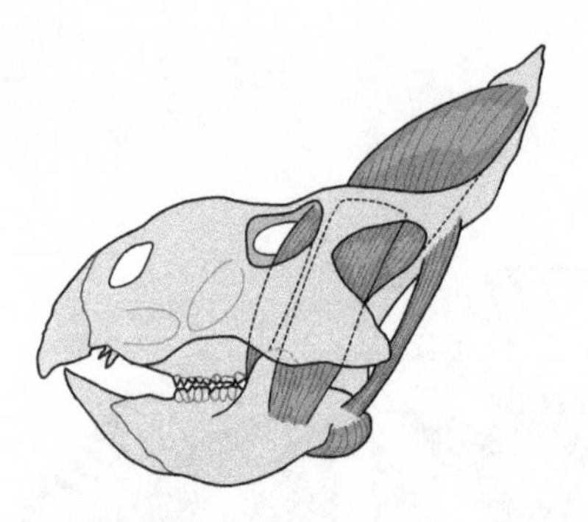

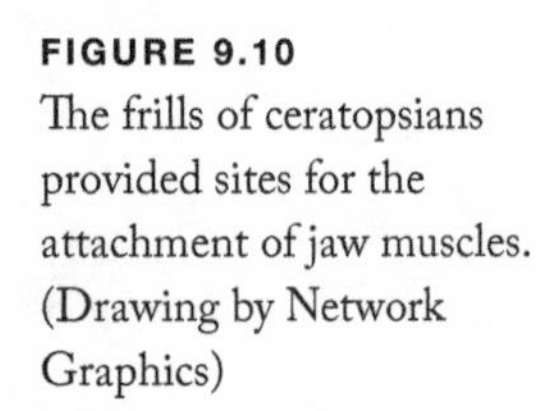

**FIGURE 9.10**
The frills of ceratopsians provided sites for the attachment of jaw muscles. (Drawing by Network Graphics)

## Function of the Horns and Frill

We usually see artistic depictions of *Triceratops* in a defensive posture, fending off the attack of a large meat-eating dinosaur with its horns and frill. But, a careful examination of ceratopsian horns and frills suggests that these were not primarily defensive structures. Indeed, the horns and frills probably performed several functions.

The large muscles that opened and closed the jaws of ceratopsians were attached mainly to the frill (figure 9.10). Different sizes and shapes of frills and their fenestrae probably reflected differences in the jaw muscles of various ceratopsians. The extremely large frills of neoceratopsians provided attachment sites for enormous jaw muscles that exerted great chewing force.

Beyond its relationship to chewing, the ceratopsian frill (and horns) must have functioned in display. Different frill sizes and shapes were probably species-specific identifiers and, within some species, may also have expressed sexual dimorphism. Analyses of variation in frill size and shape in *Protoceratops* and *Chasmosaurus* suggest that males and females of the same species differed in features of the frill, horns, and other parts of the skull (see figure 9.5). Similar kinds of differences between males and females of the same species are seen among some living mammals. Deer, in which male and female horns (antlers) differ, are a good example.

Another possible function of the ceratopsian frill was thermoregulation. Ceratopsian frills were highly vascularized; the bone contained numerous canals and grooves for blood vessels (figure 9.11). Like the plates of *Stegosaurus* (see chapter 8), the ceratopsian

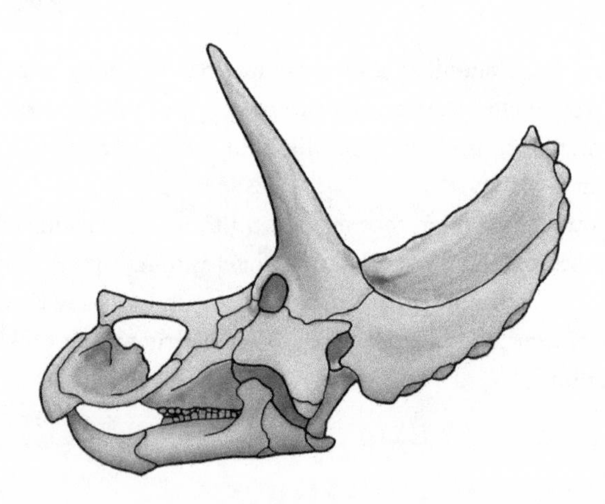

**FIGURE 9.11**
The frill of *Triceratops* was highly vascularized: it was covered with channels and grooves for blood vessels. (© Scott Hartman)

frill could have been used to spread the dinosaur's blood over a wide surface area, thus allowing for rapid heating and cooling. The frill of ceratopsians consisted of highly vascularized bone that in life was covered with huge, jaw-closing muscles. It is difficult to envision the frill as a helmet or shield used in combat. The horns might have had some defensive function, but a primary use in display seems more likely.

## Ceratopsian Evolution

Ceratopsians first appeared in Asia during the Late Jurassic, about 160 million years ago. Although the horned dinosaurs shared an ancestry with the ornithopods, that common ancestor has not yet been discovered.

Isolated bones and teeth of ceratopsians in Lower Cretaceous strata of North America are not as old as the earliest Asian ceratopsians. Indeed, *Yinlong* and *Chaoyangsaurus*, from the Upper Jurassic of northeastern China, are the oldest ceratopsians. *Liaoceratops*, from the Lower Cretaceous of northeastern China, is the earliest neoceratopsian and the same age as the psittacosaurs. *Aquilops*, from the Lower Cretaceous of Montana, and *Zuniceratops*, from the Upper Cretaceous of New Mexico, are the oldest well-known North American neoceratopsians. This suggests not only that ceratopsians originated in Asia, but that the origin of neoceratopsians must have taken place in Asia early in ceratopsian evolution.

The modest diversification of Asian ceratopsians culminated in *Protoceratops*. By 100 million years ago, ceratopsians had emigrated from Asia to western North America, where they became remarkably successful (figure 9.12). They were among the most common and most diverse of the Late Cretaceous plant-eating dinosaurs. Indeed, *Triceratops* was one of the last known dinosaurs (box 9.2).

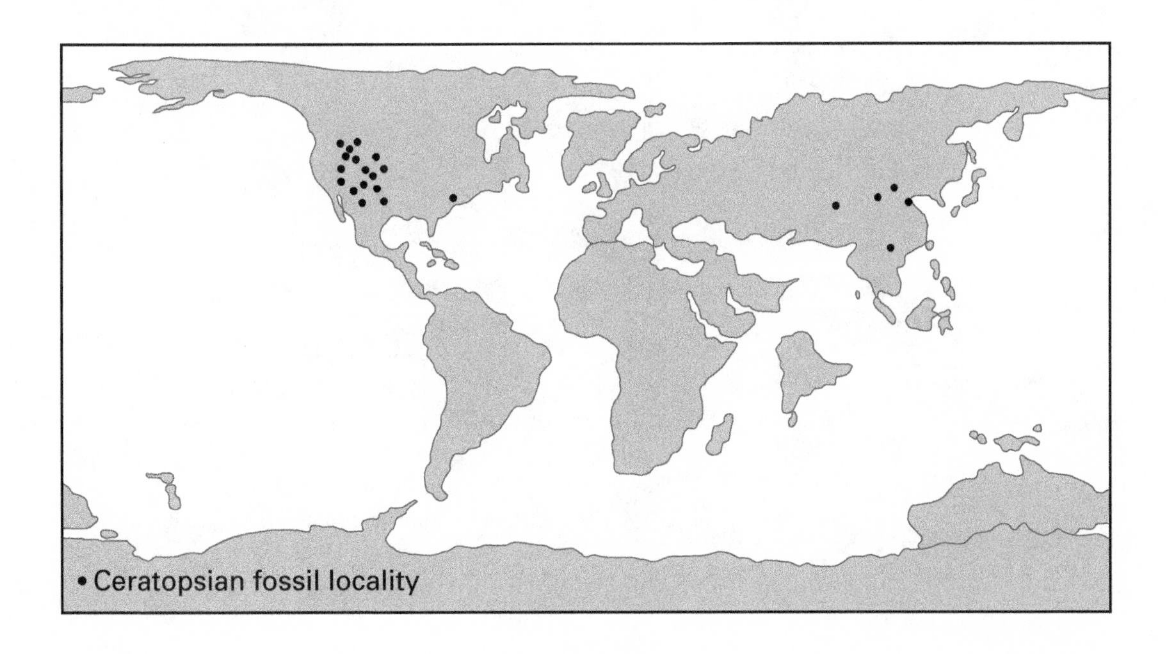


**FIGURE 9.12**
Ceratopsian fossils are known from only Asia and North America (continent position is for the Late Cretaceous). (Drawing by Network Graphics)

## Box 9.2

### How Many Species of *Triceratops*?

No one else has collected ceratopsians like John Bell Hatcher. From 1889 to 1892, Hatcher collected 32 ceratopsian skulls in eastern Wyoming and shipped them to Yale University. Most of those skulls belong to *Triceratops*, and they formed much of the basis for some of the 16 species of *Triceratops* named during the nineteenth and twentieth centuries. This great diversity of *Triceratops* species has long played a major role in the ideas about dinosaur extinction. This is because these ceratopsians were among the last dinosaurs, and their great diversity supports the notion of dinosaur prosperity just prior to their extinction.

However, paleontologists John Ostrom of Yale University and Peter Wellnhofer of the Bavarian State Museum in Munich, Germany, challenged the idea of many species of *Triceratops* at the end of the Cretaceous. These paleontologists were motivated by a desire to answer a nagging question: Did so many different species of *Triceratops* inhabit the area from Colorado to Wyoming to Montana during a mere 2 million years? Indeed, today, and at many times during the past, large terrestrial vertebrates (such as elephants) are not very speciose—one species generally covers a very wide geographic range. A good example is provided by North American bison during and since the last "Ice Age," when no more than three species were indigenous to North America.

So, how many species of *Triceratops* were there? According to Ostrom and Wellnhofer, only one! They concluded so because the variation in horn and skull shape among *Triceratops* is comparable to the variation in horn and skull shape we see today in a single species of bovid (box figure 9.2). Ostrom and Wellnhofer pointed out that, formerly, paleontologists believed every *Triceratops* skull of a different shape represented a different species. Instead it is likely that a single "herd" of *Triceratops* from Late Cretaceous Wyoming would have encompassed all the variety in skull and horn structure seen in the 16 previously named species of *Triceratops*.

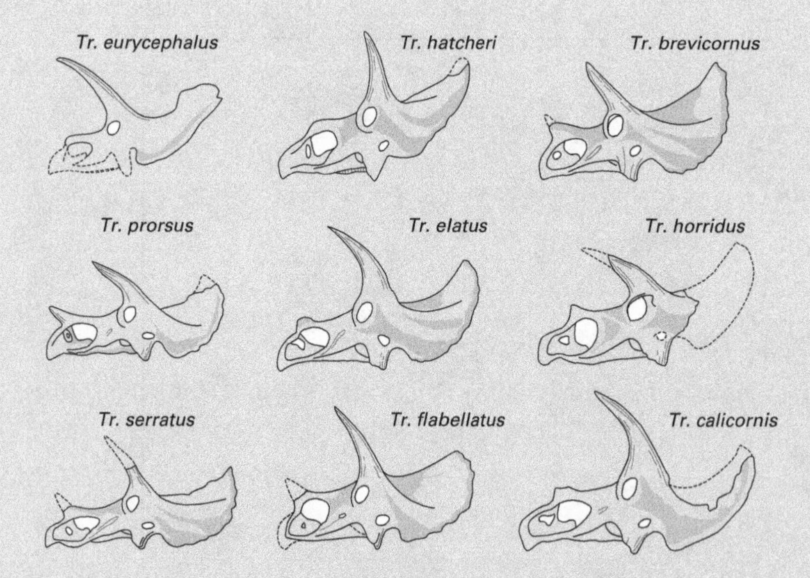


**BOX FIGURE 9.2**
Skulls of *Triceratops* from the Late Cretaceous of Wyoming show variation that led paleontologists to name many different species. (Drawing by Network Graphics)

Ostrom and Wellnhofer's reduction of the species of *Triceratops* from 16 to one is not above criticism. It would certainly be best to have a sample of *Triceratops* skulls from a single location. Such a sample would closely approximate an extinct population of these dinosaurs and thus allow an accurate assessment of skull and horn variation in a single *Triceratops* species. But, even if Ostrom and Wellnhofer are wrong, and there were as many as five different species of *Triceratops*, it is still far fewer than the 16 named by earlier paleontologists. So, our view of the prosperity of these ceratopsians just prior to dinosaur extinction needs to be revised.

## PACHYCEPHALOSAURS

The thick-headed dinosaurs, pachycephalosaurs, were bipedal ornithischians with greatly thickened bones of the skull roof. Primitive pachycephalosaurs and the dome-headed Pachycephalosauridae encompass 15 known genera from the Cretaceous of North America and Asia.

### Primitive Pachycephalosaurs

Pachycephalosaurs were long divided into two groups: the primitive, flat-headed taxa (formerly a family, Homalocephalidae) and the derived, dome-headed taxa (**Pachycephalosauridae**). Now, the primitive pachycephalosaurs are seen as an array of taxa, not a monophyletic family. However, some paleontologists argue that most or all of the flat-headed pachycephalosaurs are actually juveniles (representing early growth stages) or females of the dome-headed pachycephalosaurids (box 9.3).

The primitive pachycephalosaurs were characterized by a flat, table-like skull roof, which was of even thickness from side to side and pitted dorsally (figure 9.13). In addition, the skulls of primitive pachycephalosaurs had open supratemporal fenestrae, and short canine-like teeth were present in both the upper and lower jaws. The remaining teeth, however, were small and had sharp, serrated edges for slicing vegetation.

Well known and characteristic of the primitive pachycephalosaurs is ***Dracorex***, from the Upper Cretaceous of the United States (see box 9.3). The postcranial skeletons of primitive pachycephalosaurs are little known.

### Dome-Headed Pachycephalosaurs

Pachycephalosaurids were advanced pachycephalosaurs characterized by a prominent, dome-like thickening of the skull roof and no supratemporal openings (see box 9.3). The dome was produced by the fusion and thickening of the frontal and parietal bones. It is important to stress that it was these bones that were thickened; the volume

Box 9.3

## Ontogomorphs

For decades, paleontologists viewed the flat-headed pachycephalosaurs as a distinct family, Homalocephalidae, more primitive than the dome-headed pachycephalosaurs, the Pachycephalosauridae. However, in 2009, paleontologists John Horner and Mark Goodwin made a startling suggestion. They argued that at least some (if not all) of the flat-headed skulls were those of juveniles of the dome-headed pachycephalosaurs. In other words, during growth the skull of a pachycephalosaur changed dramatically, from flat-topped with open supratemporal fenestrae and strange nodes and horns to the massively domed skull characteristic of pachycephalosurids. Thus, as the skull became larger, the dome developed from the parietal bones, inflating and burying the fenestra, though the nodes, and horns remained.

This idea posits that the flat-headed pachycephalosur *Dracorex* is a younger growth stage of the dome-headed *Pachycephalosaurus* (box figure 9.3). Similar dramatic changes in growth have also been proposed for ceratopsids. One suggestion is that the very large ceratopsid *Torosaurus* is an advanced growth stage of *Triceratops*.

Some paleontologists have referred to these ideas as the identification of **ontogomorphs**—the development of very different anatomy (morphology) due to growth (ontogeny). The case for some dinosaur ontogomorphs appears to be strong. However, a criticism of some of the proposed ontogomorphs is that they are not based on actual growth series from a single population sample of the dinosaurs that includes both juveniles and adults. Until such samples are discovered, the cases for some ontogomorphs remain interesting possibilities requiring further documentation.

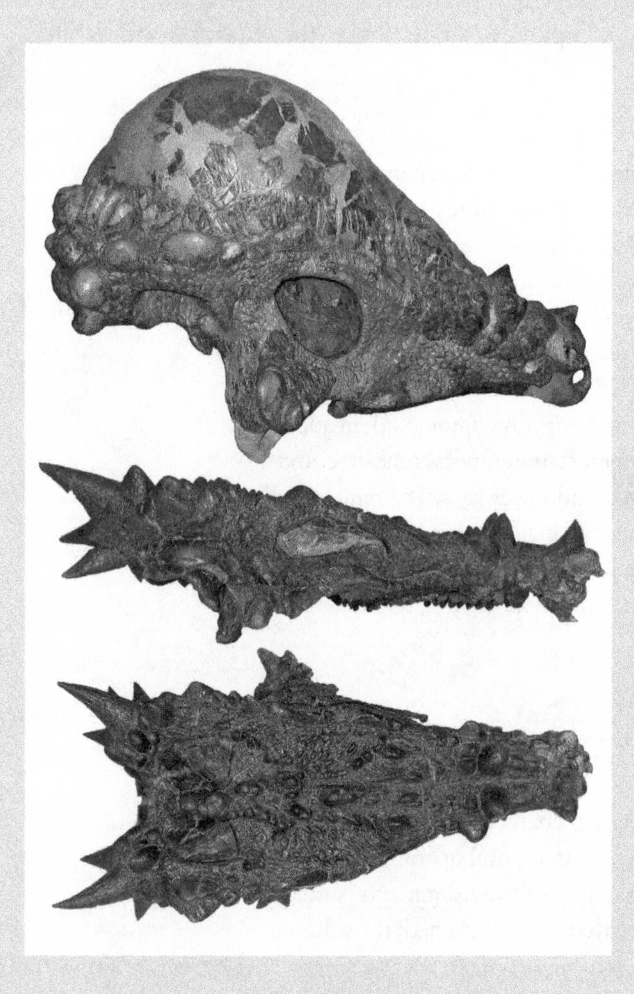

**BOX FIGURE 9.3**
Some paleontologists identify *Pachycephalosaurus* (*above*) as an ontogomorph of *Dracorex* (*center* and *below*). (Courtesy Robert M. Sullivan)

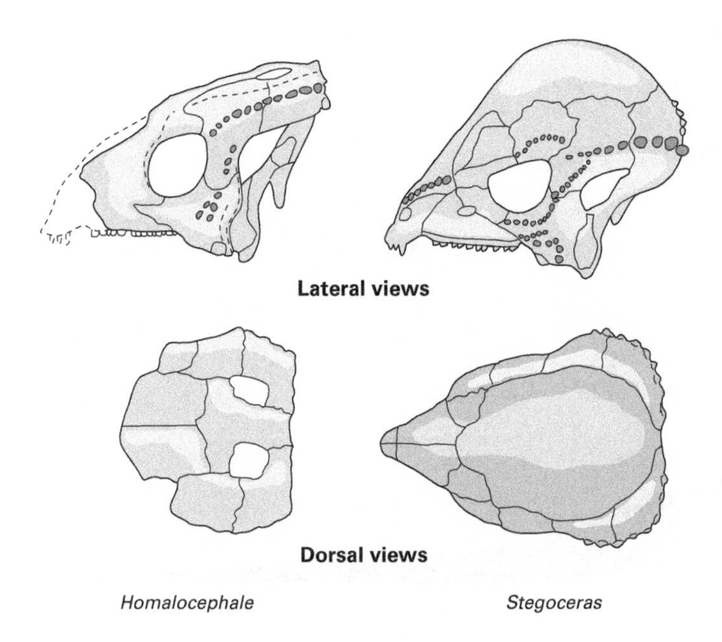


**FIGURE 9.13**
The skull of *Homalocephale* is characteristic of primitive pacycephalosaurs, whereas that of *Stegoceras* typifies the pachycephalosaurids. (Drawing by Network Graphics)

of the brain of pachycephalosaurids did not increase (figure 9.14). Unlike the flat skull roofs of primitive pachycephalosaurs, the domed skull roofs of pachycephalosaurids were smooth, not pitted.

There are at least 14 species of pachycephalosaurids known from the Cretaceous of North America and Asia. The North American genus ***Stegoceras*** is the best known and is typical of the family (figure 9.15).

The skull of *Stegoceras* had the high, smooth dome characteristic of pachycephalosaurids. In addition, a bony shelf was present around the dome. No supratemporal openings remained. The teeth of *Stegoceras* were similar to those of primitive pachycephalosaurs and indicate plant-eating.

No complete pachycephalosaurid skeleton is known, but the best known skeleton, that of *Stegoceras* (see figure 9.15), indicates that pachycephalosaurids had many of the features we associate with ornithopods. Thus, *Stegoceras* had much longer hind limbs than forelimbs and a long tail with vertebrae tightly connected by ossified tendons. It is features such as these that long justified the inclusion of the pachycephalosaurids in the Ornithopoda. Indeed, some paleontologists dismiss the apparent similarities of ceratopsians and pachycephalosaurs and regard the Pachycephalosauria as specialized, Late Cretaceous derivatives of ornithopods.

Unique features of the pachycephalosaurid postcranial skeleton are numerous. They include the long and low ilium bone of the pelvis, which contacted six to eight sacral vertebrae. Also, the joint surfaces between pachycephalosaurid vertebrae were ridged, and thus "locked" together, to stabilize the back. Finally, the head of a pachycephalosaurid was set at an angle to its vertebral column; in a normal posture, the head would have hung down so that the nose pointed at the ground and the "dome" pointed forward.

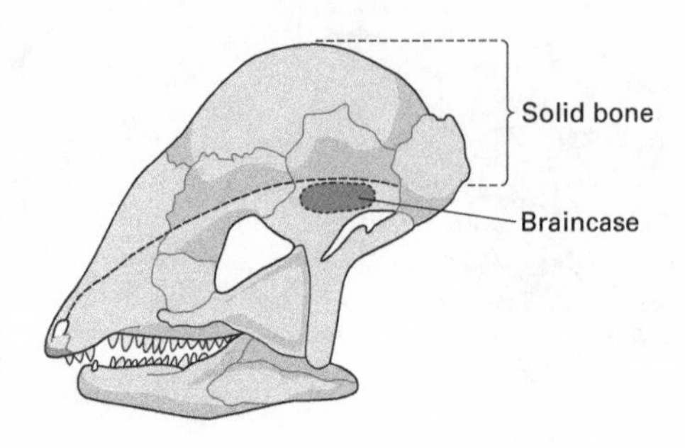


**FIGURE 9.14**
Pachycephalosaurids had greatly thickened bones in their skull roof but did not experience an increase in brain volume. (Drawing by Network Graphics)

**FIGURE 9.15**
This skeletal reconstruction of *Stegoceras* is the most complete that can be made for any pachycephalosaur. (© Scott Hartman)

## Head-Butting

The function of the peculiar, thickened skulls of pachycephalosaurids has long fascinated paleontologists. Some paleontologists argue that pachycephalosaurids used their skulls to **head-butt** against other dinosaurs, including each other (figure 9.16), much as do many living bovids. Features of the pachycephalosaurid skull and skeleton designed to resist impacts include the following:

- The greatly thickened dome of the skull, which protected the brain
- The shortened base of the skull
- The inclination of the skull relative to the vertebral column
- The widely expanded back of the skull
- The strengthening of the vertebral column, described earlier
- The reinforced upper lip of the hip socket

All of these features support the idea that pachycephalosaurids either directly butted heads or used their heads to butt the flanks of other pachycephalosaurs (or other dinosaurs). However, some paleontologists argue that head-butting by pachycephalosaurids is unlikely. Head-butting would have required the relatively small domes to

impact each other, and the shape of the bones of the pachycephalosaurid skull indicate that impact forces would have been directly transmitted to the brain, which would have produced serious fractures in cervical vertebrae.

**FIGURE 9.16**
Some paleontologists believe pachycephalosaurs butted heads to defend territory and maintain their social structure. (© Mark Hallett. Reproduced with permission of Mark Hallett Paleoart)

Today, head- and flank-butting occur in mammals such as bighorn sheep. The butting is usually performed in competition among males for territory and/or access to females. If we conclude that pachycephalosaurids butted the heads and flanks of each other and other dinosaurs, then it was likely for the same social reasons as these living mammals.

## Pachycephalosaur Evolution

The evolutionary history of the pachycephalosaurs is similar to the history of the Ceratopsia, though all bona fide pachycephalosaurs are of Late Cretaceous age. Likely originating in Asia, primitive pachycephalosaurs underwent a limited diversification and emigrated to North America, where a few taxa (or ontogomorphs) were present during the Late Cretaceous (see box 9.3). A more diverse group of advanced pachycephalosaurs (pachycephalosaurids) lived in North America, from Alberta to Texas.

The distribution of pachycephalosaur fossils indicates they preferred inland rather than coastal environments. In rocks deposited on coastal plains, pachycephalosaur fossils consist mostly of isolated and highly durable skullcaps, which may have been transported many kilometers by rivers. The most complete, well-preserved, and numerous of pachycephalosaur fossils are found in rocks deposited on inland floodplains and in deserts.

The structures for possible head-butting in the skulls and skeletons of pachycephalosaurids suggest social behavior like that of some living sheep and goats. It is tempting to speculate that pachycephalosaurids lived in herds and defended territory, maintaining their social structure by butting heads.

In North America, pachycephalosaurs existed through nearly the end of the Cretaceous. They were among the last dinosaurs.

## Summary

1. Ceratopsians and pachycephalosaurs are closely related and belong to a single group of dinosaurs, the Marginocephalia.
2. Ceratopsians consist of at least two groups, primitive ceratopsians and neoceratopsians, distinguished from other dinosaurs by features of the skull that include the presence of a frill.
3. *Psittacosaurus*, from the Lower Cretaceous of Asia, well represents a primitive ceratopsian. Neoceratopsians differed from *Psittacosaurus* in their extremely large heads, prominent frills, and pointed, sharply keeled beaks.
4. *Protoceratops* was a characteristic primitive neoceratopsian from the Cretaceous of Asia and North America with short frills and no horns.
5. Ceratopsids were advanced neoceratopsians from the Cretaceous of North America with long frills and horns.
6. Pachyrhinosaurines, with relatively short faces and frills, were primitive ceratopsids.
7. Ceratopsines, with relatively long faces and frills, were advanced ceratopsids.
8. The horns and frills of neoceratopsians functioned in display and thermoregulation, and the frills provided attachment sites for the jaw muscles. They were not primarily used in defense.
9. Ceratopsians lived during the Jurassic and Cretaceous in Asia and the Cretaceous in North America. They were most successful during the Late Cretaceous in western North America where ceratopsids were among the last dinosaurs.
10. Pachycephalosaurs were bipedal ornithischians with greatly thickened bones of the skull roof.
11. Primitive pachycephalosaurs had flat skull roofs ornamented with pits.
12. Advanced pachycephalosaurs, the Pachycephalosauridae, had smooth, domed skull roofs.
13. Pachycephalosaurid skulls and skeletons contain a variety of features that suggest these dinosaurs may have head-butted like some modern sheep and goats.
14. Pachycephalosaurs lived during the Late Cretaceous in North America and Asia.

## Key Terms

*Aquilops*
Ceratopsia
Ceratopsidae
Ceratopsinae
*Dracorex*
frill
head-butt
Marginocephalia
Neoceratopsia
ontogomorph
Pachycephalosauria
Pachycephalosauridae
Pachyrhinosaurinae
*Protoceratops*
*Psittacosaurus*
*Stegoceras*
*Triceratops*

## Review Questions

1. What evolutionary novelties are shared by ceratopsians and pachycephalosaurs to unite them as marginocephalians?
2. What features distinguish *Psittacosaurus* from neoceratopsians?
3. Describe the lifestyle of *Protoceratops*.
4. What features of *Triceratops* identify it as a typical ceratopsine? How is it different from a pachyrhinosaurine?
5. Discuss and evaluate the possible functions of the horns and frills of ceratopsians.
6. How are primitive pachycephalosaurs distinguished from derived pachycephalosaurids?
7. Why do some paleontologists think pachycephalosaurs butted heads? Why might pachycephalosaurs have done this?
8. What are the similarities and differences in the evolution of ceratopsians and pachycephalosaurs?

## Further Reading

Farke, A. A., W. D. Maxwell, R. L. Cifelli, and M. J. Wedel. 2014. A ceratopsian dinosaur from the Lower Cretaceous of western North America, and the biogeography of Neoceratopsia. *PLoS ONE* 9:e112055. (Describes *Aquilops*, an Early Cretaceous neoceratopsian from Montana)

Forster, C. A. 1996. Species resolution in *Triceratops*: Cladistic and morphometric approaches. *Journal of Vertebrate Paleontology* 16:259–270. (Argues that there are two valid species of *Triceratops*)

Galton, P. 1970. Pachycephalosaurids—dinosaurian battering rams. *Discovery* 6:23–32. (Discusses the head-butting habits of pachycephalosaurs)

Horner, J. R., and M. B. Goodwin. 2009. Extreme cranial ontogeny in the Upper Cretaceous dinosaur *Pachycephalosaurus*. *PLoS ONE* 4: e7626. (Argues that *Pachycephalosaurus* is an ontogomorph of some of the flat-headed pachycephalosaurs)

Lehman, T. M. 1989. *Chasmosaurus mariscalensis*, sp. nov., a new ceratopsian dinosaur from Texas. *Journal of Vertebrate Paleontology* 9:137–162. (Technical article describing the variation in a sample of *Chasmosaurus* from a bonebed in Texas)

Longrich, N. R., J. T. Sankey, and D. Tanke. 2010. *Texascephale langstoni*, a new pachycephalosaurid (Dinosauria, Ornithischia) from the upper Campanian Aguja Formation, southern Texas, USA. *Cretaceous Research* 31:274–284. (Describes a new pachycephalosaurid and reviews pachycephalosaurid phylogeny)

Mackovicky, P. 2012. Marginocephalia, pp. 527–549, in M. K. Brett-Surman, T. R. Holtz, Jr., and J. O. Farlow, eds., *The Complete Dinosaur*. 2nd ed. Bloomington: Indiana University Press. (Concise review of the marginocephalians)

Ostrom, J. H., and P. Wellnhofer. 1986. The Munich specimen of *Triceratops* with a revision of the genus. *Zitteliana* 14:111–158. (Discusses the taxonomy of the species of *Triceratops*)

Ryan, M. J., B. J. Chinnery-Allgeier, and D. A. Eberth, eds. 2010. *New Perspectives on Horned Dinosaurs*. Bloomington: Indiana University Press. (Collection of 36 articles on recent ceratopsian research)

Scannella, J. B., and J. R. Horner. 2010. *Torosaurus* Marsh, 1891, is *Triceratops* Marsh, 1889 (Ceratopsidae: Chasmosaurinae): Synonymy through ontogeny. *Journal of Vertebrate Paleontology* 30:1157–1168. (Argues that *Torosaurus* is an ontogomorph of *Triceratops*)

Sullivan, R. M. 2006. A taxonomic review of the Pachycephalosauridae (Dinosauria: Ornithischia). *Bulletin of the New Mexico Museum of Natural History and Science* 35:347–365. (Taxonomic review of all pachycephalosurs)

Xu, X., C. A. Forster, J. M. Clark, and J. Mo. 2006. A basal ceratopsian with transitional features from the Late Jurassic of northwestern China. *Proceedings of the Royal Society* B273:2135–2140. (Description of *Yinlong*, the oldest ceratopsian)

## Find a Dinosaur!

Horned dinosaurs are a mainstay of all dinosaur museums, and you can find a skull or skeleton (at least a replica) in just about all of the world's natural history museums. A replica of the skull of *Triceratops* is on The Mall in front of the Smithsonian's National Museum of Natural History (Washington, D.C.). Go inside the Museum to see more of *Triceratops*. At the American Museum of Natural History (New York), you can study the growth series (from babies to adults) on display of the small horned dinosaur *Protoceratops*. These fossils were collected from Cretaceous rocks in Mongolia by the fabled Central Asiatic Expeditions led by Roy Chapman Andrews in the 1920s. In front of the Yale Peabody Museum of Natural History (New Haven, Connecticut) stands a sculpture of *Torosaurus*; the real skull, named by Othniel Charles Marsh in 1891, is on display inside. In Boston, the Museum of Science has a huge (7-meter-long) *Triceratops* skeleton nicknamed "Cliff." And, there is also "Bob," a *Triceratops* on display at the Barnes County Historical Society (Valley City, North Dakota). *Triceratops* is the official state fossil of South Dakota, and at the Museum of the Black Hills Institute of Geological Research in Hill City, a replica of a fossilized skin impression of *Triceratops* is on display.

# 10

# THE DINOSAURIAN WORLD

DINOSAURS existed for about 160 million years, from the Late Triassic, 225 to 230 million years ago, to the end of the Cretaceous, 66 million years ago. This interval of time can be called the "age of dinosaurs," whereas the entire Mesozoic Era is usually termed the "age of reptiles."

Dinosaurs did not live and evolve in isolation during the age of dinosaurs. They coexisted with plants and other animals in climates and on landscapes very different from those of today. Furthermore, dinosaurs changed dramatically—through evolution and extinction—during the age of dinosaurs. This chapter examines these changes and briefly reviews the geography, climate, vegetation, and animal life that made up the dinosaurian world.

## CONTINENTAL DRIFT, SEA LEVEL, AND CLIMATE

Earth's crust is made up of rigid plates that move, thereby changing the distribution of land and sea during geological time (figure 10.1). This movement, popularly called continental drift (though not only the continents move), has a profound effect on both global sea level and climate.

The continents drift because rising heat forms long uplifted ridges on the floors of the ocean basins (see figure 10.1). These ridges are made of lava being pushed out of the Earth's interior. The new lava flows down the ridge slopes, forcing the oceanic crust away from the ridge axis and thus widening the ocean basin. The oceanic crust collides with continental crust and thus moves the continent. At the point of any crustal collision, mountain ranges form—they are the crumpled crust at the point of impact.

The process of seafloor spreading can change the volume of an ocean basin. Thus, when much heat is released from the Earth's interior, the mid-oceanic ridge is very large, and the volume of the ocean basin is reduced. The water from the basin must go

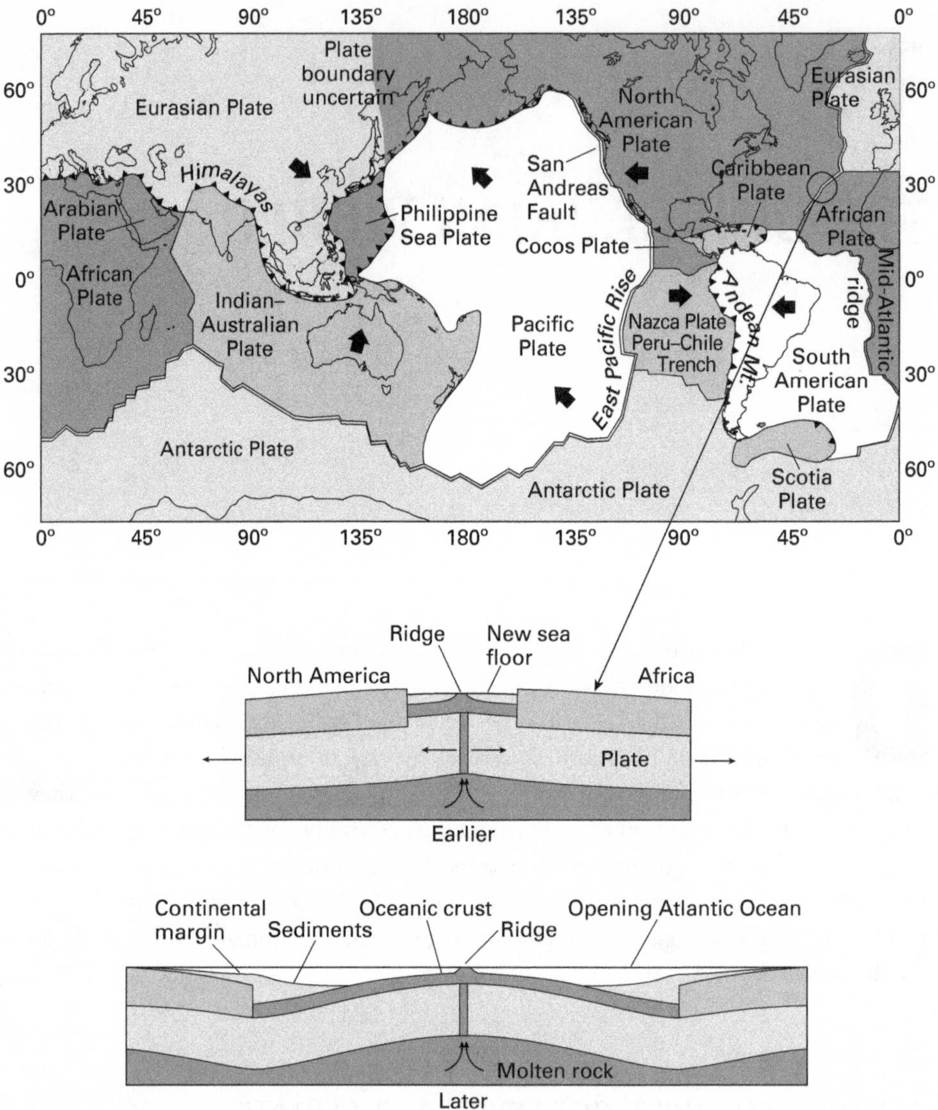


**FIGURE 10.1**
The Earth's crust is made of plates that move as ocean basins are opened (*above*). Heat and molten rock rise along a ridge along the ocean floor (*center*), causing the ocean basin to widen and pushing the continents (*below*). (Drawing by Network Graphics)

somewhere, so it is forced out onto the continents, drowning what was once dry land. In this way, at times of increased heat release, sea level is relatively high because much of the water in the ocean basins is being spilled onto the continents. The Cretaceous was such a time, with very high sea levels and much of the world's continents under seawater. In contrast, the Triassic was a time of low sea levels when most of the continents were dry land. The Jurassic was an intermediate phase between the Triassic and Cretaceous extremes.

Changing sea levels, crustal collisions, and continental movements all have a dramatic effect on world climate. When sea level is high, the continents experience greater warmth and humidity because the ocean water is storing heat from the sun. A **greenhouse** climate results. With lower sea levels, climate is generally drier and more temperate.

Crustal collisions create mountain ranges that affect wind patterns, which change climates. The configuration of the continents also affects patterns of oceanic circulation, which affects climate. Each interval of Earth history had a specific configuration of continents and, therefore, a specific climate. We see this especially during the Mesozoic, when dinosaurs lived. Triassic climates were relatively dry and strongly seasonal. In the Jurassic, climates became wetter and more **equable**, a trend culminating in the Cretaceous global greenhouse climate.

## LATE TRIASSIC: THE BEGINNING OF THE AGE OF DINOSAURS

Dinosaurs first appeared during the Late Triassic, and this interval of time, from between 225 and 230 to 201 million years ago, is known as the beginning, or dawn, of the age of dinosaurs. During the Late Triassic, dinosaurs were rare and most were small, with other reptiles dominating the landscape.

### Geography and Climate

During the Late Triassic, all the world's continents were amalgamated into a single supercontinent that geologists call **Pangea** (figure 10.2). Much earlier, during the

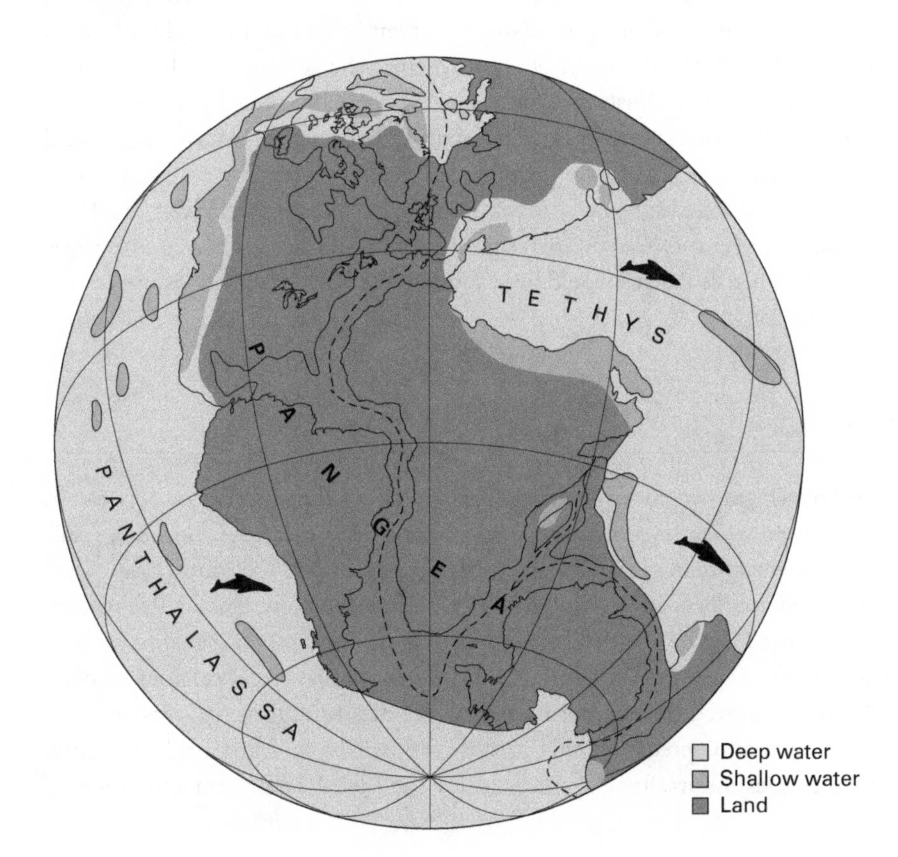


**FIGURE 10.2**
During the Late Triassic, the continents were united in one supercontinent, Pangea, surrounded by one ocean, Panthalassa. Tethys was a sea that indented the eastern margin of Pangea. (Drawing by Network Graphics)

Late Permian, more than 251 million years ago, Pangea fully coalesced from two supercontinents: **Gondwana** (now separated into South America, Africa, Antarctica, Australia, India, and Madagascar) and **Laurussia** (now separated into Europe, part of Asia, and North America). Later, during the Jurassic, Pangea separated again. When Pangea was assembled, the world had only one ocean, a sort of expanded Pacific Ocean called **Panthalassa** (see figure 10.2). The **Tethys** Sea, along the southern shore of Asia, indented Pangea between Africa and Europe. During the post-Triassic separation of Gondwana from Laurussia, the Tethys expanded progressively westward, successively creating the Mediterranean, central Atlantic, and Caribbean ocean basins.

By the end of the Triassic, Pangea had started to break up in the Caribbean and Mediterranean areas. Yet, the topography of Pangea during the Late Triassic seems to have been relatively subdued. There were relatively few large mountain ranges and few volcanoes.

There were no polar ice caps during the Late Triassic, and there is no evidence of glaciers or of cold, snowy winters. Indeed, climates were generally much warmer and less varied with latitude than they are today, a situation that continued for the entire age of dinosaurs. Recent computer modeling, however, coupled with rock and fossil evidence, indicates that Late Triassic Pangean climates were strongly **monsoonal.**

The global climate of Pangea was a true monsoon. There were only two seasons, one wet, the other dry. The abundant rainfall would have been concentrated in the summer months, and there would have been little annual temperature fluctuation. The key to the Pangean monsoon was the presence of large landmasses just north and south of the equator. During the Northern Hemisphere summer, the northern landmass would have become relatively hot, whereas the southern landmass would have been relatively cool. Moisture from the world ocean, particularly the equatorial seaway called Tethys that divided the eastern part of Pangea, would have been pulled into the Northern Hemisphere low-pressure cell, producing extensive rains, whereas the southern hemisphere high-pressure cell would have remained relatively dry. During the Southern Hemisphere summer, this process would have occurred in reverse. Thus, seasonality across Pangea would have consisted of alternating, hemisphere-wide wet and dry seasons.

## Life in the Sea

The Late Triassic seas were warm and shallow, except for the far offshore portions of Panthalassa, from which few rocks and fossils have been preserved. Clams and snails were particularly common as the dominant bottom dwellers in the Late Triassic seas, as were the ammonoid cephalopods (figure 10.3). Corals built low, mound-like reefs in many parts of the Late Triassic seafloor.

Calcareous nannoplankton, minute plant protists that secrete a calcium carbonate shell, first appeared during the Late Triassic. Sharks and heavily scaled bony fishes were successful predators in the Late Triassic seas. Long-necked marine reptiles, called **plesiosaurs**, first appeared during the Late Triassic, and **ichthyosaurs**,

**FIGURE 10.3**
Ammonoids were successful in the Mesozoic seas. These small Late Triassic ammonoids are characteristic.

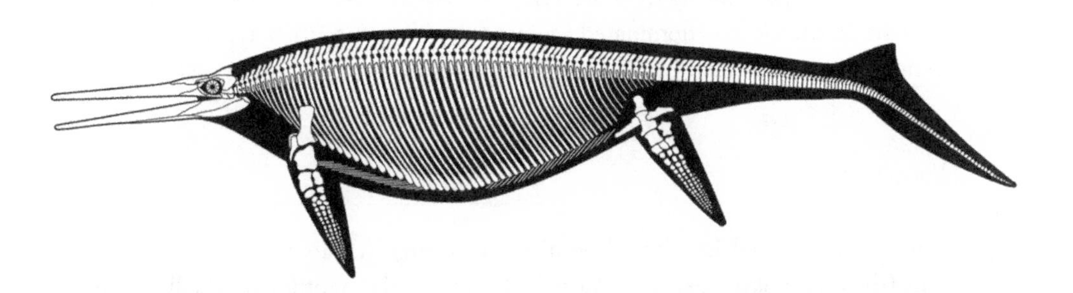

**FIGURE 10.4**
The 10-meter-long ichthyosaur *Shonisaurus* was the largest animal in the Late Triassic seas. (© Scott Hartman)

a remarkable case of evolutionary convergence of reptiles with fish-like bodies, were abundant and diverse during the Late Triassic. Indeed, the largest animal known to have lived in the Late Triassic seas was *Shonisaurus*, a 10-meter-long ichthyosaur (figure 10.4).

## Vegetation

Late Triassic vegetation was very different from present-day vegetation. Ferns and the now-extinct seed ferns dominated the understory. The trees that stood above them

**FIGURE 10.5**
The palm-like Late Triassic plant *Sanmiguelia* is thought by some paleontologists to have been an angiosperm. (Drawing by Network Graphics)

included cycads and cycadeoids, conifers, and ginkgoes. These kinds of plants, all of which survive today—although cycads and ginkgoes are very rare—belong to a group of plants having exposed seeds called **gymnosperms**.

Most paleontologists believe flowering plants, called **angiosperms**, in which the seed is covered, did not evolve until the Early Cretaceous. But there is one Late Triassic plant, ***Sanmiguelia***, that looks remarkably like a palm, which is a primitive type of angiosperm (figure 10.5). This has led some paleontologists to argue that flowering plants first evolved during the Late Triassic. However, even if this were true, the Late Triassic landscape was dominated by ferns, conifers, and other types of plants now extinct or nearly extinct.

## Vertebrates

Dinosaurs appeared in a world dominated by large (up to 2-meter-long) amphibians (the **metoposaurids**) and a variety of archosaurian reptiles, especially the meat-eating **rauisuchians** and phytosaurs and the plant-eating aetosaurs (figure 10.6). One of the largest meat-eaters of the Late Triassic was a rauisuchian, the 5-meter-long *Postosuchus*. Other plant-eaters common when dinosaurs first appeared were rhynchosaurs and dicynodonts, two groups of reptiles well adapted to eating tough, fibrous vegetation that became extinct soon after the appearance of dinosaurs. Indeed, these types of reptiles and amphibians, as well as many of the archosaurs, became extinct close to the end of the Triassic. This was one of the major extinctions of the Mesozoic (box 10.1).

Other land animals of the Late Triassic include the first turtles, pterosaurs, and mammals, groups that appeared nearly simultaneously with the dinosaurs. Indeed, the Late Triassic was an important turning point in the history of life on land, with many new types of animals appearing and several other types suffering extinction.

FIGURE 10.6
Phytosaurs (*left*) shared most Late Triassic landscapes with early dinosaurs. (© Doug Henderson)

## Dinosaurs

Dinosaurs had a nearly worldwide distribution during the Late Triassic. But, they were neither conspicuous nor abundant during the Late Triassic, except perhaps at the very end of the Triassic when they were extremely common in some fossil deposits.

Most Late Triassic dinosaurs were relatively small (up to 6 meters long) and included *Herrerasaurus*-like predators and *Lesothosaurus*-like plant-eaters, although the large prosauropods, like *Plateosaurus*, were plant-eaters as much as 8 meters long. Unlike their contemporaries, the Late Triassic dinosaurs had an upright limb posture designed to make them fast bipedal and quadrupedal walkers and runners. Some paleontologists argue that this gave the dinosaurs an edge over other reptiles and explains why they survived and prospered after the Triassic when many other types of reptiles became extinct. Others suggest that a change in vegetation and an increasingly arid climate favored the dinosaurs during the Late Triassic and Jurassic while bringing about the extinction of the other reptiles. Whichever explanation is accepted, dinosaurs did not truly dominate the land until after the end of the Triassic.

## Box 10.1

### The End-Triassic Extinctions

Paleontologists identify the extinctions at the end of the Triassic as one of the five great mass extinctions of the past 540 million years (see figure 16.1). In the sea, **ammonoid** cephalopods almost became extinct, the reef community collapsed in the Tethys, and an important group of plankton—the plant protists called radiolarians—were severely impacted. As Pangea began to split apart during the Late Triassic, extensive volcanism took place around the opening Atlantic Ocean. This produced one of the great volcanic fields of Earth history, which geologists call the Circum-Atlantic Magmatic Province (CAMP). Most paleontologists believe that CAMP volcanic eruptions produced ash and gasses that altered world ocean chemistry and likely caused the Late Triassic extinctions in the seas.

**BOX FIGURE 10.1**
The crocodile-like phytosaurs, represented by this 1-meter-long skull, suffered extinction at about the end of the Triassic.

The end-Triassic extinctions on land were much less severe. Land plants suffered no extinctions, nor did insects. Tetrapod extinctions eliminated many groups of Triassic archosaurs, such as the rauisuchians, phytosaurs, and aetosaurs discussed in chapter 4 (box figure 10.1). The large amphibians (notably the metoposaurs) also suffered substantial extinctions, as did the synapsids (mammal-like reptiles). However, a careful examination of the timing of the tetrapod extinctions shows that they were spread out over several million years. Indeed, some of the marine extinctions were similarly "spread out" over a few million years. The conclusion is that the 5 million years that encompass the Triassic–Jurassic boundary was a time of accelerated extinction rates among several groups of organisms. There was no single mass extinction.

## EARLY–MIDDLE JURASSIC: DINOSAURS ESTABLISH DOMINANCE

During the Early and Middle Jurassic, 201 to 163 million years ago, dinosaurs became the largest and most successful group of land vertebrates. But, the fossil record of the establishment of dinosaur dominance is not as complete as is the record from the Late Triassic and Late Jurassic. This is partly because vast deserts covered parts of Pangea during the Early and Middle Jurassic, especially in western North America and southern Africa, and few dinosaur fossils other than footprints are found in the resulting eolian rocks (figure 10.7). Another reason we know less about Early and Middle Jurassic dinosaurs than about earlier or later dinosaurs is simply a lack of exploration. This has begun to change, however, so that new discoveries of Early and Middle Jurassic dinosaurs have been among the most significant discoveries of the past few decades.

### Geography and Climate

By the end of the Middle Jurassic, Pangea had broken apart to the point where Laurussia and Gondwana were connected only in the western Mediterranean (figure 10.8). Nevertheless, the Pangean monsoonal circulation of the Late Triassic still influenced climate across the vast supercontinent. Highly arid climates developed in parts of Pangea during the Early and Middle Jurassic. This was especially true in western North America, which was covered by a vast sand sea during much of this time.

**FIGURE 10.7**
Eolian sandstones, cross-bedded by wind, are the most common Lower and Middle Jurassic rocks in western North America.

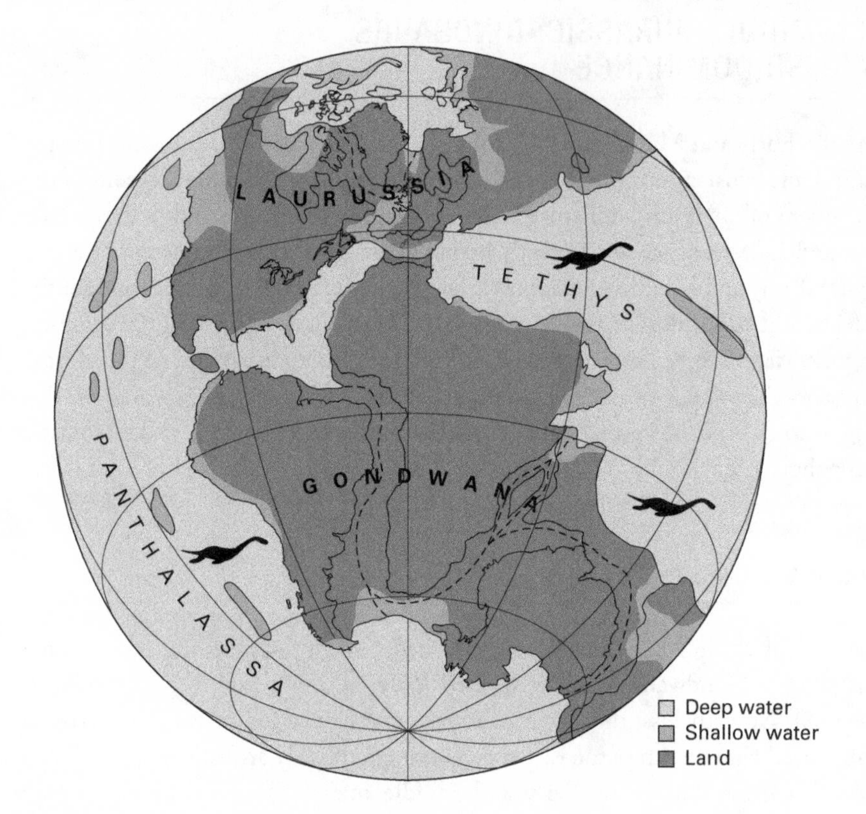


**FIGURE 10.8**
By the Middle Jurassic, Gondwana and Laurussia were almost completely separated. (Drawing by Network Graphics)

## Life in the Sea

Early and Middle Jurassic seas were populated by animals similar to those that inhabited the Late Triassic seas, except that most Jurassic animals were larger than their Triassic predecessors. Ammonoids and radiolarians recovered from the Late Triassic extinctions. The reef communities also rebounded to build the first truly large reefs of the Mesozoic. Some of the Triassic marine reptile groups became totally extinct, but ichthyosaurs and plesiosaurs persisted as particularly important marine predators. Thus, huge plesiosaurs were the largest marine predators of the Early and Middle Jurassic (figure 10.9).

**FIGURE 10.9**
Three-meter-long plesiosaurs were the largest animals in the Early and Middle Jurassic seas. (Drawing by Network Graphics)

**FIGURE 10.10**
Cycads (the stalk-like trees behind the dinosaur) were the dominant trees of the Jurassic. (© Doug Henderson)

## Vegetation

Cycads, cycadeoids, conifers, and gingkoes were the trees of the Early and Middle Jurassic, as they had been during the Late Triassic. Cycads today are rare tropical trees that somewhat resemble palms. However, cycads were the dominant small trees throughout the Jurassic (figure 10.10), and paleontologists who study fossil plants call this period in Earth history "the age of cycads." Seed ferns had become extinct by the end of the Triassic, and ferns were not as common during the Early and Middle Jurassic as they had been during the Late Triassic, possibly in part because climates worldwide may have been drier.

## Dinosaurs and Other Vertebrates

Dinosaurs dominated life on land during the Early and Middle Jurassic. The archosaurs that dominated Triassic landscapes—rauisuchians and phytosaurs, among others—were extinct, and most of the large amphibians had also disappeared. All of the large meat- and plant-eaters were dinosaurs. Examples include *Dilophosaurus*, a 6-meter-long theropod, and *Datousaurus*, a 12-to-14-meter-long sauropod (box 10.2).

### Box 10.2

### The Dinosaur National Monument of China

The most important Middle Jurassic dinosaur locality known lies just outside the city of Zigong in Sichuan Province in southern China. Chinese geologists discovered this locality in 1972, and since then, the bones of many dinosaurs and a variety of other vertebrates have been excavated from an area of more than 1,800 square meters. Many complete dinosaur skeletons were removed from the Zigong locality, but many more were left in the ground and excavated in relief for viewing by visitors to the Zigong Dinosaur Museum, built over the dinosaur locality and opened in 1987. The idea of building an exhibition facility over an ongoing dinosaur excavation was borrowed from Dinosaur National Monument in Utah, and the Zigong Dinosaur Museum may truly be called China's dinosaur national monument.

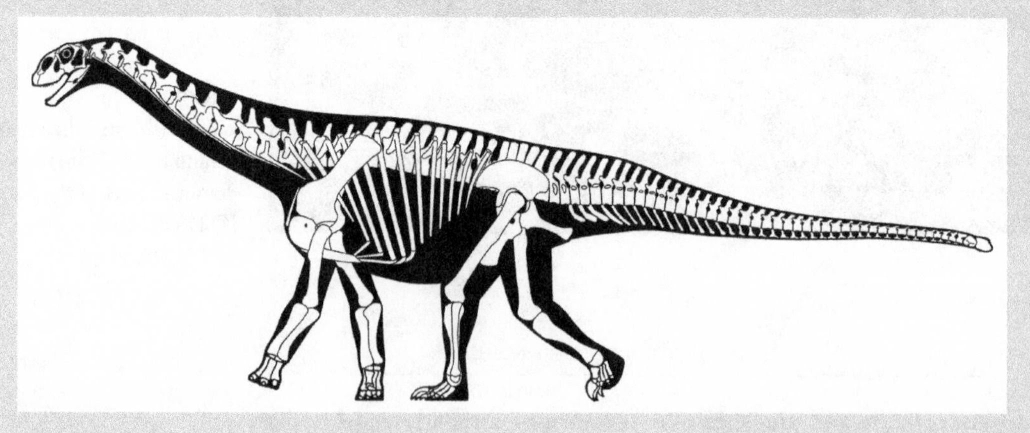

**BOX FIGURE 10.2**
The primitive sauropod dinosaur *Shunosaurus* is known from the Zigong dinosaur quarry. (© Scott Hartman)

The Middle Jurassic dinosaurs from Zigong include the oldest and most primitive stegosaur (*Huayangosaurus*) and two sauropods, *Datousaurus* and *Shunosaurus* (box figure 10.2). These sauropods were much smaller than the Late Jurassic sauropods, but the composition of the Zigong dinosaur quarry—consisting mostly of sauropods and stegosaurs—is very similar to that of the great dinosaur quarries of Late Jurassic age in the western United States. This suggests that this composition of dinosaur communities, particularly the dominance of plant-eating sauropods and stegosaurs, was a feature of both the Middle and the Late Jurassic world. The Zigong dinosaurs also indicate that stegosaurs had already evolved by the Middle Jurassic. This Chinese dinosaur locality thus provides a unique glimpse of the dinosaurs of the Middle Jurassic.

All other terrestrial vertebrates of the Early and Middle Jurassic were relatively small animals, such as mammals, pterosaurs, turtles, crocodiles, and so forth. Clearly, dinosaurs had come to rule the landscape.

## LATE JURASSIC: THE GOLDEN AGE OF DINOSAURS

Unlike the earlier part of the Jurassic, paleontologists have unearthed an extensive fossil record of dinosaurs and other organisms that lived during the Late Jurassic, 145 to 163 million years ago. This record indicates that very large land animals, the giant sauropod dinosaurs, first evolved during the Late Jurassic. This aspect of the Late Jurassic dinosaurs, as well as their wide geographic range, variety, and abundance is the reason the Late Jurassic is called the "golden age of dinosaurs."

### Geography and Climate

By the Late Jurassic, Laurussia and Gondwana were totally separated by the Tethys Sea and its westernmost extremity, the young Gulf of Mexico (figure 10.11). Tethys was tropical and full of coral reefs. World sea level was high, and much of Laurussia and significant parts of Gondwana were inundated by seawater. This created what are called **epicontinental seas**, seas situated atop continental crust, unlike today's oceans and seas, which are mainly situated atop the denser oceanic crust.

Climates during the Late Jurassic were still influenced by the monsoonal circulation pattern of earlier times, but less so as Pangea broke up and a circum-equatorial seaway (Tethys) began to influence ocean and air circulation patterns. Overall, Late Jurassic climates seem to have resembled those of the Early and Middle Jurassic but were warmer and wetter.

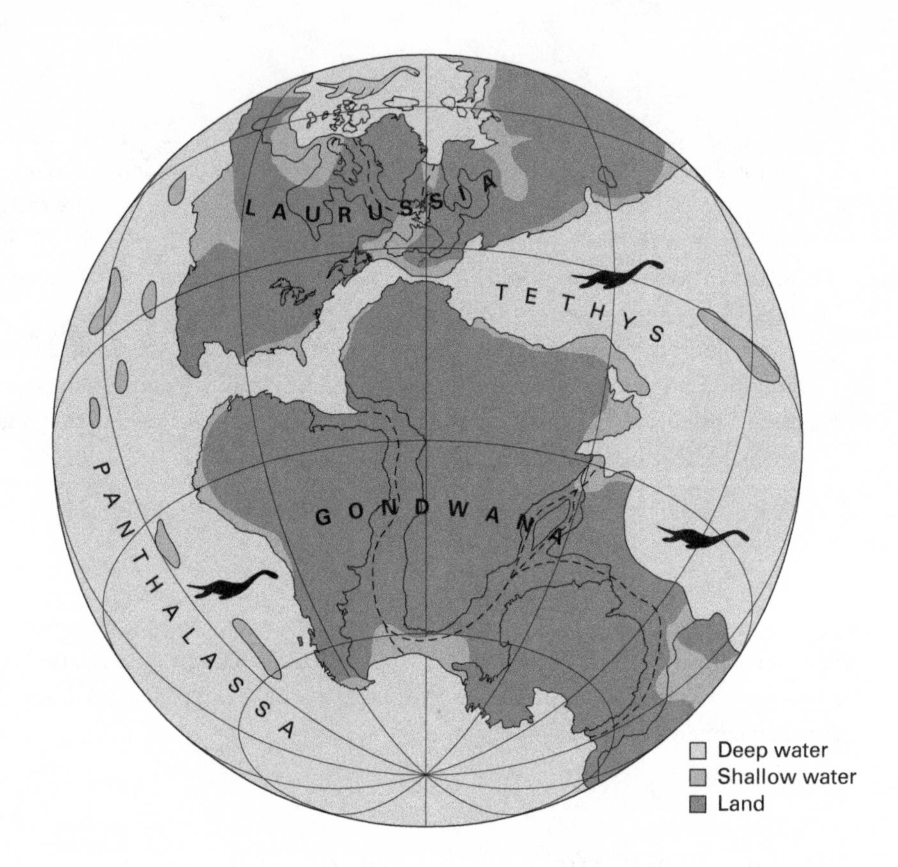


**FIGURE 10.11**
Laurussia and Gondwana were totally separated by the Late Jurassic. (Drawing by Network Graphics)

## Life in the Sea and Vegetation

Life in the sea and vegetation on land during the Late Jurassic were generally similar to those of the Early and Middle Jurassic. Some new marine organisms appeared, including oysters and sea urchins. Late Jurassic sea bottoms thus looked more modern than earlier Mesozoic seafloor communities. Extensive reefs were built on Late Jurassic seafloors, probably the result of higher sea levels. Plants adapted to warm climates lived as far poleward as 60 degrees north, as can be seen in the Upper Jurassic fossil floras found in Siberia.

## Dinosaurs and Other Vertebrates

Late Jurassic dinosaurs are well known, especially from the western United States, southern China, and eastern Africa. Huge sauropods, large stegosaurids, and small to medium-size ornithopods were the dominant plant-eaters. Megalosaurs and carnosaurs were the big meat-eaters. Almost all other land vertebrates were relatively small (some crocodiles were an exception), truly dwarfed by the gigantic dinosaurs. Dinosaur domination of the earth was complete, and the zenith of dinosaurian evolution had in many ways been reached (figure 10.12).

**FIGURE 10.12**
Gigantic sauropods, such as *Diplodocus*, lived during the Late Jurassic. (Drawing by Network Graphics)

## EARLY CRETACEOUS: A TRANSITION

The Jurassic Period did not end with significant extinctions. Nevertheless, the Early Cretaceous, 100 to 145 million years ago, was a time of profound changes in the history of life on land, for it was during the Early Cretaceous that flowering plants appeared and began their rise to dominance among land plants. This change in vegetation may lie at the core of a major transition in dinosaur evolution that took place during the Early Cretaceous.

### Geography and Climate

Worldwide sea level continued to rise from the Late Jurassic into the Early Cretaceous, inundating more of the continents. Gondwana finally fragmented, and the southern Atlantic Ocean began to form (figure 10.13).

World climates became even warmer and much wetter, producing a global greenhouse climate that persisted throughout most of the Cretaceous. Indeed, high sea levels worldwide were a primary cause of the global greenhouse climate. The world was warm, equable, and ice-free, and the pole-to-equator temperature gradient during the Cretaceous, by one estimate, was only about 20 degrees Celsius. Today, it is 41 degrees Celsius.

### Life in the Sea

The Early Cretaceous saw the diversification of new types of plankton, especially those types that are now common in the world's oceans. Notable is the rise of diatoms, silica-secreting plant protists, an important component of modern oceanic plankton. Modern types of bony fishes, the teleosts with their short jaws and symmetrical tails,

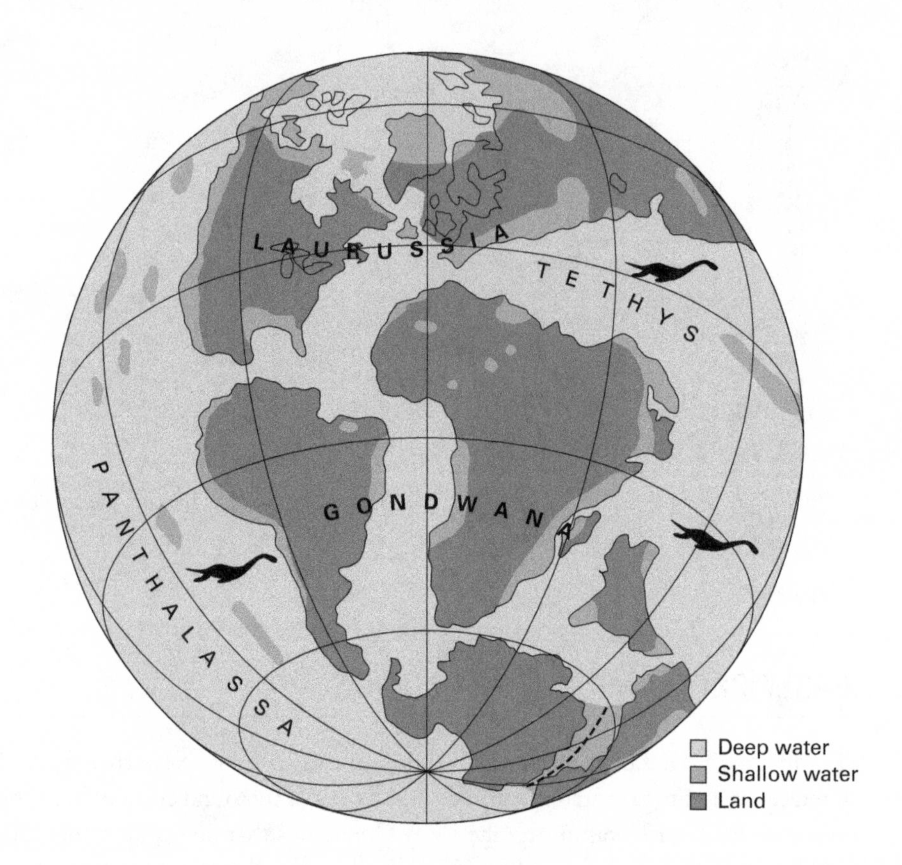


**FIGURE 10.13**
During the Early Cretaceous, Gondwana fragmented, and the South Atlantic Ocean basin opened. (Drawing by Network Graphics)

appeared. Predatory snails and crabs and clams that burrow into the seafloor began to diversify. Ammonoids were abundant and diverse. Whereas corals had been the dominant reef builders of the Triassic and Jurassic, during the Early Cretaceous, the **rudists**, a group of clams, took center stage in building reefs (figure 10.14).

## Vegetation

Although the gymnosperm floras of the Triassic and Jurassic continued to dominate the land well into the Cretaceous, flowering plants (angiosperms) were to take over rapidly (figure 10.15).

During the Early Cretaceous, conifers stole dominance from the cycads, thus ending the Jurassic "age of cycads." But, about 100 million years ago, near the end of the Early Cretaceous, angiosperms first appeared, and by the Late Cretaceous, they were more diverse than the conifers. Today, there are about 100,000 species of angiosperms and only about 550 species of conifers.

Angiosperms provide a food supply for their seeds, whereas gymnosperms, such as conifers, do not. Also, angiosperm flowers attract insects that carry their pollen to other flowers, fertilizing them. These have been the keys to angiosperm success.

**FIGURE 10.14**
During the Cretaceous, reefs, like this cliff-forming one in southwestern New Mexico, were built by rudist clams.

**FIGURE 10.15**
Flowering plants first appeared near the end of the Early Cretaceous, and by the end of the Cretaceous, as shown here, they dominated the landscape. (© Doug Henderson)

### Dinosaurs and Other Vertebrates

Dinosaurs of the Early Cretaceous differed somewhat from those of the Late Jurassic. Ornithopods (for example, *Iguanodon*) were generally larger, and sauropods were generally smaller. Nodosaurid ankylosaurs first became common at this time, whereas stegosaurs virtually disappeared. Indeed, by the end of the Early Cretaceous, almost all of the primitive ornithopods, iguanodonts, and sauropods (except those in South America) had joined the stegosaurs in practically becoming extinct. Most other land vertebrates of the Early Cretaceous were rather similar to those of the Late Jurassic, but significant changes took place among mammals. Specifically, placental and marsupial mammals first appeared near the end of the Early Cretaceous.

A major change in global climate and vegetation took place during the Early Cretaceous, so it is not difficult to understand the profound transition in dinosaurs that we see between the end of the Jurassic and the beginning of the Late Cretaceous. Dominant Late Jurassic dinosaurs—the sauropods, stegosaurs, primitive ornithopods, and iguanodontians—gave way to a totally different **dinosaur fauna**, a transition that occurred in the Early Cretaceous.

## LATE CRETACEOUS: THE LAST DINOSAURS

Dinosaurs of the Late Cretaceous, 66 to 100 million years ago, were very different from the dinosaurs of the earlier portions of Mesozoic time. New types of dinosaurs appeared during the Late Cretaceous, many of them plant-eaters possibly better adapted to feeding on angiosperms than were the plant-eating dinosaurs of the Late Jurassic. These new types of dinosaurs dominated the landscape until all dinosaurs became extinct at the end of the Cretaceous.

### Geography and Climate

By Late Cretaceous time, Gondwana had separated into its constituent continents, and Laurussia was split significantly by the enlarging of the northern Atlantic Ocean basin (figure 10.16). Sea level was at an all-time high, and most of the continents had been drowned by epicontinental seas. A huge outpouring of lava took place in India (the Deccan traps).

Climates worldwide were hot and wet like a greenhouse. Tropical and subtropical zones are estimated to have extended as far poleward as 45 degrees north and 70 degrees south. Fossil leaves of warm-adapted plants are known from Upper Cretaceous rocks in Alaska, Greenland, and Antarctica. Indeed, one paleontologist has aptly described the Late Cretaceous world as "wall-to-wall Jamaica." Near the end of the Cretaceous, however, this began to change as global sea levels fell again.

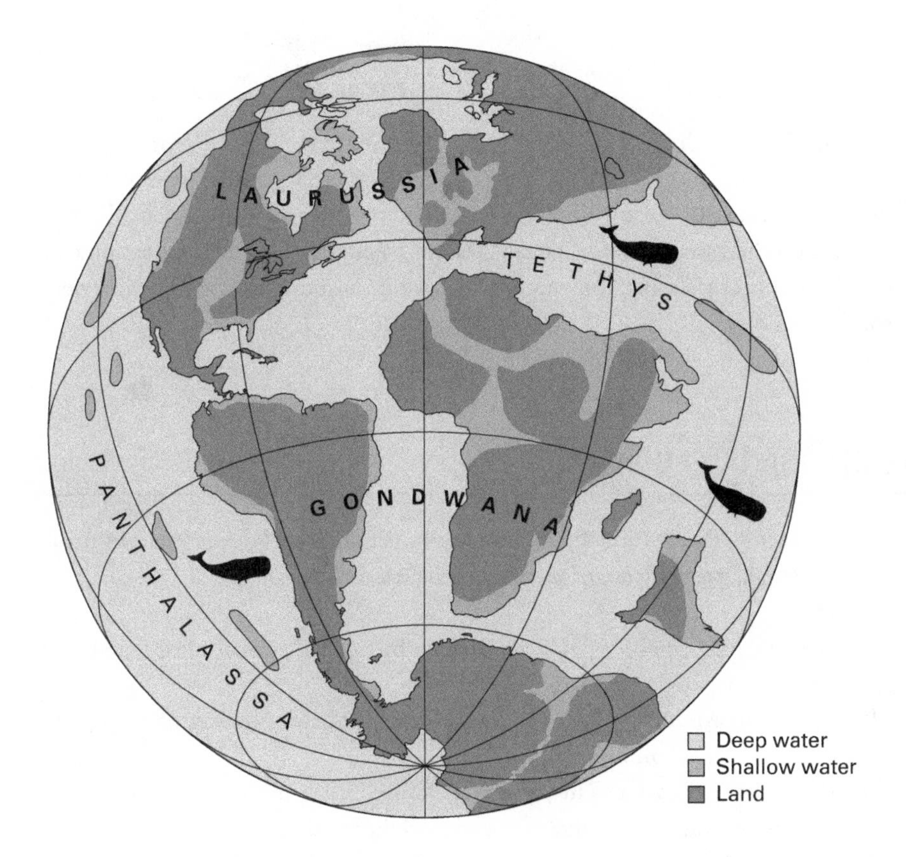


**FIGURE 10.16**
By the Late Cretaceous, Laurussia and Gondwana were almost completely fragmented, and sea levels were elevated worldwide. (Drawing by Network Graphics)

## Life in the Sea and Vegetation

The Late Cretaceous seas saw the continuing diversification of modern types of marine life—certain plankton, bony fishes, crabs, and snails—that had begun during the Early Cretaceous. It also saw the end of some very successful groups, including the ammonoids and the rudist and inoceramid (plate-like) clams. A group of marine animals unique to the Late Cretaceous was the mosasaurs, large marine lizards that were also at the top of food chains in the Late Cretaceous seas (figure 10.17).

Flowering plants diversified rapidly during the Late Cretaceous. By the end of the Cretaceous, they had virtually replaced the gymnosperms as the dominant land plants.

**FIGURE 10.17**
Mosasaurs were giant marine lizards (some more than 10 meters long) of the Late Cretaceous. (Drawing by Network Graphics)

### Dinosaurs and Other Vertebrates

Dinosaurs of the Late Cretaceous were mostly hadrosaurid ornithopods, ceratopsians, ankylosaurs, tyrannosaurids, and abelisaurid theropods. Sauropods were common only in South America. These dinosaurs coexisted with turtles, crocodiles, lizards, mammals, and other, smaller vertebrates. They were generally not as large as some of those of the Late Jurassic, but they appear to have been more diverse and were very widespread (box 10.3).

## FIVE DINOSAUR FAUNAS

If we think of each successive and distinct association of dinosaurs during the Mesozoic as a dinosaur fauna, we can recognize five such faunas:

1. The Late Triassic dinosaur fauna, dominated by relatively small meat-eaters and plant-eaters along with some large prosauropods
2. The Early and Middle Jurassic dinosaur fauna, consisting mostly of sauropods, stegosaurs, and small ornithopods
3. The Late Jurassic dinosaur fauna, generally similar to that of the Early and Middle Jurassic, but including many large sauropods and theropods
4. The transitional Early Cretaceous dinosaur fauna, in which ankylosaurs and large ornithopods dominated, whereas the numbers of sauropods and stegosaurs dwindled
5. The Late Cretaceous dinosaur fauna, dominated by hadrosaurs, ceratopsians, ankylosaurs, abelisaurids, and tyrannosaurids

These five dinosaur faunas reflect changing environments and organisms throughout the 160 million years of the age of dinosaurs. They thus indicate that much evolution and extinction of dinosaurs took place during the Mesozoic.

**Box 10.3**

### Cretaceous Dinosaurs from the Arctic and Antarctic

Dinosaur fossils were first reported from the Arctic in 1962, when they were discovered in Svalbard (Spitsbergen). By the 1990s, numerous dinosaur localities, mostly of Late Cretaceous age, were known from the Arctic, especially the North Slope of Alaska (box figure 10.3). Dinosaur fossils were also discovered in Antarctica. In 1986, scientists from Argentina discovered fragmentary remains of an ankylosaur in Upper Cretaceous rocks on James Ross Island, off the Antarctic Peninsula. Three years later, British scientists discovered fossils of an ornithopod from nearby Vega Island.

The meager record of Antarctic dinosaurs has been growing with further exploration, and it already is of importance because it documents the existence of nearly polar dinosaurs in the Southern Hemisphere during the Late Cretaceous.

Due to the drifting of the continents, reconstructions of the position of Antarctica during the Late Cretaceous suggest that the Antarctic dinosaur localities were just north of 60 degrees south then, which is only a few degrees north of their present location. There is no evidence of a polar ice cap in Antarctica during the Late Cretaceous. But, because of the tilt of Earth's axis, Antarctica would have experienced months of virtual darkness during the Late Cretaceous winters, just as it does today. This polar winter would have been a cold one, even during the Late Cretaceous when there was no Antarctic ice cap.

How did dinosaurs, which we usually think of as warm-weather animals, survive in Antarctica during the Late Cretaceous? Two possibilities exist. One is that contrary to popular belief, dinosaurs were not just warm-weather animals, but were warm-blooded, like living mammals and birds, and thus could endure very cold temperatures. The evidence for warm-blooded dinosaurs, however, is debatable (see chapter 14). A more likely solution to the puzzle of Antarctic dinosaurs may be that dinosaurs did not live there year round, but undertook seasonal migrations, as birds do today.

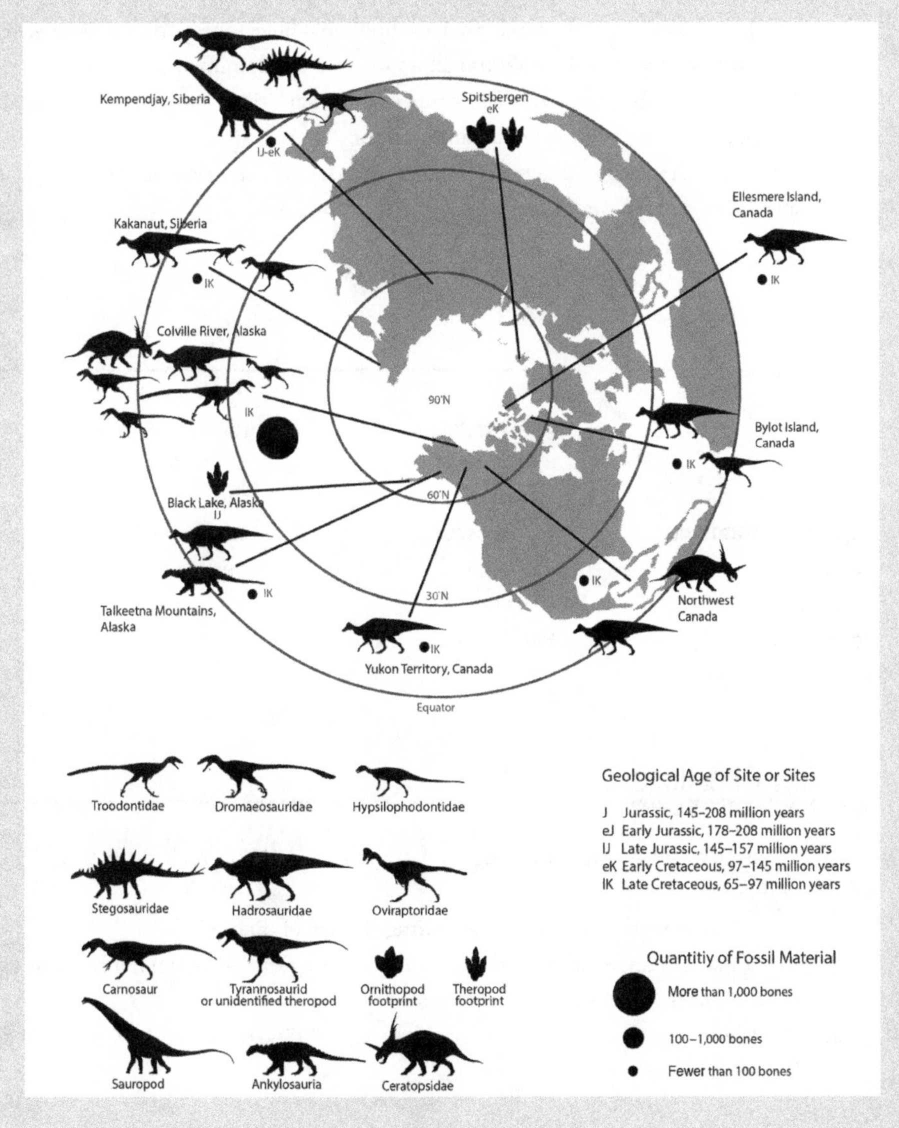


**BOX FIGURE 10.3** Dinosaur fossils are known from numerous localities in the Arctic. (© Scott Hartman)

## Summary

1. Dinosaurs appeared during the Late Triassic when all the continents were united in a single supercontinent, Pangea.
2. Climates on Late Triassic Pangea were warm and monsoonal, and the vegetation was dominated by ferns and gymnosperms.
3. Dinosaurs were not the dominant land vertebrates of the Late Triassic; other archosaurs were.
4. Various archosaurs and many other land vertebrates became extinct at about the end of the Triassic.
5. Dinosaurs established themselves as the dominant land vertebrates by the end of the Middle Jurassic.
6. Early and Middle Jurassic plant-eating dinosaurs were mostly prosauropods, sauropods, and stegosaurs.
7. The Late Jurassic was the golden age of dinosaurs; huge sauropods, stegosaurs, primitive ornithopods, megalosaurs, and allosaurs were the dominant dinosaurs.
8. During the Early Cretaceous, dinosaurs were in transition, as vegetation changed and the world climate became wetter and warmer.
9. The Late Cretaceous dinosaurs were mostly hadrosaurs, ceratopsians, ankylosaurs, abelisaurs, and tyrannosaurids.

## Key Terms

ammonoid
angiosperm
dinosaur fauna
epicontinental sea
equable
Gondwana
greenhouse
gymnosperm
ichthyosaur
Laurussia
metoposaurid
monsoonal
Pangea
Panthalassa
plesiosaur
rauisuchian
rudist
*Sanmiguelia*
Tethys Sea

## Review Questions

1. Describe world geography during the Late Triassic and how it changed during the age of dinosaurs.
2. How did the world climate change during the age of dinosaurs?
3. What major changes in vegetation took place during the age of dinosaurs? How might these changes have influenced dinosaur evolution?
4. What were the five dinosaur faunas of the age of dinosaurs, and how did they differ from each other?

## Further Reading

Behrensmeyer, A. K., J. D. Damuth, W. A. DiMichele, R. Potts, H.-D. Sues, and S. L. Wing. 1992. *Terrestrial Ecosystems Through Time.* Chicago: University of Chicago Press. (Provides a detailed review of Mesozoic terrestrial vegetation, animal life, and climate on pp. 327–372)

Colbert, E. H. 1965. *The Age of Reptiles.* New York: Norton. (Somewhat outdated, but otherwise an excellent review of Mesozoic life)

Currie, P. J., and E. B. Koppelhaus, eds. 2005 *Dinosaur Provincial Park: A Spectacular Ancient Ecosystem Revealed.* Bloomington: Indiana University Press. (A collection of 28 technical articles on one of the world's greatest Late Cretaceous dinosaur ecosystems)

Foster, J. 2007. *Jurassic West: The Dinosaurs of the Morrison Formation and Their World.* Bloomington: Indiana University Press. (Everything you want to know about the most famous Late Jurassic dinosaurs and their environment)

Fraser, N. 2006. *Dawn of the Dinosaurs: Life in the Triassic.* Bloomington: Indiana University Press. (Very readable review of the Triassic world and its biota)

Gee, C. T., ed. 2010. *Plants in Mesozoic Time: Morphological Innovations, Phylogeny, Ecosystems.* Bloomington: Indiana University Press. (A collection of 14 technical articles presenting recent research on Mesozoic plants)

Godefroit, P., ed. 2012. *Bernissart Dinosaurs and Early Cretaceous Terrestrial Ecosystems.* Bloomington: Indiana University Press. (A collection of 33 technical articles on Bernissart and much more about Early Cretaceous dinosaurs and ecosystems)

Hallam, A. 1965. *Jurassic Environments.* Cambridge: Cambridge University Press. (Provides an overview of the Jurassic world)

Lucas, S. G. 2013. The end-Triassic extinction, pp. 475–486, in N. MacLeod, J. D. Archibald, and P. S. Levin, eds., *Grzimek's Animal Life Encyclopedia.* Vol.2, *Extinction.* Detroit: Gale Cengage Learning. (Review of the end-Triassic extinctions and their possible causes)

Novas, F. E. 2009. *The Age of Dinosaurs in South America.* Bloomington: Indiana University Press. (Comprehensive review of the Mesozoic history of an entire continent)

Russell, D. A. 1989. *An Odyssey in Time: The Dinosaurs of North America.* Toronto: University of Toronto Press. (A colorful review of the geological and environmental context of North American dinosaurs)

Stanley, S. M., and J. A. Luczaj. 1989. *Earth System History.* New York: Freeman. (An introductory-level college textbook that provides an overview of fossils and the history of life)

Sues, H.-D., and N. C. Fraser. 2010. *Triassic Life on Land: The Great Transition.* New York: Columbia University Press. (Describes the Triassic rise of modern terrestrial ecosystems)

# 11

# DINOSAUR HUNTERS

THE history of the discovery of dinosaurs is full of adventure. Fantastic discoveries, exotic places, physical hardships, danger, and dynamic personalities are just some of the elements of this history. Sometimes it is easy to lose sight of the fact that what we know about dinosaurs is the result of about two centuries of dinosaur-fossil collecting by amateurs and professionals worldwide. It is their discoveries, beginning in the 1820s, that have shaped scientific views about dinosaurs. Indeed, if we examine both the history of dinosaur discoveries and the history of scientific ideas about dinosaurs, there is a notable correlation. New discoveries have led to new ideas, with the result that ideas about dinosaurs have changed dramatically since the 1820s.

This chapter does not present a comprehensive history of the discovery and collection of dinosaurs. That history merits a book all its own and is well told in the books listed as further readings at the end of this chapter. Instead, the focus here is on how dinosaur discoveries have shaped scientific ideas about the "terrible lizards."

## EARLIEST DISCOVERIES

The scientific study of dinosaurs began in England during the 1820s, although it is clear that many cultures, such as the ancient Chinese, had encountered dinosaur bones long before then. The first dinosaur to be described scientifically was the theropod ***Megalosaurus*** (see chapter 5), named by British geologist and naturalist **William Buckland** (1784–1856) in 1824. However, two years earlier, British country doctor **Gideon Mantell** (1790–1852) discovered dinosaur teeth and bones in Sussex. In 1825, Mantell named those fossils ***Iguanodon*** because they resembled the teeth of a living iguana (*odont* is Greek for "tooth") (figure 11.1).

**FIGURE 11.1**
These lithographs of teeth of *Iguanodon* are from Gideon Mantell's original article, published in 1825. (From G. Mantell. 1825. Notice on the *Iguanodon*, a newly discovered fossil reptile, from the sandstone of Tilgate Forest, in Sussex. *Philosophical Transactions of the Royal Society of London* 115:179–186)

Neither Buckland nor Mantell knew the fossils they described as dinosaurs, but they did recognize them as the remains of large, extinct reptiles. Similar fragmentary fossils of reptiles continued to be found in Britain through the 1830s. These included a partial skeleton of an armored reptile, named ***Hylaeosaurus*** by Mantell in 1833, as well as other fragments that formed the basis for the names *Cetiosaurus*, *Poekilopleuron*, and *Thecodontosaurus*.

At that time, the foremost authority on fossil reptiles in Britain was **Richard Owen** (1804–1892), a comparative anatomist who worked for most of his career at the Royal College of Surgeons and, later, the British Museum of Natural History, both in London (figure 11.2). In 1842, Owen published a comprehensive review of British fossil

**FIGURE 11.2**
Richard Owen, who coined the word "Dinosauria" in 1842. (Drawing by M. Tardivel)

reptiles in the *Report of the British Association for the Advancement of Science*. There, Owen coined the term **Dinosauria**, from the Greek roots *deinos*, meaning "terrible" (Owen actually meant "fearfully great"), and *sauros*, meaning "lizard" or "reptile." Owen included *Megalosaurus*, *Iguanodon* and *Hylaeosaurus* in the Dinosauria, but other British fossil reptiles known to him and subsequently shown to be dinosaurs were excluded. For example, Owen identified the theropod *Poekilopleuron* as a crocodile, and the sauropod *Cetiosaurus* as a gigantic marine crocodile. Owen characterized dinosaurs as having teeth set in bony sockets, large sacra composed of five fused vertebrae, ribs with two heads, a complex shoulder girdle, long hollow limb bones and mammal-like feet. In fact, Owen saw several of the features of dinosaurs as more mammal-like than reptile-like, and he even speculated that dinosaurs had hearts and respiratory systems very similar to those of living mammals.

Yet, despite Owen's comparisons of dinosaurs to mammals, his work and that of his contemporaries produced a very reptilian image of the dinosaurs. This image, the first comprehensive scientific view of the dinosaurs, emerged in the 1850s through Owen's collaboration with artist and sculptor **Benjamin Waterhouse Hawkins** (1807–1894). Between 1852 and 1854, this collaboration resulted in various paintings by Hawkins and, most notably, several life-size sculptures of dinosaurs for the grounds of the Crystal Palace exhibition center at Sydenham, now a London suburb where they still stand (figure 11.3). Indeed, as a publicity stunt just before the sculptures were unveiled, Hawkins organized a dinner for 20 scientists held inside the hollow body of the life-size *Iguanodon*.

The Owen–Hawkins collaboration produced images of dinosaurs as bulky, lizard- and toad-like brutes. These images differ greatly from current conceptions of dinosaurs and are demonstrably wrong. But, to be fair to Owen, Hawkins, and their contemporaries, we need to remember the context that produced these incorrect restorations of

**FIGURE 11.3**
Benjamin Waterhouse Hawkins created these images of dinosaurs. (Courtesy Robert M. Sullivan)

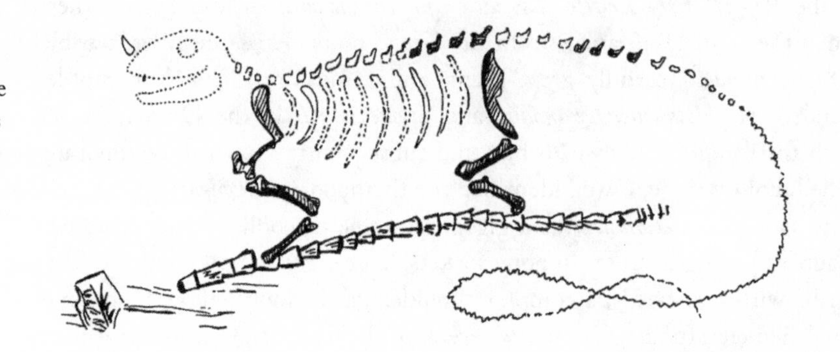

**FIGURE 11.4**
Mantell reconstructed the skeleton of *Iguanodon* with the thumb spike on the dinosaur's nose. (From G. Mantell. 1834. Discovery of the bones of the *Iguanodon* in a quarry of Kentish Rag [a limestone belonging to the Lower Greensand Formation] near Maidstone, Kent. *Edinburgh New Philosophical Journal* 17:200–201)

dinosaurs. First, let us not forget that not a single complete skeleton of a dinosaur had yet been discovered in 1854. Thus, many aspects of dinosaur size, shape, and appearance were necessarily highly speculative. In fact, the most complete dinosaur skeleton available to Owen was a very incomplete *Iguanodon* that had been found in a quarry at Maidstone in Kent. Mantell had studied this fossil and used it to reconstruct a complete skeleton of *Iguanodon* (figure 11.4). A spike of bone that was part of this skeleton was placed by Mantell on the nose of the dinosaur and also appears there on Hawkins's sculpture. But, the discovery of complete *Iguanodon* skeletons at Bernissart, Belgium, in 1878, revealed this to be a thumb, not a nose, spike.

Because Owen and his contemporaries lacked complete dinosaur skeletons, they filled in the gaps from their knowledge of living reptiles, especially lizards and crocodiles. This knowledge was also heavily colored by an image of reptiles as stupid and sluggish brutes, prejudices that are still with us (see box 14.1).

The incomplete knowledge of dinosaurs and these prejudices about living reptiles inspired the Owen–Hawkins restorations of dinosaurs during the 1850s. What emerged was the concept of ponderous, dim-witted, and terrible lizards, or "dinosaurs as reptiles."

## COMPLETE SKELETONS

The first dinosaurs brought to scientific attention in North America were teeth found in 1855 in what is now Montana. They were studied by **Joseph Leidy** (1823–1891), a Philadelphia anatomist who is considered the founder of vertebrate paleontology in the United States (figure 11.5). Leidy was familiar with British publications on dinosaurs, and he recognized the similarities of the teeth from Montana to those of *Megalosaurus* and *Iguanodon*. The names Leidy coined for the Montana teeth included ***Deinodon*** (terrible tooth) and ***Trachodon*** (rough tooth), and these were the first North American dinosaurs to be described scientifically.

But a far more significant discovery was in store for Leidy much closer to home. In 1868, he examined a partial skeleton of a dinosaur found by a farmer digging in a marl pit near Haddonfield, New Jersey. This was the most complete skeleton of a dinosaur yet discovered, and Leidy christened it ***Hadrosaurus*** (heavy lizard).

**FIGURE 11.5**
Joseph Leidy is shown here with the femur of *Hadrosaurus*. (Courtesy Ewell Sale Stewart Library, Academy of Natural Sciences of Philadelphia)

The Haddonfield skeleton had nearly complete forelimbs and hind limbs, and Leidy realized that the dinosaur must have been an upright biped, not a sprawling quadruped like the Owen–Hawkins *Iguanodon*. This change in dinosaur posture was soon reflected in a life-size model of *Hadrosaurus* in New York's Central Park, constructed in 1868 by Benjamin Waterhouse Hawkins.

The skeleton of *Hadrosaurus* may have forced some rethinking of the posture and appearance of dinosaurs by 1870. But, a true revolution in scientific understanding was to take place between 1870 and 1900 because of the discovery of many more nearly complete dinosaur skeletons.

These skeletons came from western North America during what has been called the "**great dinosaur rush.**" The many dinosaur skeletons were primarily collected for and studied by two paleontologists: **Edward Drinker Cope** (1840–1897) and **Othniel Charles Marsh** (1831–1899) (figure 11.6). Cope was born into a wealthy Quaker family, was a protégé of Leidy (though largely self-taught), and worked and lived mainly in Philadelphia, where he used his family fortune to support his paleontological research. Marsh was trained primarily in Europe and was the first professor of paleontology at Yale University, where he founded the Peabody Museum of Natural History. It was named after his wealthy uncle, George Peabody, who underwrote many of Marsh's paleontological endeavors.

Neither Cope nor Marsh collected many dinosaurs themselves. Instead, they employed collectors who scoured much of the American West, from the Dakotas

**FIGURE 11.6**
Edward Drinker Cope (*left*) and Othniel Charles Marsh (*right*), scientific rivals, studied most of the dinosaur fossils discovered during the "great dinosaur rush" of the nineteenth century. ([*left*] Courtesy Department of Library Services, American Museum of Natural History, New York [negative no. 312408]; [*right*] courtesy John Ostrom, Yale Peabody Museum of Natural History)

to New Mexico, discovering and collecting dinosaurs and many other fossils, then shipping them back to Cope in Philadelphia and Marsh in New Haven for preparation, study, and description. For a variety of reasons, a tremendous personal rivalry arose between Cope and Marsh, as each tried to outdo the other in paleontological discoveries. This rivalry, which began in the early 1870s, lasted until Cope's death in 1897. It benefited paleontology by fueling the discovery of many new fossils, but it also led Cope and Marsh to sometimes publish their research too hastily and make many unnecessary mistakes.

Tremendous new dinosaur discoveries, particularly by Marsh's collectors, brought to science many new types of dinosaurs and complete skeletons. Thus, Marsh first described some of the best-known dinosaurs, Jurassic giants such as *Stegosaurus*, "*Brontosaurus*," and *Allosaurus*, from the dinosaur quarries worked by his collectors at **Como Bluff** in Wyoming. Indeed, beginning in 1891, Marsh published remarkably accurate reconstructions of the nearly complete skeletons of many of the dinosaurs his collectors had discovered (figure 11.7). Furthermore, Marsh proposed a comprehensive classification of the dinosaurs in 1882 (see box 5.1). His terminology, reconstructions, and ideas vitally shaped scientific conceptions about dinosaurs. His collectors, some

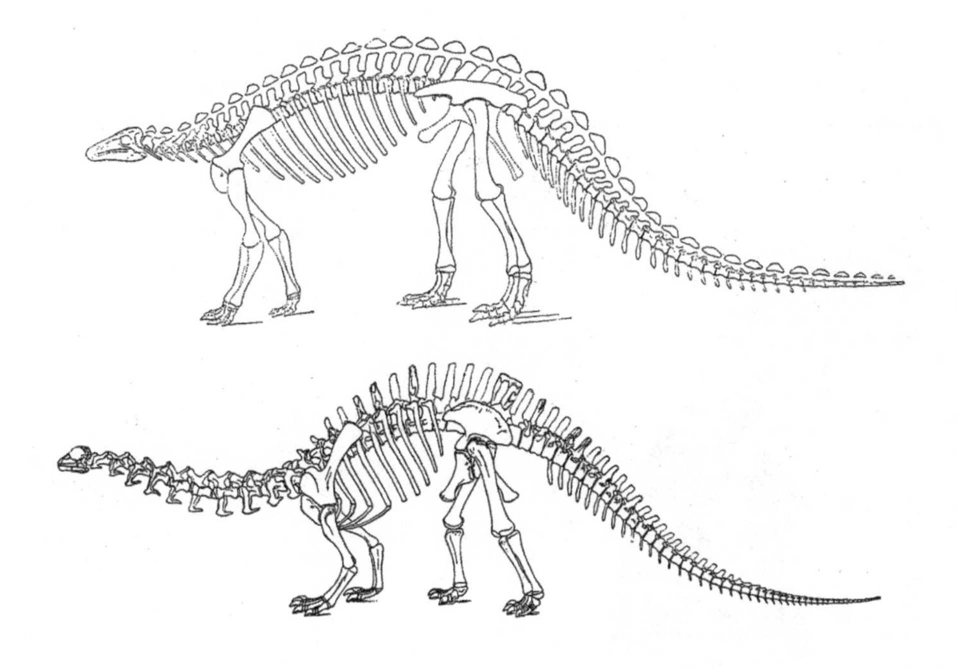

**FIGURE 11.7**
Marsh published many very accurate reconstructions of dinosaur skeletons, two of which are shown here. (From E. H. Colbert. 1984. *The Great Dinosaur Hunters and Their Discoveries*. Mineola, N.Y.: Dover)

of whom went on to become prominent paleontologists themselves, also discovered many new types of dinosaurs (box 11.1).

Cope devoted less attention to dinosaurs than did Marsh, but his collectors' discoveries were also very significant. And, Cope was more willing to speculate on the appearance and habits of the dinosaurs than was Marsh. So, we find Cope introducing the idea that sauropods were aquatic in 1897 (see figure 6.18) and directing artist **Charles R. Knight** (1874–1953) to draw two active and agile, fighting theropods (see figure 17.7). Indeed, the scientific legacy of Cope and Marsh's scientific ideas about dinosaurs is well summed up in the artwork of Charles R. Knight.

Knight painted and sculpted dinosaurs largely in collaboration with paleontologist **Henry Fairfield Osborn** (1857–1935), a protégé of Cope who founded the Department of Vertebrate Paleontology at the American Museum of Natural History. After Cope and Marsh died, the "great dinosaur rush" continued, partly financed by the American Museum of Natural History. **Barnum Brown** (1873–1963) collected dinosaurs for the American Museum all over the American West. And, one of Cope's collectors, **Charles Hazelius Sternberg** (1850–1943), collected dinosaurs freelance for many museums, especially in Alberta, Canada, and New Mexico (figure 11.8). Indeed, it would be fair to say that Brown and Sternberg collected more dinosaurs than anyone else in history.

Primarily through the artistry of Charles R. Knight, the dinosaurs available to Osborn and his colleagues at the turn of and early years of the twentieth century produced an image of dinosaurs fundamentally different than the Owen–Hawkins collaboration. Multiple skeletons allowed for the discovery of accurate, upright postures and body shapes in dinosaurs. Huge, bipedal theropods and strange, armored ankylosaurs, stegosaurs, and ceratopsians were not extrapolations from a handful of bones,

## Box 11.1

### John Bell Hatcher

**John Bell Hatcher** began his paleontological career as one of Othniel Charles Marsh's hired collectors and went on to become an outstanding paleontologist in his own right (box figure 11.1). Born in 1861 in Virginia, Hatcher was raised on a farm. As a young man he worked in a coal mine to save money for a college education.

**BOX FIGURE 11.1**
John Bell Hatcher.

In 1884, Hatcher received a bachelor's degree in geology from Yale College and was immediately hired by Marsh to collect fossils. Although Hatcher collected fossils for Marsh until 1893, the relationship between the two men was rough. They squabbled often over Hatcher's salary and over the nature and schedule of Hatcher's duties. Yet, despite the bickering, Hatcher made truly amazing discoveries for Marsh, many of fossil mammals in what are now the states of Wyoming and South Dakota. His great dinosaur discoveries were made during the field seasons of 1889 to 1892, during which he collected the skulls and skeletons of 50 (!) ceratopsians in eastern Wyoming (see box 9.2). Most of these were of *Triceratops* and even now represent much of what we know about that dinosaur.

After Hatcher left Marsh's employ, his dinosaur-collecting days ended. He participated in fossil-mammal-collecting expeditions to Argentina from 1896 to 1899 and shortly thereafter took a position at the Carnegie Museum of Natural History. But, Hatcher had been a sickly child, and as an adult he suffered from rheumatoid arthritis and various other maladies. His death, from typhoid fever in 1904 at the age of 42, cut short the career of one of America's most promising paleontologists.

**FIGURE 11.8**
Barnum Brown (*left*) and Charles Hazelius Sternberg (*right*) were the greatest dinosaur collectors of all time. ([*left*] Courtesy Department of Library Services, American Museum of Natural History, New York [negative no. 37243])

but real animals based on the discovery of many articulated skeletons. Yet, despite the huge increase in knowledge of dinosaurs since 1854, the image of dinosaurs as ponderous, dim-witted brutes remained (figure 11.9). True, there were exceptions, like the Cope–Knight fighting theropods (see figure 17.6), but dinosaurs as reptiles, with all the supposed negative reptilian attributes, remained the predominant scientific view.

**FIGURE 11.9**
Charles R. Knight's classic painting of a *Triceratops* engaging in battle with a *Tyrannosaurus*. (Courtesy Field Museum of Natural History, Chicago [negative no. 59442])

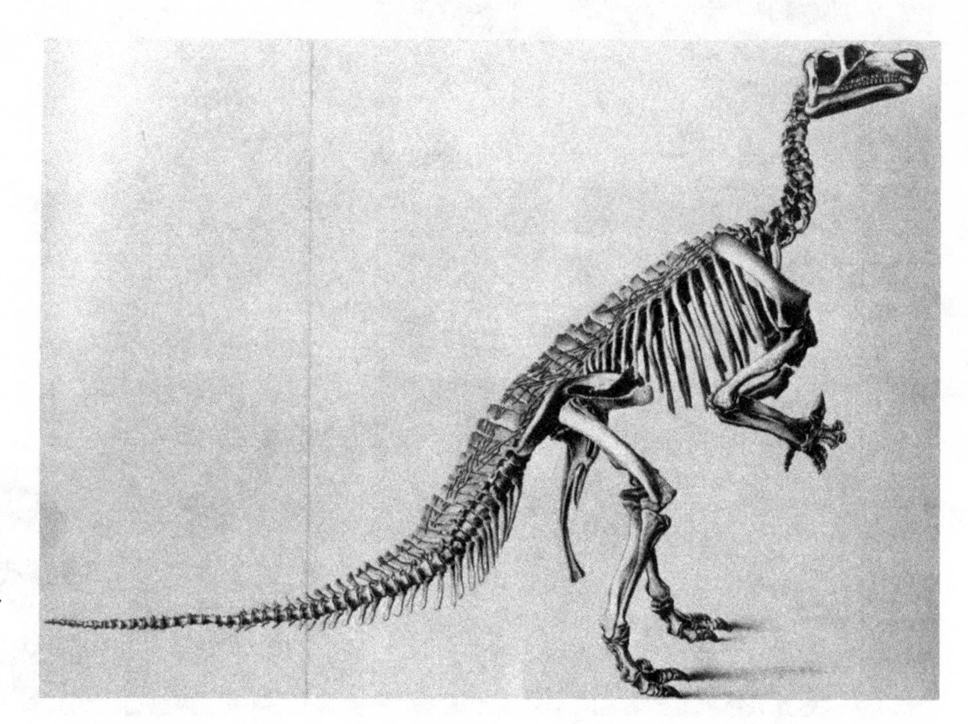

**FIGURE 11.10**
Louis Dollo's reconstruction of the skeleton of *Iguanodon* was based on more than two dozen complete skeletons from Bernissart. (From L. Dollo. 1882. Première note sur les dinosauriens de Bernissart. *Bulletin du Musée royal d'histoire naturelle de Belgique* 1:55–74)

It would be incorrect to think that all the great discoveries and new ideas about dinosaurs between the 1850s and the early 1900s took place just in North America. Dinosaurs continued to be collected in Europe, and in 1878, complete skeletons of *Iguanodon* were discovered in **Bernissart, Belgium** (see box 7.2). Belgian paleontologist **Louis Dollo** (1857–1931) studied these fossils and provided a reconstruction of *Iguanodon* (figure 11.10) fundamentally different from the early attempts of Mantell, Owen, and Hawkins.

Ideas about other aspects of dinosaur science were not confined to North America, either. For example, in 1887, British paleontologist **Harry G. Seeley** (1839–1909) proposed a classification of dinosaurs that divided them into two groups: Saurischia and Ornithischia. This classification (see chapter 2) differed from those of Cope and Marsh but is the one still used today (figure 11.11).

In many ways, the science of dinosaur paleontology came into its own during the second half of the nineteenth century. The discovery of a variety of dinosaurs, and

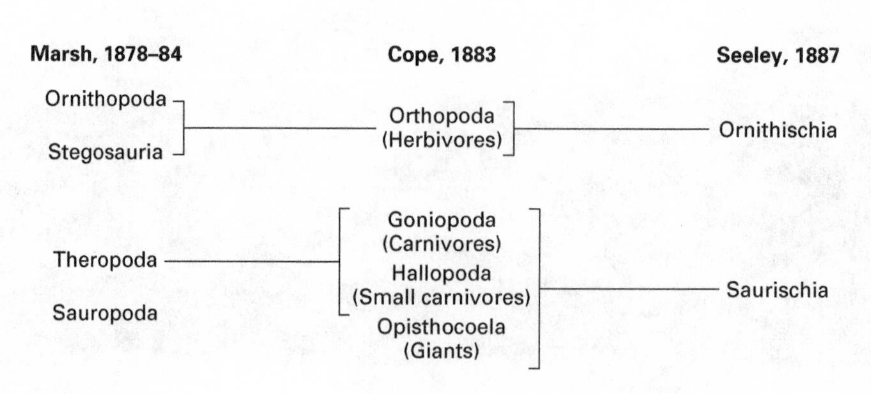


**FIGURE 11.11**
Three different classifications of dinosaurs, by Marsh, Cope, and Seeley, were proposed in the 1880s. (Drawing by Network Graphics)

especially of articulated, essentially complete dinosaur skeletons, made this possible. But, the emphasis remained on the reptilian nature of the dinosaurs even though a more thorough knowledge of their anatomy was available.

## TWO GREAT EXPEDITIONS

It is fair to say that scientific opinions about dinosaurs did not change significantly between the turn of the twentieth century and World War II. Much time was still spent analyzing the products of the great dinosaur discoveries of the previous half century, and new discoveries continued to be made, such as Barnum Brown's discovery of *Tyrannosaurus rex* in Montana in 1902. Whatever new ideas arose in dinosaur science between 1900 and 1945 were largely the result of two great expeditions.

The first of these took place between 1907 and 1912 at **Tendaguru Hill** in what was then German East Africa and is now Tanzania. These expeditions were led by German paleontologists **Werner Janensch** (1878–1969) and Edwin Hennig (1882–1977) and brought thousands of bones of Jurassic dinosaurs, including several nearly complete skeletons, back to Berlin (figure 11.12). The Tendaguru dinosaurs included articulated skeletons of a new dinosaur, then the largest land animal of all time, the gigantic sauropod *Brachiosaurus*. The Tendaguru dinosaurs included several types, such as *Brachiosaurus*, also known from the western United States. This provided a dinosaur-basis for correlation (see box 3.2) and demonstrated a broader geographic distribution for some dinosaurs than had previously been suspected. Most importantly, the extensive dinosaur collections from Tendaguru demonstrated that significant dinosaur collections could be found in parts of the world outside of Europe and western North America.

The second great dinosaur expedition (actually a series of expeditions) of the first half of the twentieth century further emphasized that great dinosaur finds were waiting outside of Europe and North America. This series of expeditions, the **Central Asiatic Expeditions** of the American Museum of Natural History, led by **Roy Chapman Andrews** (1884–1960), collected fossils in the **Gobi Desert** of China and Mongolia during the 1920s and 1930s. Although the expeditions' primary purpose was not to collect dinosaurs (box 11.2), they unearthed a rich lode of them in the Cretaceous beds of the Gobi, including the primitive ceratopsian *Protoceratops* and the small theropods *Oviraptor* and *Velociraptor*, among others. Most famous of the dinosaur discoveries of the Central Asiatic Expeditions, however, were the nests of dinosaur eggs discovered at Bayn-Dzak, Mongolia (figure 11.13).

These expeditions, like the German efforts at Tendaguru, brought to light new types of dinosaurs and new information about dinosaur biology and distribution, all of which significantly added to our scientific understanding of dinosaurs. New analyses of dinosaurs, such as Swedish paleontologist Carl Wiman's 1923 interpretation of the hollow tube on the head of the hadrosaurid *Parasaurolophus* as a resonating chamber, also appeared in the first half of the twentieth century. But, the overall image of dinosaurs did not change from that of the turn of the century—the view of dinosaurs as reptiles as put forth by Cope, Marsh, and Osborn, and as painted by Charles R. Knight—remained substantially intact.

**FIGURE 11.12**
The skeleton of *Brachiosaurus* (or *Giraffatitan*), on display in Berlin, was brought back from East Africa by a German expedition in the early twentieth century. (Courtesy Museum für Naturkunde, Berlin)

## THE CALM BEFORE THE STORM?

Curiously, dinosaur paleontology was relatively quiet between the 1940s and the early 1970s. Dinosaurs continued to be collected during this time, and some outstanding discoveries were made, such as the skeletons of hundreds of *Coelophysis* found in Upper Triassic strata in northern New Mexico by the American Museum of Natural History in the late 1940s. Other Triassic dinosaurs were discovered in Brazil and Argentina, and dinosaur collecting became a worldwide activity, with important discoveries made in such far-flung locales as southern Africa, India, and China. Between

## Box 11.2

### The Story Behind the Central Asiatic Expeditions

The principal purpose of the Central Asiatic Expeditions, which collected fossils in the Gobi Desert in 1922, 1923, 1925, 1928, and 1930, was not to collect dinosaurs. Instead, the expeditions were initiated because Henry Fairfield Osborn believed that the origin of humankind was to be found in Asia.

**BOX FIGURE 11.2**
Roy Chapman Andrews. (Courtesy Department of Library Services, American Museum of Natural History, New York [negative no. 410927])

Roy Chapman Andrews, the leader of the Central Asiatic Expeditions, first went to China in 1916 to collect living mammals for the American Museum of Natural History (box figure 11.2). The idea of collecting fossils in Asia came to Andrews only in 1919, when he met Swedish geologist Johan Gunnar Andersson (1874–1960). Andersson was employed by the Chinese government as a mining geologist and adviser. As an adjunct to his official duties, which began in 1914, Andersson amassed an extensive collection of archeological materials and mammal fossils in eastern China. This collection must have impressed Andrews, who later discussed it with Osborn. Osborn was also aware of a fossil human tooth bought from a Chinese druggist and described by German paleontologist Max Schlosser (1854–1932) in 1903. Schlosser boldly suggested, largely on the basis of this fossil, that the origin of humankind was to be found in Asia, a suggestion heartily endorsed by Osborn. Thus, Schlosser's and Andersson's fossils and theories about human origins and the bold, adventurous spirit of Andrews provided the basis for the Central Asiatic Expeditions. Ironically, no human fossils were found, but a large number of other fossil mammals, dinosaurs, and dinosaur eggs were discovered.

**FIGURE 11.13**
Dinosaur eggs were discovered by the Central Asiatic Expeditions at the "flaming cliffs" of Bayn-Dzak, Mongolia. (Courtesy Department of Library Services, American Museum of Natural History, New York [negative no. 410767])

1947 and 1949, Soviet paleontologists led expeditions back to Mongolia and uncovered many new Cretaceous dinosaurs. These and other discoveries helped to fill in scientific understanding of the dinosaurs and the world they inhabited, but they did not fundamentally alter the prevailing, turn-of-the-century view of dinosaurs as reptiles.

Proof of this is best seen in the work of two men: Czech artist **Zdeněk Burian** (1905–1981) and American paleontologist **Edwin H. Colbert** (1905–2001). Burian's paintings of dinosaurs, many painted during the 1950s, show the heavy influence of Charles R. Knight and well reflect turn-of-the-century ideas about dinosaur biology and behavior. Colbert, an outstanding researcher on dinosaurs and other reptiles, wrote many popular books on dinosaurs from the 1940s through the 1960s. Several of his books featured Knight's paintings and popularized the scientific image of dinosaurs as reptiles.

## THE DINOSAUR RENAISSANCE

Although major new ideas about dinosaurs did not emerge during the 1940s, 1950s, or 1960s, dinosaur science was not dead. Instead, those years were a period when new ideas about dinosaurs were incubating, ready to hatch during the early 1970s.

During the 1960s, the joint **Polish–Mongolian Paleontological Expeditions** revisited the Gobi Desert. Again, new dinosaurs were to be had, especially small and very bird-like theropods. Also during the 1960s, field crews from Yale University, led by paleontologist **John Ostrom**, explored the Lower Cretaceous dinosaur-bearing strata along the Montana–Wyoming border. Their most significant discovery was also a small theropod, ***Deinonychus***.

The skeletons of the small theropods from Mongolia and of *Deinonychus* suggested very active and agile animals, quite different from the ponderous, dim-witted brutes of the dinosaurs-as-reptiles image. Furthermore, in the 1970s, Ostrom argued persuasively in several scientific publications that dinosaurs were the ancestors of birds (see chapter 15).

The ideas about highly active and agile theropods and dinosaurs as bird ancestors were clearly not based only on the discoveries and research of the 1960s. They could also be found in speculations published by some paleontologists as far back as the 1870s. However, in the 1970s, one bold paleontologist, **Robert Bakker**, challenged paleontological orthodoxy by arguing the case for warm-blooded and active dinosaurs that were much more bird-like in their biology and behavior than paleontologists had previously envisioned. Since Bakker's initial proposals, over the past 40 years, paleontologists have used a wealth of new information on the biology of living reptiles, mammals, and birds, as well as new techniques, many computer-aided, to analyze dinosaur fossils.

The renaissance in the study and interpretation of dinosaur fossils continues today. Renewed interest in collecting dinosaurs has resulted in discoveries in Antarctica, Alaska, Australia, and just about every other place that dinosaur fossils might be had. New interpretations of dinosaur phylogeny, rooted in cladistic analysis, prevail. Spirited scientific debate about dinosaur metabolism (see chapter 14) and the causes of dinosaur extinction (see chapter 16) are ongoing. Discoveries of fossilized nests and baby dinosaurs in Montana (see chapter 13) and a revitalized interest in dinosaur footprints (see chapter 12) have produced a wealth of informed speculation about dinosaur behavior. This new research has reshaped our entire scientific view of dinosaurs. The "dinosaurs-as-reptiles" view of ponderous, dim-witted brutes has given way to the view of dinosaurs as fast, active, and agile animals, more akin to living birds, their apparent descendants, than to living reptiles. This view of dinosaurs as bird ancestors has also been brought to life by many talented artists, some of them well-schooled in dinosaur science.

The **dinosaur renaissance**, however, has not been without its excesses. The conceptions of galloping brontosaurs and super-intelligent theropods have little or no supporting evidence. And, no matter how fast, active, agile, and bird-like some dinosaurs may have been, it is impossible to view a 50-ton sauropod as anything other than a huge, slow-moving, powerful, and not particularly brainy animal. New discoveries and new analyses continue to support an image of the dinosaurs, not as the ponderous reptiles of the turn-of-the-century paleontologist, but as bird-like—the most distinctive animals to have ever lived.

## CHANGING IDEAS IN DINOSAUR SCIENCE

The history of dinosaur discoveries presented here emphasizes the interplay of discoveries and scientific ideas about dinosaurs. It reveals three distinct concepts of dinosaurs from the 200 or so years during which they have been studied scientifically (figure 11.14).

**FIGURE 11.14**
These three views of *Iguanodon* reflect changing scientific ideas about dinosaurs over the past 200 years. (Drawing by Network Graphics)

The first concept, of the 1850s, was based on a very incomplete knowledge of dinosaur anatomy and on a view of living reptiles as slow-moving and stupid. Dinosaurs were seen as huge, ponderous, dim-witted reptiles.

The second concept had emerged by the turn of the twentieth century and was founded on a far more extensive knowledge of dinosaur anatomy. Nevertheless, the idea of dinosaurs as reptiles, however unique as reptiles they might have been, remained.

The third, and current concept of dinosaurs emerged during the 1970s. Dinosaurs are now seen as fast, active, agile, bird-like animals. This view is based on new discoveries of bird-like theropods, on a new understanding of the biology of living animals, and on the recognition of dinosaurs as the ancestors of birds.

Which concept is the correct one? Certainly, it is as easy to dismiss the 1850s view of dinosaurs now as it was in 1900. And, the early twentieth century concept of dinosaurs is clearly wrong for theropods and in several other ways. Today's concept of dinosaurs stems from the analysis of much more information than earlier concepts were based on, so it seems closer to the truth than the older concepts. But, what will our concept of dinosaurs be in the next century?

## Summary

1. Dinosaurs first came to scientific attention in Britain during the 1820s.
2. Richard Owen coined the word "Dinosauria," meaning "terrible lizards," in 1842.
3. The first restorations of dinosaurs, in the 1850s, reflected terribly incomplete knowledge of dinosaur anatomy and prejudices about living reptiles. Dinosaurs were portrayed as ponderous, sluggish, dim-witted brutes.
4. The "great dinosaur rush" of the 1870s and 1880s in western North America brought to science many more dinosaurs, including nearly complete skeletons, described

principally by two American paleontologists: Edward Drinker Cope and Othniel Charles Marsh.

5. By the beginning of the twentieth century, many dinosaurs could be reconstructed with fair accuracy, but scientific views of dinosaurs still stressed their slow-moving and slow-witted reptilian nature.
6. Two great dinosaur-collecting expeditions of the first half of this century—to East Africa from 1907 to 1912 and to the Gobi Desert during the 1920s and 1930s—brought more dinosaurs to scientific attention, but they did not alter the prevailing view of dinosaurs as reptiles.
7. Between 1940 and 1970, still more dinosaur discoveries were made, but they too failed to alter early twentieth century views of dinosaurs as reptiles.
8. A new concept of dinosaurs as fast, active, agile, and bird-like emerged in the 1970s, following new discoveries of small theropods, a new understanding of the biology of living animals, and the recognition of dinosaurs as the ancestors of birds.

## Key Terms

Roy Chapman Andrews
Robert Bakker
Bernissart, Belgium
Barnum Brown
William Buckland
Zdeněk Burian
Central Asiatic Expeditions
Edwin H. Colbert
Como Bluff
Edward Drinker Cope
*Deinodon*
*Deinonychus*
dinosaur renaissance
Dinosauria
Louis Dollo
Gobi Desert
"great dinosaur rush"
*Hadrosaurus*
John Bell Hatcher
Benjamin Waterhouse Hawkins
*Hylaeosaurus*
*Iguanodon*
Werner Janensch
Charles R. Knight
Joseph Leidy
Gideon Mantell
Othniel Charles Marsh
*Megalosaurus*
Henry Fairfield Osborn
John Ostrom
Richard Owen
Polish–Mongolian Paleontological Expeditions
Harry G. Seeley
Charles Hazelius Sternberg
Tendaguru Hill
*Trachodon*

## Review Questions

1. What were the first dinosaurs described in Europe? In North America? Who described them?
2. Who introduced the term "Dinosauria," and what does it mean?
3. What provided the basis for the Owen–Waterhouse dinosaur sculptures?
4. How did the "great dinosaur rush" influence scientific understanding of dinosaurs?
5. What was the early twentieth century scientific image of dinosaurs, and why did that image remain unchanged until the 1970s?
6. How did the two great dinosaur-collecting expeditions of the first half of the twentieth century contribute to our knowledge of dinosaurs?
7. What is the current concept of dinosaurs, and what is it based on?

## Further Reading


Brinkman, P. D. 2010. *The Second Jurassic Dinosaur Rush: Museums and Paleontology in America at the Turn of the Twentieth Century*. Chicago: University of Chicago Press. (Discusses the early history of dinosaur collecting and science in the United States)

Colbert, E. H. 1984. *The Great Dinosaur Hunters and Their Discoveries*. Mineola, N.Y.: Dover. [Republication of *Men and Dinosaurs* (1968)] (The most extensive history of dinosaur collecting through the 1960s)

Desmond, A. J. 1979. *The Hot-Blooded Dinosaurs*. New York: Dial. (An excellent history of changing ideas about dinosaurs during the 1970s)

Gallenkamp, C. 2001. *Dragon Hunter: Roy Chapman Andrews and the Central Asiatic Expeditions*. New York: Viking. (Biography of Andrews with heavy focus on the Central Asiatic Expeditions)

Kielan-Jaworowska, Z. 1969. *Hunting for Dinosaurs*. Cambridge, Mass.: MIT Press. (The story of the Polish–Mongolian paleontological expeditions)

Mair, G. 2003. *African Dinosaurs Unearthed: The Tendaguru Expeditions*. Bloomington: Indiana University Press. (Complete history of the dinosaur discoveries and excavations from Tendaguru Hill in Tanzania)

Ostrom, J. H., and J. S. McIntosh. 1966. *Marsh's Dinosaurs*. New Haven, Conn.: Yale University Press. (The history of collecting at Como Bluff and many previously unpublished lithographs of Jurassic dinosaurs prepared for Othniel Charles Marsh)

Owen, R. 1842. Report on British fossil reptiles. *Report of the British Association for the Advancement of Science* 11:60–204. (The word "Dinosauria" is coined on p. 103)

Sternberg, C. H. 1985. *Hunting Dinosaurs in the Bad Lands of the Red Deer River, Alberta, Canada*. Edmonton: NeWest. [Republication of the second edition (1932) of the original book] (The story of Sternberg's work in Canada and many other locales)

Weishampel, D. B., and N. M. White, eds. 2003. *The Dinosaur Papers, 1676–1906*. Washington, D.C.: Smithsonian Institution Press. (Reprints and analyzes many of the early and classic scientific articles about dinosaurs)

# 12

# DINOSAUR TRACE FOSSILS

WHEN we think of dinosaur fossils, what first comes to mind are the complete skeletons that are the mainstays of the world's great natural history museums. To collect dinosaurs does mean unearthing such skeletons, or at least finding a skull or some bones. But, there is also an important record of dinosaurs preserved in their skin impressions, footprints, eggs, gizzard stones (gastroliths), tooth marks, and feces (coprolites) (figure 12.1). Such "trace fossils" (though, technically speaking, only tracks are truly trace fossils) provide important evidence of dinosaur behavior and distribution that significantly augments the information gleaned from studying their bones (body fossils). The past few decades have witnessed a resurgence in the study of dinosaur trace fossils, particularly footprints. In this chapter, I evaluate the behavioral and distributional significance of dinosaur footprints, eggs, gastroliths, tooth marks, and coprolites. Dinosaur skin impressions are discussed in chapter 13.

## DINOSAUR FOOTPRINTS

Dinosaur footprints have been discovered in Upper Triassic through Upper Cretaceous rocks on all continents. Indeed, in many regions, dinosaur footprints are often the only dinosaur fossils known. These footprints provide important information about the posture, gait, foot structure, speed, and social behavior (see chapters 6 and 13) of dinosaurs that cannot be learned from a study of their skeletons alone. In order to interpret dinosaur footprints, we must first understand how they are identified and learn some associated terminology.

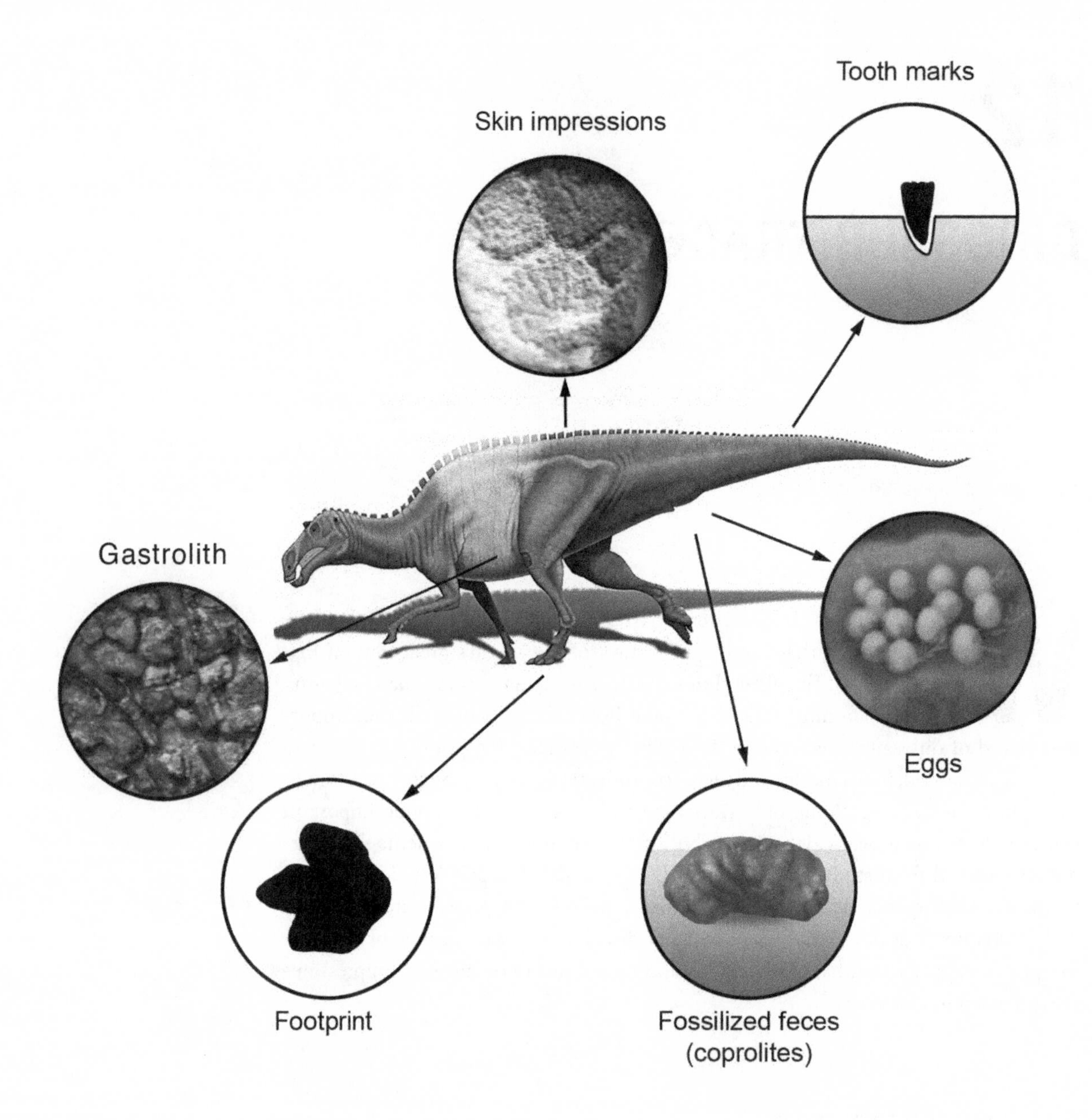


**FIGURE 12.1**
Dinosaur trace fossils include skin impressions, tooth marks, eggs, coprolites, footprints, and gastroliths. (© Scott Hartman)

## Understanding Dinosaur Footprints

A dinosaur **footprint (track)** results from the interaction between the structure of a living dinosaur's foot and the land surface (substratum) upon which it walked. A series of consecutive footprints of an individual dinosaur is called a **trackway**, and this is an obvious example of "fossilized behavior." To understand this behavior, we must first measure the trackway and each footprint, thus calculating the foot length, **stride**, **pace** length, and **pace (or step) angle** (figure 12.2). These measurements, and the shapes of the footprints themselves, provide the basis for identifying the dinosaur that made them.

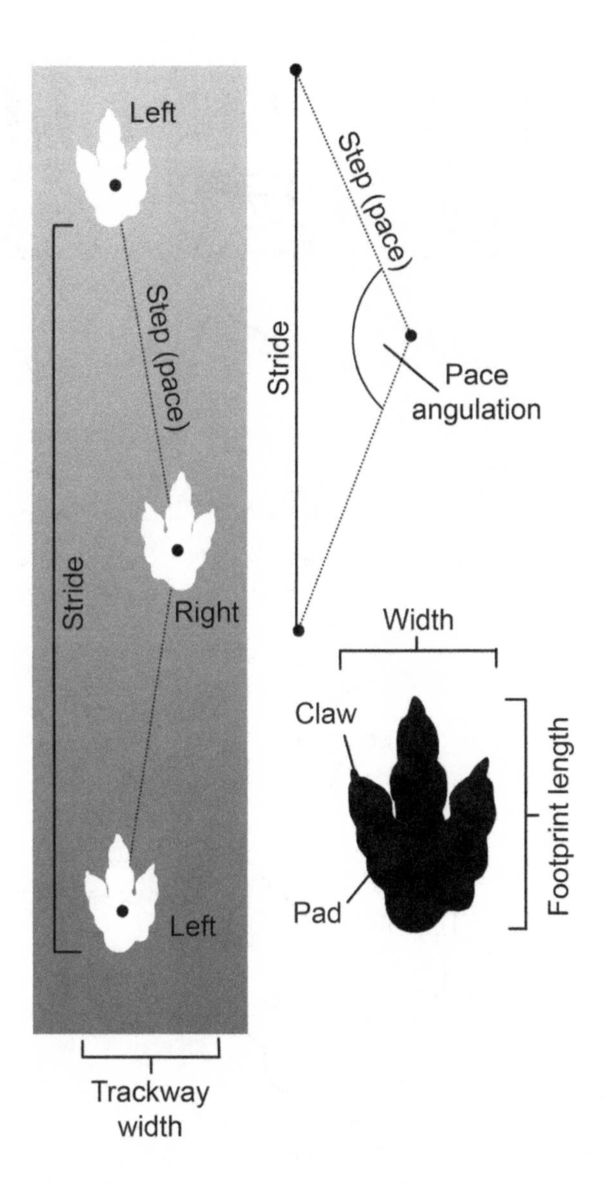


**FIGURE 12.2**
A sequence of footprints is called a trackway. Pace, stride, and pace angle can usually be measured from a dinosaur trackway. (© Scott Hartman)

Identifying dinosaur footprints begins by matching them with the known foot structures of dinosaur skeletons (figure 12.3). For example, the three-toed footprints of some theropod and ornithopod dinosaurs seem rather similar, but theropod footprints can be easily distinguished by their long, slender toes, claws, and lack of "heel" impressions, features absent in ornithopod footprints (figure 12.4).

The problem with matching dinosaur footprints to skeletons is that many footprints are known for which no matching skeletons exist. There is also the problem of the distinctiveness of dinosaur feet and footprints. For example, many types of hadrosaur dinosaurs are distinguished by little more than their different skull shapes. As far as paleontologists can tell, the feet of these different hadrosaurs were of essentially

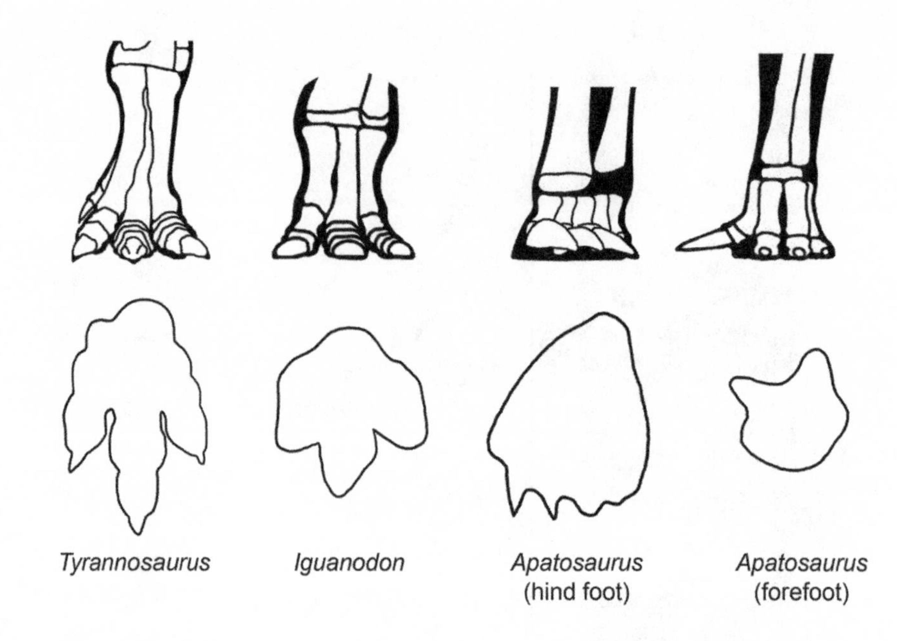


**FIGURE 12.3**
Dinosaur foot skeletons can sometimes be matched to footprint shapes. (© Scott Hartman)

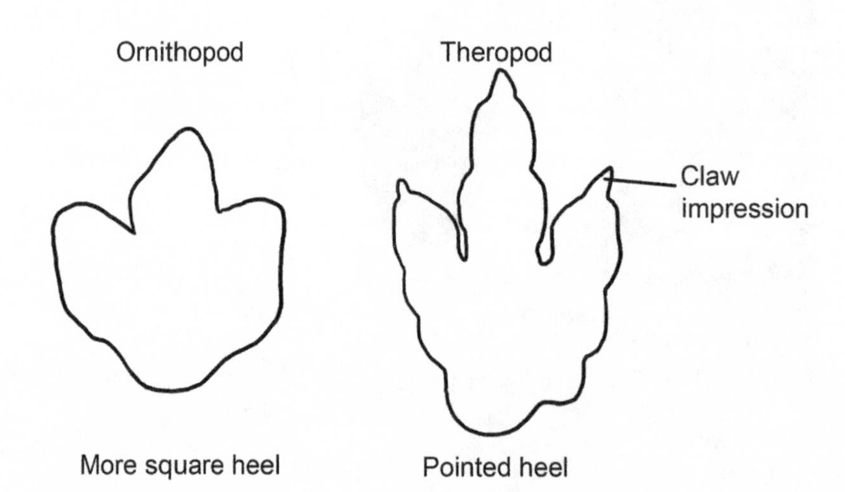


**FIGURE 12.4**
Ornithopod and theropod footprints are somewhat similar but can be distinguished by the features shown here. (© Scott Hartman)

one uniform structure. This means that hadrosaur footprints are of no use in distinguishing the different types of hadrosaurs that are distinguished only by skull shape. Despite this, distinctive dinosaur footprint types may be recognized in certain deposits, even though the identity of the footprint-maker remains unknown.

Because of the need to discuss different types of dinosaur footprints in a concise way, distinctive types of dinosaur footprints are given Latinized scientific names similar to, but not the same as, the names assigned to dinosaur body fossils. These names, however, were sometimes created without regard to the various factors that affected the shape of the footprints (box 12.1).

Box 12.1

## A Theropod Footprint by Any Other Name

Latinized scientific names are given not just to body fossils, but to trace fossils as well. This is especially true of dinosaur footprints, for which hundreds of different names have been proposed. Unfortunately, many paleontologists who proposed names for types of dinosaur footprints failed to consider variation in footprint size and shape during an individual dinosaur's life span. They also failed to consider variation in the substratum upon which the dinosaur walked, as well as variation in a dinosaur's speed and gait. Indeed, in modern tracks, these sources of variation can produce rather different-looking footprints even though only one type of animal, or even a single individual, made the footprints (box figure 12.1A). Variation in footprint shape based on such nonmorphological features is called **extramorphological variation**.

**BOX FIGURE 12.1A**
These seagull footprints show remarkable variation in shape even though a single group of gulls made them.

*(continued)*

The result of the failure to consider extramorphological variation has been a plethora of names for dinosaur footprints, many of them synonyms. Perhaps no footprints have suffered more from such naming than those of theropods. Such footprints belong to bipedal dinosaurs with three toes bearing long claws that touched the substratum. They range in age from Late Triassic to Late Cretaceous and have been described from all continents except Antarctica. A variety of names are still applied to them, even when the sources of variation discussed above are considered (box figure 12.1B). Such names are useful in discussing theropod-footprint distribution, but paleontologists should be cautious and take into account the many possible sources of variation before naming theropod and other dinosaur footprints.

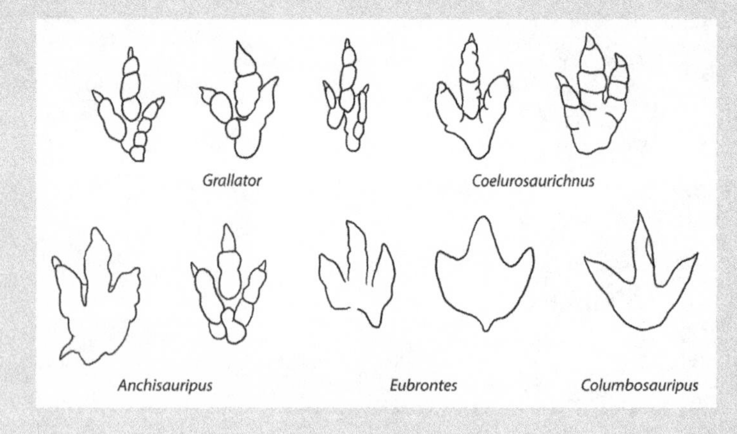


**BOX FIGURE 12.1B**
Various names, some shown here, have been applied to theropod footprints. (© Scott Hartman)

The most important of these factors is that a footprint represents the interaction of a foot with the substratum. In other words, the type of substratum, and where in the substratum the footprint is preserved, can very much affect the footprint's shape (figure 12.5). It is particularly important to recognize that a dinosaur footprint may be the original footprint on the surface of the substratum or one of a series of "underprints" in all the sediment layers underneath the surface that were disturbed by the weight of the dinosaur. The underprints can look quite different from the surface footprint, so we need to be careful to distinguish footprints from underprints when naming dinosaur footprints.

## Interpreting Dinosaur Footprints

Dinosaur footprints document the former presence of dinosaurs in places, time intervals, and environments where dinosaur body fossils are often absent. Footprints also provide important evidence about the posture and gait of dinosaurs. For example, the footprints of theropods confirm the erect posture and bipedality of these dinosaurs that is inferred from studying their skeletons, whereas the footprints of ankylosaurs and ceratopsians may suggest more upright forelimb postures and gaits than have been inferred by studying their limb structures (figure 12.6). This has been a source

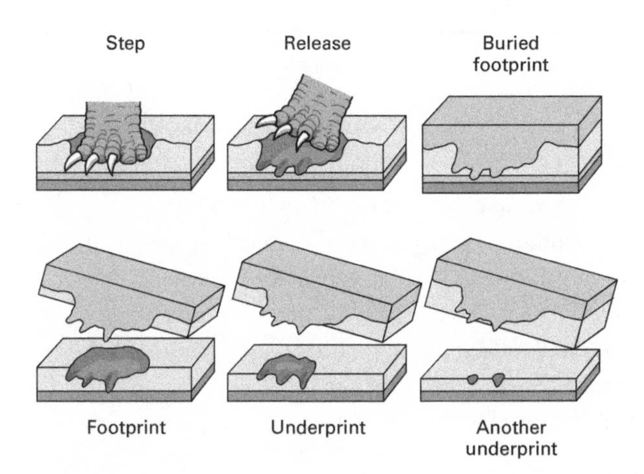


**FIGURE 12.5**
A dinosaur footprint forms during the interaction between foot and substratum. Not only may the footprint on the surface of the substratum be preserved, but "underprints" may also be fossilized. (Drawing by Network Graphics)

of debate over the postures and gaits of these dinosaurs (see box 9.1). Perhaps the most important piece of information paleontologists derive from studying dinosaur footprints is an estimate of dinosaur speed (box 12.2). Footprints are the only reliable information from which to obtain such estimates.

**FIGURE 12.6**
These trackways of an ankylosaur and a ceratopsian suggest more erect forelimb postures than do studies of their bones. (© Scott Hartman)

Box 12.2

## Speed Estimates from Dinosaur Footprints

Dinosaur speed can be estimated from footprints using a method developed by British researcher Robert McNeill Alexander. Here's how it's done. A stride is the distance from one point on a footprint to the same point on the next print of the same foot (see figure 12.2). When dinosaurs walked, they took short strides; when dinosaurs ran, they took longer strides. Long-legged dinosaurs, of course, took longer strides than smaller ones, whether walking or running. So, we can't simply use stride length to estimate dinosaur speeds, because a longer-legged dinosaur, such as *Tyrannosaurus*, would take longer strides than a shorter-legged dinosaur, such as *Ornithomimus*, when both dinosaurs were walking (or running). Instead, we eliminate the effect of leg length on speed in order to use stride length to estimate a dinosaur's speed. This is most easily done by dividing the length of the dinosaur's stride by the length of its leg to arrive at the relative stride of the dinosaur:

$$\text{Relative stride} = \text{length of stride (in meters)} \div \text{leg length (in meters)}$$

In this calculation, leg length is the distance from the hip joint to the ground when the dinosaur is in a normal, standing posture. Note that relative stride is dimensionless (not a length in meters) because dividing stride length in meters by leg length in meters eliminates the dimension.

But how do we determine the leg lengths of dinosaurs that left only fossil footprints? Measuring the leg lengths of complete dinosaur skeletons and then trying to match them to footprints might seem the most direct approach. This, however, is seldom possible because determining which skeletons correspond to which footprints is never obvious. Furthermore, many footprints cannot be matched to the feet of known dinosaur skeletons. Nevertheless, measurements of skeletons of a wide variety of dinosaurs indicate that their feet are about one-fifth as long as their legs. In other words, a *Tyrannosaurus* with a foot (or footprint) length of 0.64 meters would have had a leg length of $5 \times 0.64$ m $= 3.2$ m. If the footprints of this *Tyrannosaurus* showed a stride of 4.16 meters, then its relative stride would be its stride length divided by its leg length: $4.16$ m $\div 3.2$ m $= 1.3$.

Now that we can calculate the relative stride of any dinosaur, we need a way to convert relative stride into an estimate of speed of movement. To do so, we need to examine the relationship between relative stride length and speed in living animals, such as mammals. This is done by measuring the relative strides of mammals clocked at different speeds when walking and running. When measuring speed, however, we face the problem that larger animals move faster than smaller ones, even if they have the same relative strides. This is similar to the stride and leg-length problems discussed earlier.

To resolve this problem, we need to divide the actual speed of the mammal by some dimension of body size, in this case the square root of leg length × gravitational acceleration. (The use of this dimension of body size is dictated by the laws of physics.) This gives us a value called **dimensionless speed**:

$$\text{Dimensionless speed} = \text{actual speed (in meters per second)} \div \text{leg length (in meters)} \\ \times \text{gravitational acceleration (in meters per second squared)}$$

(Use the square root key on your calculator to calculate the denominator.) Dimensionless speed thus allows us to estimate equivalent speeds despite the size (dimension) of an animal.

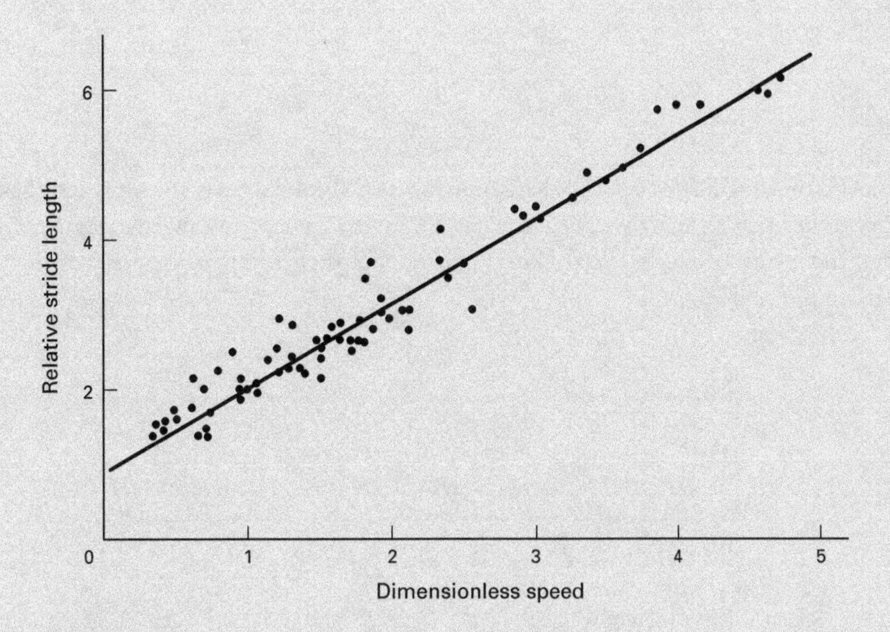


**BOX FIGURE 12.2**
This plot of relative stride length against dimensionless speed is based on data from living mammals and ostriches. Note that both axes of the plot are dimensionless. (Drawing by Network Graphics)

The relationship between relative stride and dimensionless speed for living mammals produces a trend that accords with intuition: the longer the relative stride, the faster the dimensionless speed (box figure 12.2). The relationship between relative stride and dimensionless speed also allows us to estimate dinosaur speeds simply by finding the dimensionless speed of a living mammal that corresponds to the relative stride length of the dinosaur. For the relative stride length of the *Tyrannosaurus* mentioned earlier, 1.3, the dimensionless speed is 0.4. To estimate the actual speed of the dinosaur, we can plug this value into the equation for dimensionless speed and solve the equation for actual speed:

$$\text{Dimensionless speed} = \text{actual speed} \div \sqrt{(\text{leg length} \times \text{gravitational acceleration})}$$

$$\text{Actual speed} = \text{dimensionless speed} \times \sqrt{(\text{leg length} \times \text{gravitational acceleration})}$$

Note that gravitational acceleration = 10 meters per second squared. So:

$$\text{Actual speed} = 0.4 \times \sqrt{3.2 \text{ m} \times 10 \text{ m/s}^2}$$

$$\text{Actual speed} = 2.26 \text{ m/s (about 4.6 miles per hour)}$$

This is a brisk walk for a human. Applying this method to a variety of dinosaur trackways produces estimated speeds that range from walks to fast runs of as much as 43 kilometers per hour (a 4-minute mile is equivalent to 25 kilometers per hour) by some theropods (box table 12.2). Some trackways thus document very fast dinosaurs, which helps to dispel the old idea that dinosaurs were slow, lumbering behemoths. Most trackways, however, record normal walking speeds—presumably because this was the typical mode of progression among dinosaurs.

(*continued*)

Other ways of estimating dinosaur speeds have been devised. These methods are not strictly based on footprints because they rely on body weight estimates, limb angulation, and other factors requiring skeletal information. The method presented here relies simply on footprints and measurements easily derived from them (length of foot and leg).

BOX TABLE 12.2

| Taxon | Number of trackways | Height at hip (m) | Velocity (km/h) |
|---|---|---|---|
| Sauropods | 2 | 1.5–3.0 | 4 |
| Ornithomimid | 10 | <1.0 | 6 |
| Theropods | 7 | 0.8–1.3 | 4–8 |
| Theropods | 8 | 1.5–2.3 | 3–9 |
| Theropods | 2 | 1.5–2.1 | 5–10 |
| Theropods | 17 | 0.8–1.5 | 5–16 |
| Theropods | 3 | 0.3–1.2 | 2–11 |
| Theropods | 3 | 0.6 | 7–8 |
| Theropods | 2 | 0.4–0.6 | 5–8 |
| Theropods | 15 | 1.2–1.9 | 6–43 |
| Ornithopods | 2 | 0.2–0.6 | 2–3 |
| Ornithopods | 3 | 1.0–2.6 | 4–6 |
| Ornithopods | 10 | <1.0 | 16 |

Estimates of dinosaur speeds from footprints range from slow walks to very fast running.

## Footprint Myths

Although much important information comes from studying dinosaur footprints, many myths and misconceptions about dinosaurs are also based on footprints. Foremost among these are supposed human footprints associated with dinosaur footprints (box 12.3) and the idea of sauropods swimming in the sea.

## Box 12.3

### Did Humans Walk with Dinosaurs?

Most of us have heard claims of human footprints found together with dinosaur footprints. Such claims contradict what we know from the body fossil record: dinosaurs became extinct 66 million years ago, more than 60 million years before humans evolved. Most of these claims come from one of the great dinosaur footprint localities, the limestone bed of the Paluxy River, just west of the town of Glen Rose in Somervell County, Texas.

There, thousands of footprints of Early Cretaceous theropod and sauropod dinosaurs have been studied by paleontologists since the first investigations by Roland T. Bird of the American Museum of Natural History in the 1930s. However, the first supposed human footprints associated with these dinosaur footprints were found in 1910. Since that time, many "human footprints" have been discovered here and studied, especially by so-called "creation scientists" intent on overturning the standard paleontological interpretation of Earth history. Indeed, the "human footprints" from near Glen Rose have figured prominently in books and films by creationists who assert that humans and dinosaurs lived side by side.

During the 1980s, qualified paleontologists carefully re-examined the "human footprints" from near Glen Rose. They found that most of these footprints were not footprints at all, but erosional features only remotely similar to human footprints. Others were actually eroded bipedal dinosaur footprints. And, some of the "human footprints" were manmade—chiseled in rock—and are not even accurate replicas of a human foot (box figure 12.3). Indeed, we now know that several residents of Glen Rose have, over the years, supplemented their incomes by carving human-like footprints for sale to unsuspecting passersby.

Careful scientific scrutiny reveals that no authentic human footprints are associated with the dinosaur footprints near Glen Rose or anywhere else. There remains absolutely no evidence that humans walked with the dinosaurs.

**BOX FIGURE 12.3**
This "human footprint" from near Glen Rose is actually a human carving and constitutes an unfortunate and unsuccessful attempt to manipulate the nature and significance of the fossil record. (Courtesy Glen K. Kuban)

**FIGURE 12.7**
The original interpretation of a swimming sauropod (*above*) has been reanalyzed to show that most of the footprints discovered are actually underprints (*below*). (Drawing by Network Graphics)

The idea of swimming sauropods is based on footprints of Early Cretaceous age along the Paluxy River near Glen Rose, Texas, studied by paleontologist Roland T. Bird during the 1930s and 1940s. In a 1941 article, Bird analyzed the trackway of a single sauropod that consists almost entirely of shallow forefoot impressions. Bird explained the lack of hind-foot impressions as a result of the sauropod's hindquarters being buoyed up in fairly deep water so that only its front feet touched the bottom (figure 12.7).

Recent careful re-examination of this sauropod trackway, however, reveals that most of the sauropod footprints are, in fact, underprints and that a hind-foot impression is also present. This undermines the evidence of swimming sauropods based on the Texas footprints.

## DINOSAUR EGGS

Dinosaur **eggs** were first discovered in France in 1869, but the most famous dinosaur eggs are those discovered in Mongolia in 1923 (figure 12.8). Today, dinosaur eggs are best known from the Upper Cretaceous of Asia, the western United States, and Argentina, though they have also been described from France, Spain, India, and other locales, the oldest being of Late Triassic age.

Dinosaur eggshells were composed of organic matter and an inorganic mineral, crystalline calcium carbonate, as are the eggshells of living reptiles and birds. The well-organized, interlocking crystals of calcium carbonate made the dinosaur eggshell rigid and thus gave it a good chance of being fossilized. The fact that relatively few dinosaur eggs older than Late Cretaceous age are known has led some paleontologists to speculate that many primitive dinosaurs may have laid soft-shelled eggs lacking an extensive mineral matrix.

The shape of dinosaur eggs varies considerably, ranging from spherical to almost cylindrical (figure 12.9). Some kinds of dinosaur had eggs with characteristic mineral textures, and some fossilized eggs contain well-preserved embryos or are closely associated with the skeletons of hatchling and/or adult dinosaurs. These circumstances allow most dinosaur eggs to be identified as those of a particular type of dinosaur.

**FIGURE 12.8**
This nest of dinosaur eggs was discovered in Mongolia in 1923. Originally identified as *Protoceratops* eggs, they are now thought to be the eggs of a small theropod. (Courtesy Department of Library Services, American Museum of Natural History, New York [negative no. 410765])

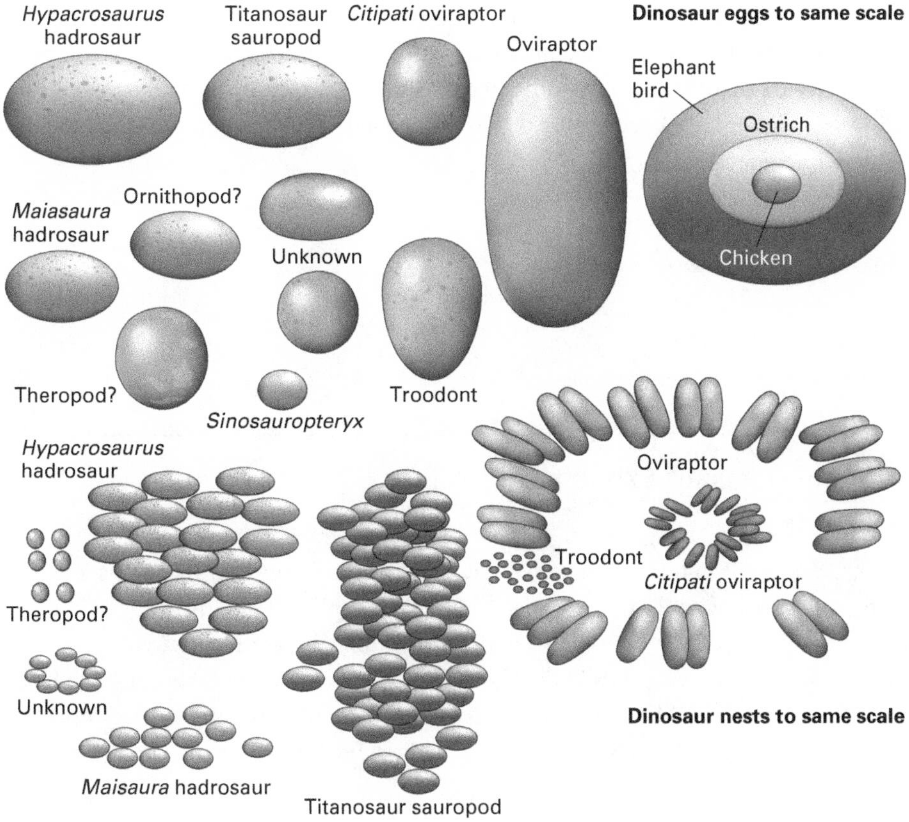

*Hypacrosaurus* hadrosaur
Titanosaur sauropod
*Citipati* oviraptor
Dinosaur eggs to same scale
Oviraptor
Elephant bird
Ostrich
Chicken
Ornithopod?
*Maiasaura* hadrosaur
Unknown
Theropod?
*Sinosauropteryx*
Troodont
*Hypacrosaurus* hadrosaur
Oviraptor
Troodont
*Citipati* oviraptor
Theropod?
Unknown
*Maisaura* hadrosaur
Titanosaur sauropod
Dinosaur nests to same scale

**FIGURE 12.9**
The shape and size of dinosaur eggs vary considerably. (Drawing by Network Graphics)

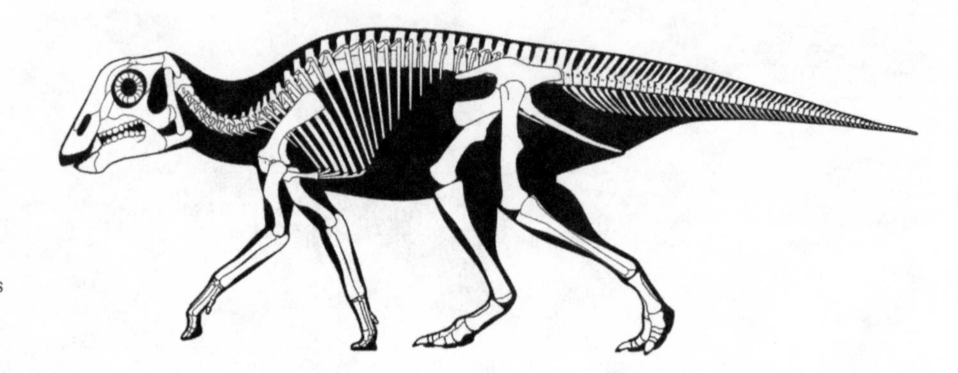

**FIGURE 12.10**
This drawing of the skeleton of an embryonic hadrosaur (*Maiasaura*) is based on bones found inside dinosaur eggs. (© Scott Hartman)

The largest dinosaur eggs, those of sauropods, were more than 30 centimeters long. Eggs are known from most dinosaur groups, including theropods, prosauropods, sauropods, ceratopsians, and ornithopods, and paleontologists are certain that all dinosaurs reproduced by laying eggs.

The eggs of some dinosaurs were laid in hollow mounds or shallow pits. The sizes of the clutch varied, and as many as 20 eggs in one clutch have been documented for the hadrosaur *Maiasaura*. Some clutches appear to have been laid in **nests** that also contained hatchlings. A few dinosaur eggs discovered still contain fossilized embryonic dinosaurs inside them (figure 12.10).

Dinosaur eggs present paleontologists with various types of information. Perhaps most significant is that most, if not all, dinosaurs reproduced by laying eggs. Nests with eggs have also been a source for interpreting the parenting behavior of some dinosaurs (see chapter 13). Also, dinosaur eggs provide information on dinosaur distribution where dinosaur bones or other trace fossils are not available. Finally, analysis of dinosaur eggs of Late Cretaceous age has figured in the speculation about the cause of dinosaur extinction (box 12.4).

Box 12.4

## Brittle Eggshells: Cause of Dinosaur Extinction?

One of the greatest accumulations of dinosaur eggs is in the Upper Cretaceous strata of the Nanxiong Basin of southeastern China. Here, Chinese paleontologists have collected about 20,000 eggshell fragments and about 300 complete eggs, including some found in 24 complete or nearly complete nests. These eggs have been assigned Latin names and have been found in association with very few dinosaur bones (box figure 12.4). The eggs are from the uppermost Cretaceous sediments in the Nanxiong Basin, however, and they must have been laid by dinosaurs during the last few million years before dinosaur extinction. Indeed, in the strata immediately above the dinosaur eggs are found fossils of some of the earliest Tertiary mammals.

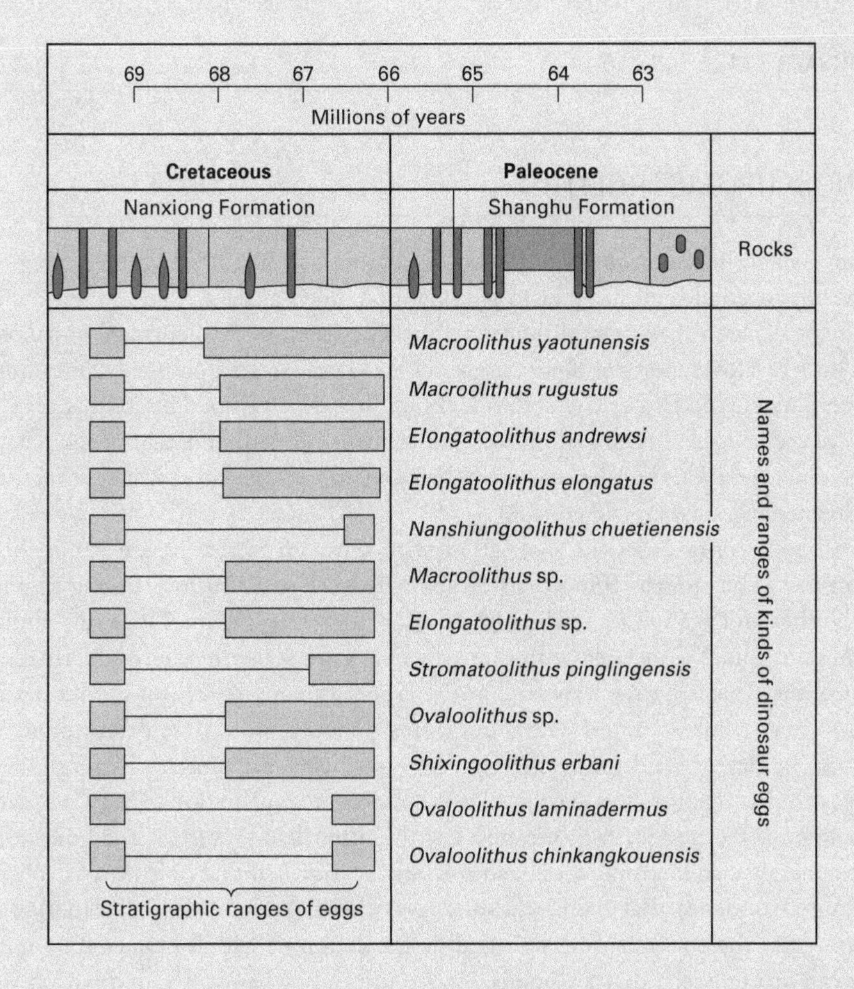


**BOX FIGURE 12.4** Fossil dinosaur eggs are abundant in the Upper Cretaceous of the Nanxiong Basin in southeastern China, as shown by the range bars of the different kinds of eggs in the left portion of the diagram. (Drawing by Network Graphics; data from Z. Zikui et al. 1991. Extinction of the dinosaurs across the Cretaceous–Tertiary boundary in Nanxiong Basin, Guangdong Province. *Vertebrata PalAsiatica* 29:1–20)

Chinese paleontologists and geochemists recently conducted an extensive analysis of the dinosaur egg record from the Nanxiong Basin. They found that many of the youngest eggs, those laid just before the extinction of the dinosaurs, had shells that were thin, brittle, and enriched in trace elements, such as manganese and strontium. In addition, these fossil eggshells have an unusual ratio of the two common isotopes of the oxygen atom, $^{16}O$ and $^{18}O$.

The Chinese scientists considered these abnormalities to be the result of dry climates and an excess supply of trace elements in southeastern China at the end of the Cretaceous. They maintained that plant-eating dinosaurs ingested an abnormally high amount of these trace elements, and the dinosaurs that preyed upon them thus also ate the trace elements in high concentrations. The trace elements adversely affected the formation of eggshells by female dinosaurs, making the shells so brittle that they fractured easily and failed to protect the dinosaur embryos inside. In addition, some of the embryos may have died well before hatching because of the excessive amounts of trace elements. The deaths of so many embryonic dinosaurs resulted in greatly reduced dinosaur populations, producing a major extinction, according to the Chinese scientists.

This explanation of dinosaur extinction relies on too long a chain of reasoning. Perhaps the weakest link in the chain is the assumption that the high levels of trace elements in the dinosaur eggshells were concentrated when the eggshells were formed in the bodies of female dinosaurs. It is more likely that the trace elements were introduced and concentrated during the fossilization of the dinosaur eggs. This might also explain the oxygen isotope ratios in the eggshells. And, even though the youngest Cretaceous eggshells are thinner than the older ones, it is not clear that they were too thin to protect the developing dinosaur embryos. Thus, it is easy to question the conclusion that the dinosaur eggs of the Nanxiong Basin provide evidence of a specific cause of dinosaur extinction. Nevertheless, the dinosaur eggs from the Upper Cretaceous of southeastern China do represent an exceptional record, one deserving of much further analysis.

## DINOSAUR GASTROLITHS

Many living birds and some reptiles swallow stones and hold them in a crop region (gizzard) to grind food in order to aid digestion. Such gizzard stones, when found with fossils, are called **gastroliths** (from the Greek *gastro* for "stomach" and *lithos* for "rock"). Most reports of dinosaur gastroliths are of isolated polished stones from Upper Jurassic and Lower Cretaceous rocks in the western United States that may or may not actually have been gastroliths. Many different kinds of dinosaur skeleton have been discovered with polished stones in their abdominal cavities, and these stones are undoubtedly gastroliths (figure 12.11).

Polished stones found in association with dinosaur skeletons are confidently identified as gastroliths. The stones presumably were polished by grinding against each other and against food while in the dinosaur's crop region. However, similar stones not found associated with dinosaur bones are sometimes also identified as gastroliths. This is particularly true in the Upper Jurassic Morrison Formation of the western United States, where numerous polished stones are often found in clumps in clay or sandstone where no dinosaur bones are present. Regurgitation of gastroliths by dinosaurs has been suggested as an explanation of why no fossil bones are associated with the stones. On the other hand, perhaps the bones did not mineralize and become fossilized at those locations, so that only the gastroliths remain as evidence that a dinosaur once was there. A more likely explanation is simply that these stones were polished by the action of wind or water (box 12.5). We can be confident that a stone is a gastrolith only when it is found within the abdominal cavity of an articulated fossil skeleton. If stones are not found associated with dinosaur bones, we cannot confidently term them gastroliths; instead they can be called **gastromyths**.

Today, living animals swallow stones to relieve hunger pangs, to serve as ballast while swimming, or to aid in grinding and crushing food. All three possibilities have been invoked to explain why some dinosaurs evidently swallowed stones. Relief of hunger pangs is difficult to document in a dinosaur, and because dinosaurs were generally not aquatic animals, it seems unlikely they swallowed the stones for ballast. If anything, it is most probable that dinosaur gastroliths aided digestion much as the gizzard stones do in many of today's birds.

## DINOSAUR TOOTH MARKS

Tooth marks on dinosaur bone are direct evidence of predation, scavenging, and fighting (figure 12.12). Well-known examples include the bones (vertebrae and ribs) of theropods, sauropods, and hadrosaurs with broken theropod teeth embedded in them. Most of these examples provide direct evidence of predator–prey interaction. It is important to determine whether the bone around the tooth mark healed (was remodeled) before the dinosaur died. If it was not healed, then the tooth mark likely is evidence of scavenging.

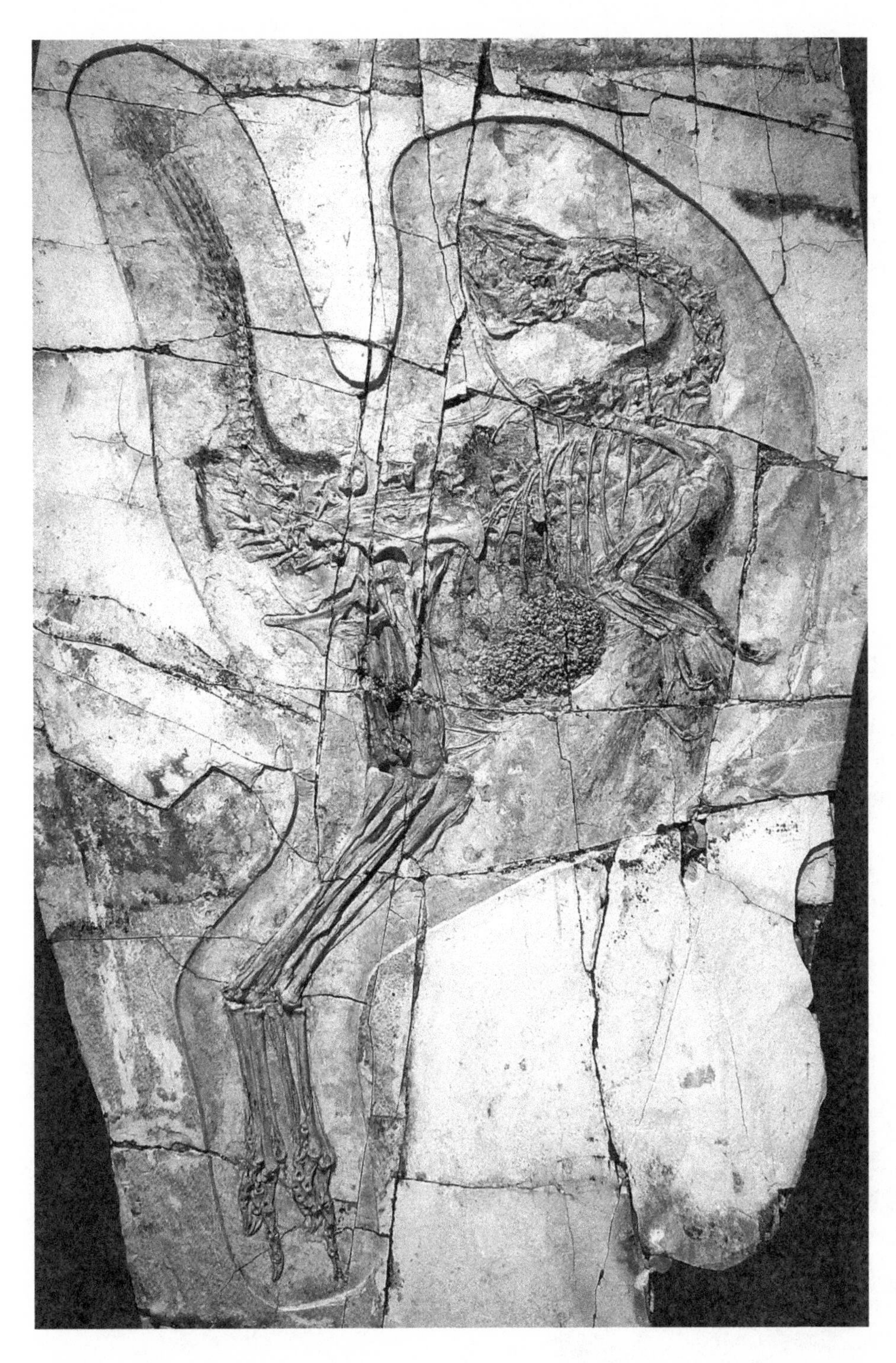

**FIGURE 12.11**
The polished stones inside this *Caudipteryx* skeleton are considered gastroliths. (Photograph © Stephen Czerkas)

Tooth marks provide important evidence of theropod feeding and bone utilization. A recent study suggested that theropods did not consume bone, as do some modern predators and scavengers, such as hyenas. The study also concluded that theropods preferentially hunted juvenile dinosaurs.

## Box 12.5

### Sauropod Gastromyths

A gastrolith is a hard object (usually a stone) of no caloric value that is (or was) swallowed and retained in the digestive tract of an animal. Paleobotanist George Wieland (1865–1953) coined the term in 1906 to apply to stones swallowed by recent and extinct vertebrates, but in particular those associated with plesiosaur fossils. In the century that followed, gastroliths have been well documented in only a few dinosaurs, notably the theropod *Caudipteryx* (see figure 12.11), the ceratopsian *Psittacosaurus*, and the prosauropod *Massospondylus*. Very few bona fide examples of sauropod gastroliths are known, though many claims of sauropod gastroliths have been made. Indeed, several paleontologists have argued that the tiny-headed sauropods, with their relatively weak chewing apparatus, needed "in-gut-mastication" by gastroliths to process food.

A case of supposed sauropod gastroliths that did not survive critical scrutiny is that of Late Jurassic "*Seismosaurus*" (actually a large species of *Diplodocus*). The partial skeleton of this dinosaur, consisting mostly of the hip and parts of the dorsal and caudal vertebral series, was excavated in northern New Mexico during the 1980s. About 240 polished siliceous (mostly quartzite) stones between about 40 and 100 millimeters in diameter were found next to, and in some cases, touching bones of the partial skeleton (box figure 12.5). Few of these stones were within or close to the dinosaur's abdominal cavity, all were well rounded, and some were imbricated (tilted) on cross-beds. The bones and stones were in a sandstone body that represents a Late Jurassic river channel. The case that the stones are gravel pebbles in a fluvial deposit, not gastroliths, is incontrovertible.

The stones associated with the bones of "*Seismosaurus*" are highly polished, and this fits a long-held idea that high polish is characteristic of gastroliths. However, diverse examples of bona fide gastroliths include stones that are not highly polished. Instead, their surfaces are extensively pitted and rilled because of stone-to-stone abrasion. High polish does not diagnose gastroliths.

But, old misconceptions die hard, even (or possibly especially) in science. Many paleontologists still believe that high polish is characteristic of a gastrolith and that sauropods had gastroliths. Indeed, a web search of the word "gastrolith" uncovers a world of misinformation. Ironically, about 50 years ago, in 1966, paleontologist Jack Dorr stated the only well-founded way to identify gastroliths: "No smooth, rounded, or even highly polished stones can be identified positively as gastroliths unless they are found within the fossilized skeleton of an animal in an area formerly occupied by the digestive tract." It's really that simple!

**BOX FIGURE 12.5**
These polished quartzite stones were misinterpreted as gastroliths of the sauropod dinosaur "*Seismosaurus*."

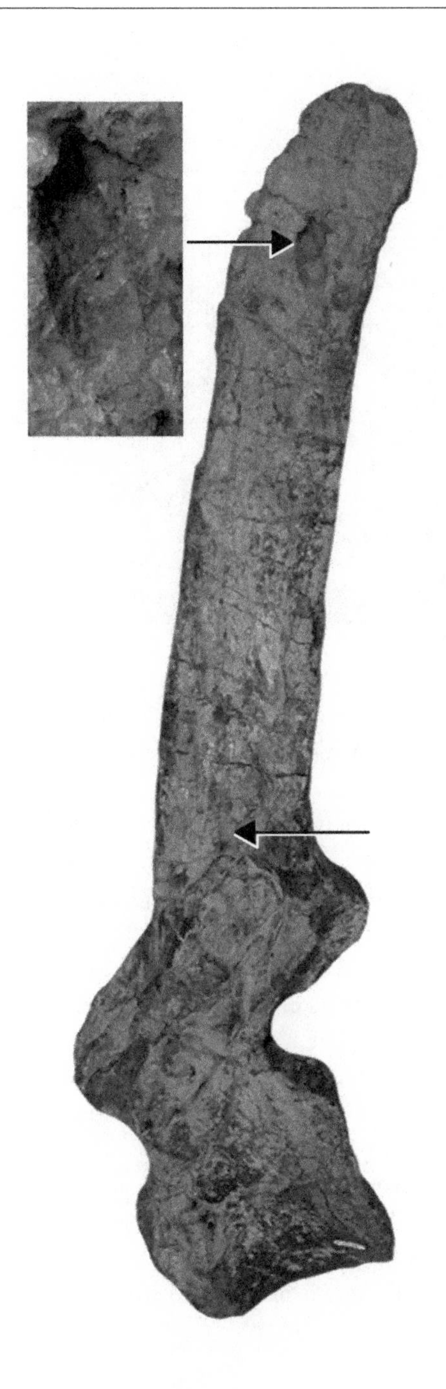

**FIGURE 12.12**
This Late Cretaceous hadrosaur vertebra has two tooth marks (*arrows*). The upper tooth mark (*inset*) is 3 centimeters long and appears to be a postmortem bite mark.

## DINOSAUR COPROLITES

The term **coprolite**, from the Greek roots *copros* (feces) and *lithos* (rock), is applied to fossilized feces. Coprolites are known from a wide variety of animals of different ages, ranging from crustaceans to extinct humans. They provide paleontologists with direct evidence of an extinct animal's diet. But, their use is greatly limited because we are unable, except in rare cases, to link confidently a given coprolite to the animal that

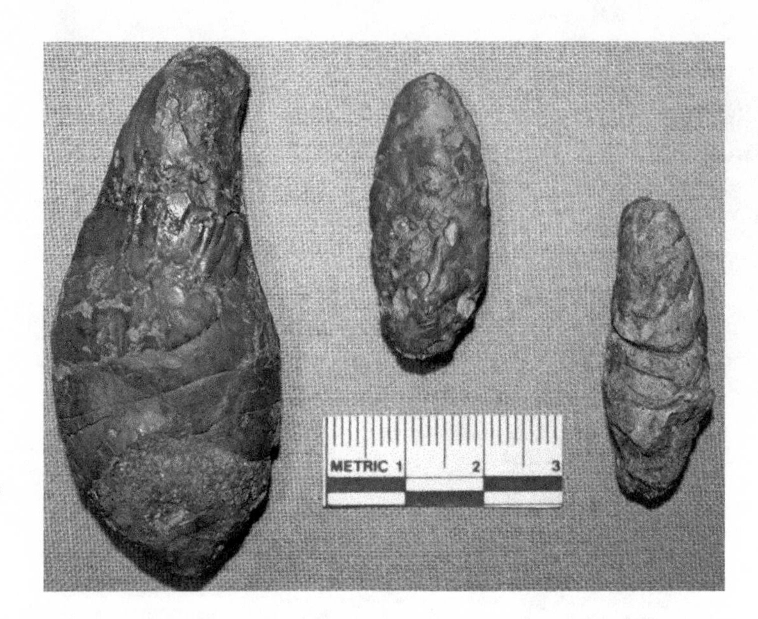


**FIGURE 12.13**
Were any of these three types of coprolite from the Upper Triassic of New Mexico produced by a dinosaur?

produced it. This limitation especially affects dinosaurs. Few bona fide dinosaur coprolites have been identified, and until relatively recently, there has been little analysis of them.

Coprolites from a Late Triassic fossil locality in New Mexico provide a good example of the problems encountered in trying to interpret them. This locality produces bones and teeth of dinosaurs and a variety of other reptiles, large amphibians, and fishes. Three types of coprolites are present (figure 12.13).

The most common coprolites are small (up to 2 centimeters long), cigar-shaped, and do not contain noticeable bone fragments or fish scales. The next most common type also lacks bone and scale fragments, but have spiral grooves on their surfaces. The third, least common, type of coprolite, is much larger than the other two types and contains bone and scale fragments. What kind of animal produced each type of coprolite, and are any of the coprolites from dinosaurs?

The spiral grooves on the second type of coprolite are typical of fishes. The third type obviously belongs to a fish-eater, but that could be any of several different kinds of reptiles or large amphibians known from the locality. What animal produced the first type of coprolite is unknown, and there is no way to be certain that any of the coprolites was produced by a dinosaur.

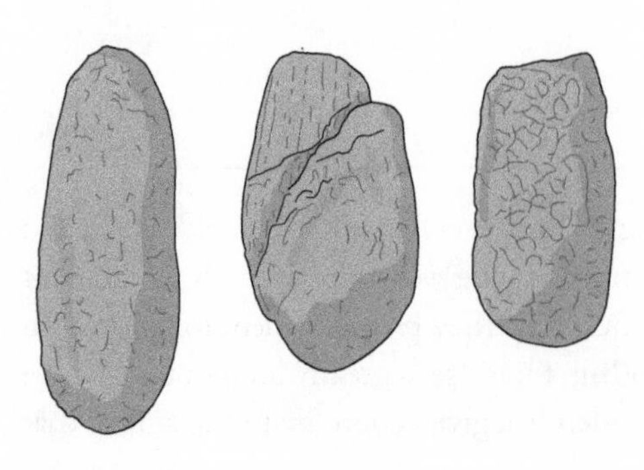

**FIGURE 12.14**
These dinosaur coprolites are from the Cretaceous of India. (Drawing by Network Graphics)

What we need is a fossil locality where only dinosaur bones are associated with coprolites, or coprolites so large that only dinosaurs could have produced them. Such localities are known and contain coprolites as large as 29 centimeters long (figure 12.14). A coprolite nearly 44 centimeters long, with an estimated original volume of more than 2 liters, found in the Upper Cretaceous of Saskatchewan, Canada, has been attributed to *Tyrannosaurus rex*, mostly because of its large size. Indeed, known dinosaur coprolites are mostly of predatory dinosaurs. This is because the feces of these dinosaurs were rich in phosphatic minerals from the undigested bones of their prey and thus fossilized more readily than did the feces of plant-eating dinosaurs, which lacked such minerals. Coprolites of herbivorous dinosaurs are known, such as those of presumed hadrosaurs from the Upper Cretaceous of Montana. They are packed with conifer debris, which provides direct evidence of the feeding preferences of at least some herbivorous dinosaurs.

At present, few dinosaur coprolites have been identified, and relatively little analysis of them has been undertaken. This means that collecting and studying dinosaur coprolites remains a largely untapped field of research that may teach us more about the behavior and diets of dinosaurs.

## Summary

1. Dinosaur trace fossils include skin impressions, footprints, eggs, gastroliths, tooth marks, and coprolites.
2. Dinosaur footprints provide important information about dinosaur posture, gait, speed, behavior, and distribution.
3. Estimates of dinosaur speeds from trackways range mostly from walks to fast runs of as much as 43 kilometers per hour for theropods.
4. There are many myths surrounding dinosaur footprints, including the idea that humans walked side by side with dinosaurs and that sauropods swam in the sea.
5. Dinosaur footprints receive scientific names, but attention is not often paid to several sources of variation when naming dinosaur footprints.
6. Dinosaur eggs are known from most kinds of dinosaur and indicate that dinosaurs laid hard-shelled eggs like those of living birds.
7. Dinosaur gastroliths can only be identified with certainty when stones are found associated with dinosaur skeletons. Dinosaurs probably used gastroliths primarily to crush and grind food.
8. Dinosaur coprolites have been little analyzed, largely because it is difficult to identify the kind of dinosaur that produced a specific coprolite.

## Key Terms

coprolite
dimensionless speed
egg
extramorphological variation
footprint (track)
gastrolith
gastromyth
nest
pace
pace (or step) angle
stride
trackway

## Review Questions

1. What are the principal types of dinosaur trace fossils?
2. What kinds of information can paleontologists obtain from dinosaur footprints?
3. How is the speed of a dinosaur estimated from a trackway?
4. Estimate the speed of a dinosaur with a stride length of 1 meter and a leg length of 1.2 meters.
5. Which dinosaurs laid eggs?
6. What types of information can paleontologists obtain from dinosaur eggs?
7. How are dinosaur gastroliths identified?
8. Why might some dinosaurs swallow stones?
9. Why is it difficult to interpret dinosaur coprolites?

## Further Reading

Alexander, R. M. 1989. *Dynamics of Dinosaurs and Other Extinct Giants.* New York: Columbia University Press. (Chapter 3 explains how to estimate dinosaur speeds from dinosaur footprints)

Bird, R. T. 1941. Did *Brontosaurus* ever walk on land? *Natural History* 53:60–67. (This is the original article on swimming sauropods)

Carpenter, K. 1999. *Eggs, Nests, and Baby Dinosaurs.* Bloomington: Indiana University Press. (A comprehensive synthesis of all aspects of dinosaur reproduction)

Carpenter, K., K. F. Hirsch, and J. Horner, eds. 1994. *Dinosaur Eggs and Babies.* Cambridge: Cambridge University Press. (Collection of technical articles on dinosaur eggs)

Chin, K. 2012. What did dinosaurs eat? Coprolites and other direct evidence of dinosaur diet, pp. 589–601, in M. K. Brett-Surman, T. R. Holtz, Jr., and J. O. Farlow, eds., *The Complete Dinosaur.* 2nd ed. Bloomington: Indiana University Press. (Reviews tooth marks, stomach contents, and coprolites as evidence of dinosaur diets)

Chin, K., T. T. Tokaryk, G. M. Erickson, and L. C. Calk. 1998. A king-sized theropod coprolite. *Nature* 393:680–682. (Discusses a coprolite of *Tyrannosaurus rex*)

Hone, D. W. E., and O. W. M. Rauhut. 2010. Feeding behavior and bone utilization by theropod dinosaurs. *Lethaia* 43:232–244. (Review of theropod tooth marks and other data relevant to their feeding)

Hunt, A. P., J. Milàn, S. G. Lucas, and J. A. Spielmann, eds. 2012. Vertebrate coprolites. *Bulletin of the New Mexico Museum of Natural History and Science* 57:1–387. (Comprehensive volume of research on vertebrate coprolites)

Lockley, M. 1991. *Tracking Dinosaurs: A New Look at an Ancient World.* Cambridge: Cambridge University Press. (A very readable introduction to the study and interpretation of dinosaur footprints)

Lockley, M. G., and A. P. Hunt. 1995. *Dinosaur Tracks and Other Fossil Footprints of the Western United States.* New York: Columbia University Press. (Thoroughly reviews all dinosaur tracks found in the western United States)

Lockley, M. G., and A. Rice. 1990. Did "Brontosaurus" ever swim out to sea? Evidence from brontosaur and other dinosaur footprints. *Ichnos* 1:81–90. (Challenges the idea of swimming sauropods introduced in Bird's 1941 article)

Lucas, S. G. 2000. The gastromyths of "*Seismosaurus*," a Late Jurassic dinosaur from New Mexico. *Bulletin of the New Mexico Museum of Natural History and Science* 17:61–67. (Critiques previous claims of gastroliths in sauropods)

Martin, A. J. 2014. *Dinosaurs Without Bones: Dinosaur Lives Revealed by Their Trace Fossils.* New York: Pegasus. (A very readable review of all dinosaur trace fossils)

Milne, D. H., and S. D. Schafersman. 1983. Dinosaur tracks, erosion marks and midnight chisel work (but no human footprints) in the Cretaceous limestone of the Paluxy River bed, Texas. *Journal of Geological Education* 31:111–123. (Thoroughly debunks the "human footprints" associated with Cretaceous dinosaur footprints)

Mossman, D. J., and W. A. S. Sarjeant. 1983. The footprints of extinct animals. *Scientific American* 248:74–85. (A popular introduction to the study of fossil footprints)

Thulborn, R. A. 1990. *Dinosaur Tracks.* London: Chapman & Hall. (A comprehensive, technical review of dinosaur footprints)

Wings, O. 2015. The rarity of gastroliths in sauropod dinosaurs: A case study in the late Jurassic Morrison Formation, western USA. *Fossil Record* 18:1–16. (Reviews the minimal evidence of gastroliths in Late Jurassic sauropods of the western United States)

## Find a Dinosaur!

Tracking dinosaurs? Then follow them to the Raymond M. Alf Museum of Paleontology (Claremont, California), one of the finest footprint museums anywhere. Here, you can see many dinosaur tracks on display, as well as the tracks of other extinct reptiles and ancient birds and mammals. The Beneski Museum of Natural History of Amherst College (Amherst, Massachusetts) displays the remarkable footprint collection of Edward Hitchcock, made in the nineteenth century. The collection includes many footprints of Early Jurassic theropod dinosaurs, which were among the first dinosaur tracks known to science.

You can also track dinosaurs outdoors. Dinosaur Valley State Park (Glen Rose, Texas) contains some very famous Early Cretaceous dinosaur tracks, mostly made by sauropods and the theropods that were tracking them. The St. George Dinosaur Discovery Site (St. George, Utah) reveals tracks similar to Hitchcock's, mostly of early theropods. Just west of Denver, Colorado, Dinosaur Ridge provides guided tours of the spectacular Early Cretaceous dinosaur tracksites preserved in Colorado's front range. At Clayton Lake State Park in northeastern New Mexico, a boardwalk traverses a sandstone surface with up to 800 tracks made by Early Cretaceous beach-going ornithopods. And, less formally, you can tour an Early Jurassic dinosaur tracksite 8 kilometers west of Tuba City, Arizona, with a local Navajo guide. These outdoor dinosaur tracksites provide remarkable access to the fossil tracks in their natural setting.

# 13

# DINOSAUR BIOLOGY AND BEHAVIOR

CHAPTERS 5 through 9 presented a survey of the dinosaurs and many conclusions about their biology and behavior. This chapter summarizes some of that information and explores other aspects of dinosaur biology and behavior. The study of dinosaur soft-tissue (nonskeletal) anatomy and behavior is full of speculation, some of it sensational and unwarranted. Here, I shall take a cautious approach to these subjects, presenting reasonable speculation and avoiding science fiction.

## DINOSAUR BIOLOGY

Many aspects of dinosaur biology, especially skeletal anatomy, have already been reviewed in this book, and chapter 14 discusses the complex subject of dinosaur metabolism. The focus of this chapter is on four topics not addressed in detail elsewhere in this book: dinosaur external appearance, weight, growth, and longevity.

### External Appearance

Fossils of the integument (scales, feathers, scutes, and **skin impressions**) are known for most of the major dinosaur taxa. They indicate that the skin of most dinosaurs was covered with scales similar to the scales of some living reptiles (figure 13.1). However, recent discoveries also indicate that many theropods had feathers or a feather-like body covering (see chapter 15).

Skin impressions of hadrosaurs are particularly well known and indicate a body covered in scales. A recent study of the skin impressions of *Saurolophus*, a Late Cretaceous hadrosaur, known from Canada and Mongolia, distinguished the two species of the genus by differences in scale shape and pattern. This is an exciting result because

**FIGURE 13.1**
This close-up fossilized skin impression of a hadrosaurid shows reptilian scales only a few millimeters wide. Many dinosaurs were covered with such scales.

today herpetologists distinguish many living species of reptiles (especially lizards) based on scale patterns. This can now be done with *Saurolophus*, and a better record of skin impressions may allow such distinctions among other dinosaur species.

The color of dinosaurs is totally a matter of conjecture, because the pigment in their skin and scales did not fossilize. However, color patterns have been preserved in some dinosaurs, and some indicate a striped coloration, a common form of camouflage in living animals.

Classically, paleontologists and artists who create paintings or sculptures of dinosaurs used the color patterns of living reptiles, especially lizards, as a guide to the probable colors of dinosaurs. However, most artists now paint dinosaurs with very flamboyant, bright color patterns unlike those of most living reptiles and more like those of birds. These flashy dinosaurs are eye-catching, but no serious student of dinosaurs views the coloration attributed to any dinosaur as anything but speculation.

The posture and overall body shape of a dinosaur is determined by analyzing its skeleton. Skeletal anatomy is a guide to the size, shape, and configuration of the muscles and provides an understanding of how the dinosaur moved (figure 13.2). Of course, this is old news to the readers of this book. What is not old news, and what is not so certain, is how fat or thin a given dinosaur was, whether it had flaps of skin on its neck or head as do some modern lizards and birds, and other features of dinosaur external anatomy that are difficult to predict from skeletal anatomy. Fortunately, some exceptionally preserved dinosaur fossils provide direct knowledge of some dinosaur soft tissues (box 13.1).

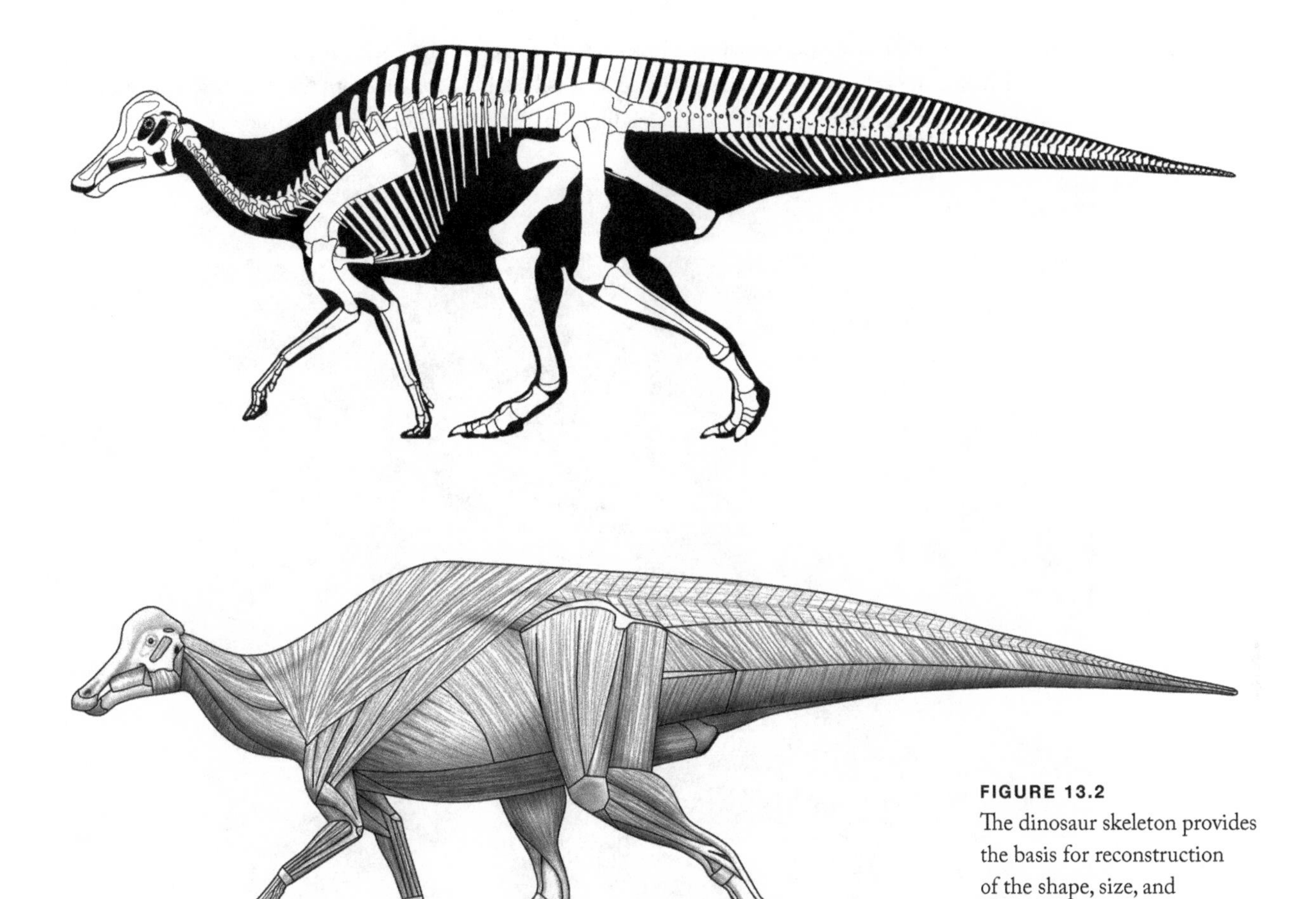

**FIGURE 13.2**
The dinosaur skeleton provides the basis for reconstruction of the shape, size, and arrangement of muscles. (© Scott Hartman)

## Box 13.1

### The Biology of *Psittacosaurus*

We now know a tremendous amount about the biology of the ceratopsian dinosaur *Psittacosaurus*, much more than we know about most dinosaurs. This is because of its exceptional fossil record, which includes some remarkable recent discoveries from the Lower Cretaceous of northeastern China. Some highlights include the following:

1. Studies of the growth of *Psittacosaurus* indicate that it grew faster than modern reptiles but slower than living mammals or large dinosaurs.
2. These studies also indicate maturity was reached at about nine years of age.
3. An adult skeleton found together with 34 juvenile skeletons has been interpreted as a nest with evidence of a high degree of parental care.
4. Analysis of skeletons of psittacosaurs killed in a volcanic mudflow suggests that groups of closely related dinosaurs lived together as a sort of "family."

(*continued*)

5. Soft-tissue preservation of one exceptional psittacosaur fossil shows a bushy tail with quill-like spines (box figure 13.1). This dinosaur otherwise had scaly skin, and its fossil retains a color pattern that indicates dark and light bands (possibly for camouflage) and countershading (lighter coloring on the ventral side of the body).

**BOX FIGURE 13.1**
This exquisitely preserved skeleton of *Psittacosaurus*, from the Lower Cretaceous of northeastern China, retains skin impressions and shows quill-like projections around the tail. (Courtesy of Naturmuseum Senckenberg, Frankfurt)

It is certain that paleontologists' and artists' views of the external appearance of dinosaurs have evolved as new ideas about dinosaur biology and behavior have emerged (see chapter 11). Older ideas of sluggish, cold-blooded dinosaurs produced restorations of flabby and lethargic creatures. New ideas of fast, warm-blooded dinosaurs produce renderings of sleek and agile animals. We can see that ideas about dinosaur external appearance not only involve careful inferences from dinosaur skeletal anatomy but also require speculation about **coloration** and **soft-tissue anatomy** that cannot be directly inferred from skeletal anatomy. Furthermore, paleontological conceptions of dinosaur biology and behavior have always shaped perceptions of dinosaur external appearance.

## Weight

Many popular dinosaur books list weights of dinosaurs. These **weight estimates** are one way to state the size of a dinosaur. Indeed, most dinosaur weight estimates emphasize the very large size of the dinosaurs. How are dinosaur weights estimated?

Several methods are used. Many methods are mathematical in nature. One of these measures the **cross-sectional area** of weight-bearing limb bone (figure 13.3). The more weight a limb bone bears, the larger its cross-sectional area, and an equation that predicts weight borne from the cross-sectional area of a limb bone (usually the femur) can be calculated for living vertebrates. The cross-sectional area of the dinosaur limb bone can be plugged into such an equation.

What is determined, of course, is not the weight of the dinosaur but only the weight supported by the limb bone. A further adjustment upward of this value must be made in order to estimate the dinosaur's body weight. This adjustment depends on the dinosaur's posture and shape, leading to some uncertainty in the weight estimate. Also uncertain is just how applicable the cross-section-to-weight-borne equation is to extinct animals much larger, and presumably heavier, than the living animals from which the equation was originally determined. Because of these uncertainties, the cross-sectional area method is not the ideal method of estimating a dinosaur's weight.

**Scale models** of dinosaurs have long been a common way to estimate weights. This is because estimating dinosaur weights from scale models is relatively straightforward once a model of known scale (say one-fiftieth the length of the dinosaur) is available. The volume of the model is calculated by displacement of water. That volume is then

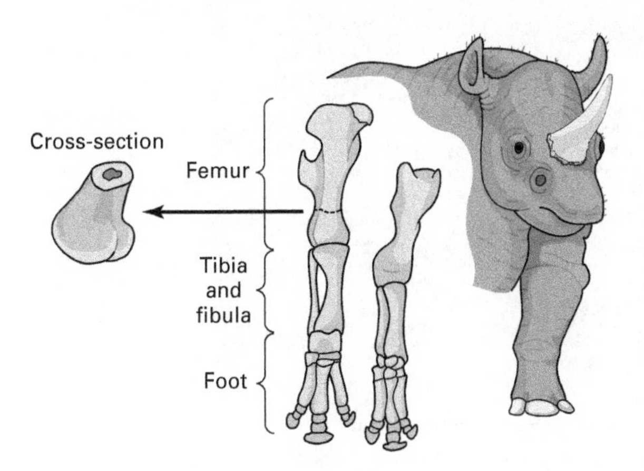


**FIGURE 13.3**
The cross-sectional area of a limb bone is related to the weight that bone must bear. (Drawing by Network Graphics)

multiplied by the cube of the linear scale of the model so that it becomes the volume of a full-size dinosaur identical in shape to the model. This volume of the "real" dinosaur is then multiplied by 0.9 kilograms per liter, which is the mass of a liter of living crocodile, to arrive at a mass (weight) in kilograms (box 13.2).

## Box 13.2

### Your Own Estimates of Dinosaur Weights

Most paleontologists estimate the weights of dinosaurs by using scale models. The more accurate the scale model, the more accurate the weight estimate. You can test this method by using available scale models, such as the plastic dinosaurs you can purchase in a toy store. Here's how to do it.

First, you need the following: solid, waterproof plastic scale models of dinosaurs; a water supply; a calculator; a ruler; and a measuring cup from the kitchen or a graduated cylinder from a chemistry lab. The more precisely calibrated the measuring cup or graduated cylinder, the more accurate will be your estimate.

Start by measuring the length of the dinosaur scale model. If your ruler is not metric, convert this measurement into centimeters (1 inch = 2.54 centimeters). Then, determine the actual length of the full-size dinosaur. Now, divide the length of the actual dinosaur by the length of the model. For a 6.7-meter-long *Stegosaurus* and a 10.2-centimeter-long model, this value is approximately 66, so the model is 1⁄66 the size of the dinosaur.

Now submerge the model in water to measure its volume (box figure 13.2). Fill the measuring cup or graduated cylinder to a set level, say 5 ounces, or 100 milliliters if the cylinder is metric. Drop in the dinosaur toy.

**BOX FIGURE 13.2**
The volume of a dinosaur scale model can be determined by measuring the amount of water it displaces. This volume can then be used to estimate the weight of a full-size dinosaur.

If it doesn't sink, gently push it with your finger or a pen point until it is submerged. Read the new level of the water; the difference between it and the original water level is the volume of the scale model. If you must push the model to submerge it, make sure you do not put your finger or the pen point into the water any more than necessary, because that will artificially increase the estimate of the model's volume.

If your estimate of the volume of the model is in fluid ounces, you need to convert it to milliliters (1 fluid ounce = 29.6 milliliters). The 10.2-centimeter-long toy *Stegosaurus* displaces 0.9 ounces, or about 27 milliliters.

Now, cube the scale of the model to obtain its cubic scale. The *Stegosaurus* model is 1⁄66 of the length of the actual dinosaur, so it is $1/66 \times 66 \times 66$ (or $66^3$) = 1/287,496 the volume of the actual dinosaur. If we multiply the inverse of this value by the volume of the model, we get the volume (in milliliters) of a life-size replica of the toy dinosaur.

So,

$$287{,}496 \times 27 = 7{,}762{,}392 \text{ mL}$$

Dividing this value by 1,000 converts the volume to liters:

$$7{,}762{,}392/1{,}000 = 7{,}762 \text{ L}$$

A liter of water weighs 1 kilogram, and the average living crocodile weighs 0.9 kilograms per liter. If we apply this to our *Stegosaurus,* the weight of a dinosaur with a volume of 7.762 liters should be as follows:

$$7{,}762 \text{ L} \times 0.9 \text{ kg/L} = 6{,}986 \text{ kg}$$

A metric ton is 1,000 kilograms, so this is about 7 metric tons. One kilogram = 2.2 pounds, so this is 15,369 pounds, or about 7.7 tons.

This method is simple, and it forms the basis of many published dinosaur weight estimates. However, these estimates are only as accurate as the models themselves (figure 13.4). Because we are often uncertain exactly how "fleshy" or "lean" dinosaurs were, dinosaur model-making is an imprecise art. Dinosaur weight estimates are full of uncertainty, which is why this book uses skeletal lengths, not weight estimates, to convey the size of a given dinosaur.

## Growth and Longevity

Living reptiles grow throughout their lives, although the rate at which they grow decreases as they age (figure 13.5). This type of growth, called **indeterminate growth,** contrasts with the **determinate growth** of mammals and birds. Determinate growth

**FIGURE 13.4**
Different scale models yield different weight estimates of the same type of dinosaur.

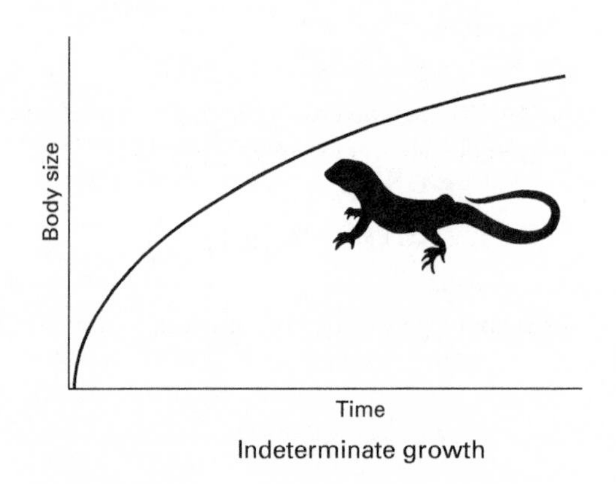


Maturity
Body size
Time
Determinate growth

**FIGURE 13.5**
Living reptiles undergo indeterminate growth, whereas mammals and birds undergo determinate growth. (Drawing by Network Graphics)

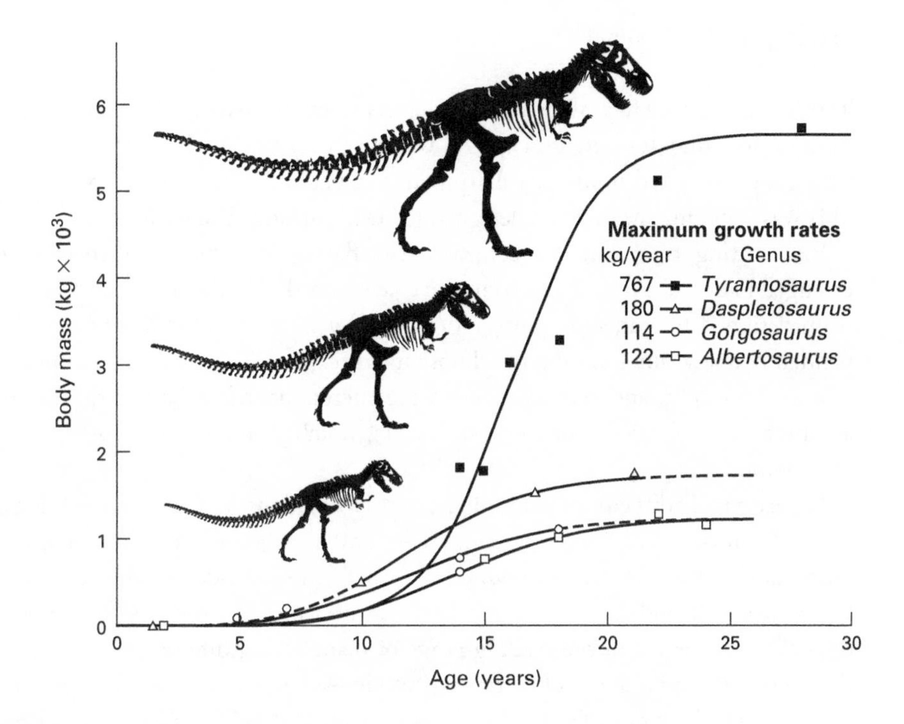


**FIGURE 13.6**
Dinosaurs grew rapidly and underwent determinate growth. (Drawing by Network Graphics)

means that after an animal grows to adulthood it stops growing. (In general, rates of determinate growth are faster than those of indeterminate growth.) Another factor affecting growth rate is metabolism; on average, warm-blooded vertebrates grow at least 10 times faster than do cold-blooded vertebrates.

What type of growth did dinosaurs have, and how long did an individual dinosaur typically live? These questions were long difficult to answer directly from dinosaur fossils because fossils provide little direct evidence of the age, in years, of an individual dinosaur. However, recent studies of dinosaur growth based on new techniques that analyze bone microstructure and the spacing of growth lines in bone have given new insight. These studies suggest relatively rapid and determinate growth in dinosaurs and that no dinosaur had a lifespan of more than 100 years (figure 13.6). Dinosaurs experienced growth rates faster than those of living reptiles and even as fast as some living birds. However, according to the new analyses, the dinosaurs thought of as ancestral to birds, the small theropods, actually had relatively slow growth rates, as was the case with *Archaeopteryx*. The giant dinosaurs were able to grow so large because their growth rate accelerated through life. In other words, a late spurt of growth produced the giant dinosaurs.

## DINOSAUR BEHAVIOR

Although chapters 5 through 9 discussed many aspects of dinosaur behavior, it is useful to summarize them here and to discuss some subjects mentioned briefly at greater length. Important aspects of dinosaur behavior were feeding and locomotion, reproduction and parenting, attack and defense, and social (group) behavior.

## Feeding and Locomotion

Tooth and jaw structure allow paleontologists to easily distinguish plant-eating from meat-eating dinosaurs (figure 13.7). As we repeatedly saw in chapters 4 and 9, meat-eating dinosaurs had numerous sharp, serrated, blade-like teeth set in powerful jaws. This was a feeding mechanism designed to stab, tear, and slice flesh.

Plant-eating dinosaurs, in contrast, had flatter, leaf-shaped teeth, sometimes arranged in dental batteries, set in massive jaws and skulls. This was a feeding mechanism designed to tear, slice, pulp, and/or grind vegetation. Herbivory evolved independently many times among the dinosaurs: three times in ornithopods, once in the sauropodomorphs, and several times in the theropods. Most herbivorous dinosaurs ate high-fiber plants. And, some dinosaurs, notably some theropods, were likely omnivorous.

It is extremely difficult to determine exactly what kinds of plants or animals a given kind of dinosaur ate. Stomach contents—conifer twigs and needles in hadrosaur mummies, a lizard in *Compsognathus*, and so forth—provide some direct evidence. Coprolites even indicate that some meat-eating dinosaurs resorted to cannibalism (box 13.3). Inferences of the **feeding range** of plant-eating dinosaurs—how far above the ground the dinosaur could crop vegetation—suggest some specific types of plant food that may have been favored by different kinds of dinosaurs (figure 13.8). But, preserved gut contents are not known for most dinosaurs, and inferences about feeding do little to narrow the range of possible plant foods. So it remains difficult to identify the specific food items most dinosaurs ate.

One of the principal reasons most animals, including dinosaurs, move (locomote) is to obtain food. Dinosaur **locomotion** has been discussed at various points in this book, and we can now draw some general conclusions.

Dinosaur skeletons indicate that most dinosaurs were ground-dwelling walkers and runners. However, there is now evidence that some small theropods were likely arboreal (lived in trees) or at least capable of climbing trees. Other than some hadrosaurs, no compelling argument can be presented that any dinosaur was aquatic (lived in the water). Indeed, as we saw in chapter 6, earlier ideas that sauropods were aquatic do not stand up to a critical analysis of sauropod anatomy. Therefore, dinosaurs stand out as a remarkable group of almost wholly ground-dwelling animals.

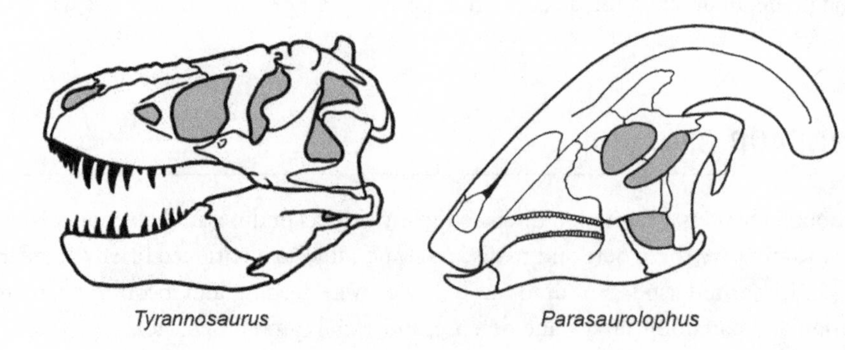


**FIGURE 13.7**
The teeth, jaws, and skull structure of meat-eating *Tyrannosaurus* contrast with those of plant-eating *Parasaurolophus*. (© Scott Hartman)

## Box 13.3

### Dinosaur Cannibals

**Cannibalism** (eating one's own kind) is common among some living predators, such as sharks and crocodiles. Evidence of cannibalism exists for some predatory dinosaurs. Most impressive are skeletons of the small theropod *Coelophysis* found at the Upper Triassic Ghost Ranch dinosaur quarry in northern New Mexico. One of these skeletons has coprolites in the pelvic area full of the bones of juvenile *Coelophysis* (box figure 13.3).

**BOX FIGURE 13.3**
These tiny juvenile bones of *Coelophysis* (scale bar in millimeters) came out of a coprolite of an adult *Coelophysis*.

Today, cannibalism occurs in many predatory reptiles and mammals. Most often, a reptile or mammal will eat younger individuals of its species when hungry, simply because they are easy to capture. This may seem cruel, but it may actually make sense in situations where the population density of young reptiles or mammals is very high and/or other food is difficult to catch.

The large adult individual of *Coelophysis* from Ghost Ranch with small juvenile *Coelophysis* in its coprolites fits the modern pattern of reptilian cannibalism. Clearly, cannibalism has a long antiquity, at least 210 million years, and may have been common among theropod dinosaurs.

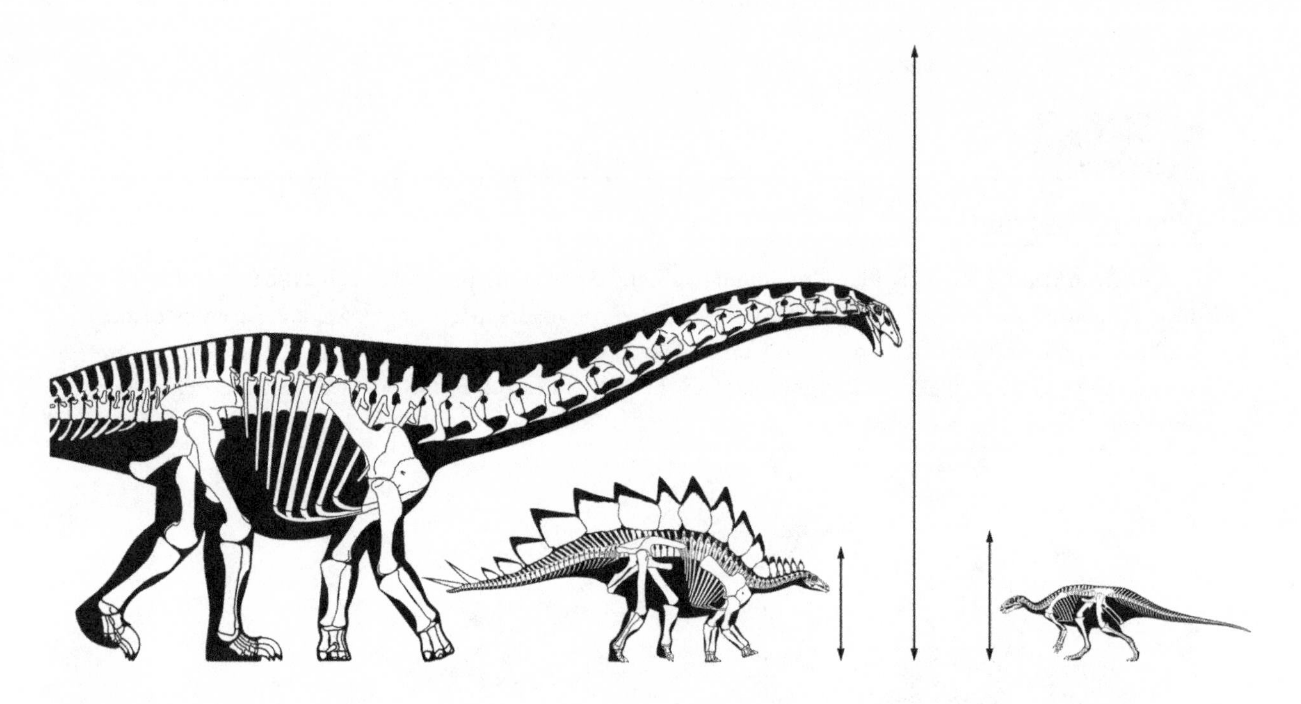

**FIGURE 13.8**
Feeding range may provide some clues to the types of plants eaten by some dinosaurs. (© Scott Hartman)

## Reproduction and Parenting

Chapter 12 discussed the evidence suggesting that dinosaurs reproduced by laying eggs. **Clutches** of eggs are particularly well known for several kinds of dinosaurs.

Very significant for interpretations of dinosaur reproductive and parenting behavior have been Late Cretaceous egg sites in Montana. There, at a place called "Egg Mountain," a number of clutches of eggs attributed to two dinosaurs—the hadrosaur ***Maiasaura*** and the primitive ornithopod *Orodromeus*—have been discovered (figures 13.9 and 13.10). These clutches are in oval to subcircular, crater-like depressions with raised rims. It is almost certain the eggs were not exposed to the air after being laid, but were covered with a thin layer of soil or vegetation (figure 13.11).

Numerous skeletons of hatchling dinosaurs have been found around these nests, as have some bones and footprints of adult dinosaurs of the same species as the hatchlings. This provides circumstantial (though not incontrovertible) evidence of parental care.

There is very suggestive evidence of parental care of young dinosaurs in the hadrosaurid *Maiasaura* (figure 13.12), but what of other dinosaurs? Certainly the **nests** and hatchlings of some theropods could have been cared for, as were those of *Maiasaura*. But evidence of such care is inferred, not actually fossilized, except in rare cases like the *Oviraptor* that died brooding its nest (see box 5.4). Other dinosaurs, such as the cannibalistic ***Coelophysis***, probably did not care for their young beyond the amount of care characteristic of living crocodilians (see box 13.3). But new discoveries in the Lower Cretaceous of northeastern China of the ceratopsian *Psittacosaurus*

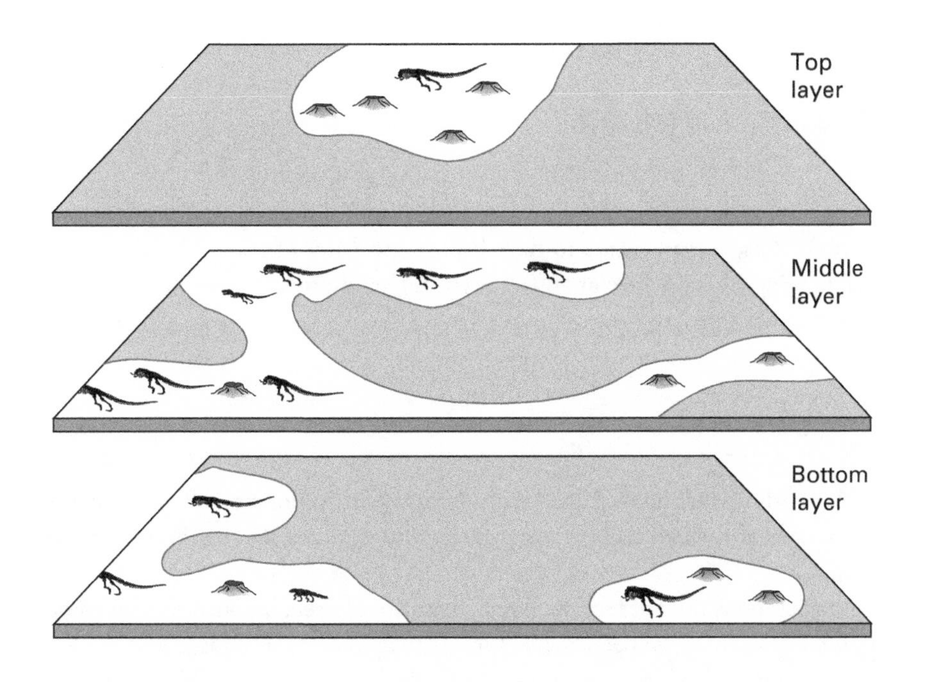


**FIGURE 13.9**
Nests and dinosaur bones have been discovered at "Egg Mountain" in Montana. (Drawing by Network Graphics)

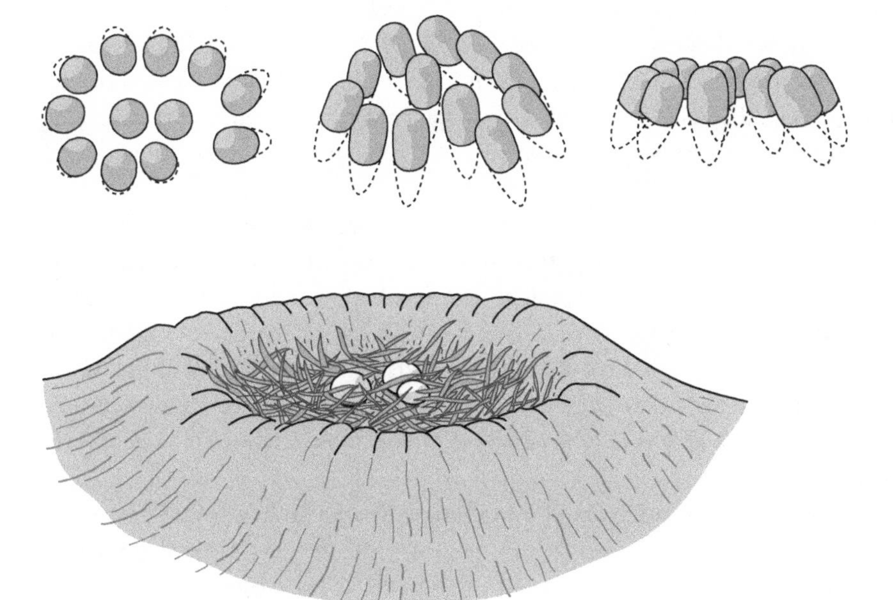

**FIGURE 13.10**
These three views (*top*, *oblique*, and *side*) are of an *Orodromeus* clutch from "Egg Mountain." Each egg is about 20 centimeters long. (Drawing by Network Graphics)

**FIGURE 13.11**
This restoration of a *Maiasaura* nest includes a layer of vegetation that may have covered the eggs. (Drawing by Network Graphics)

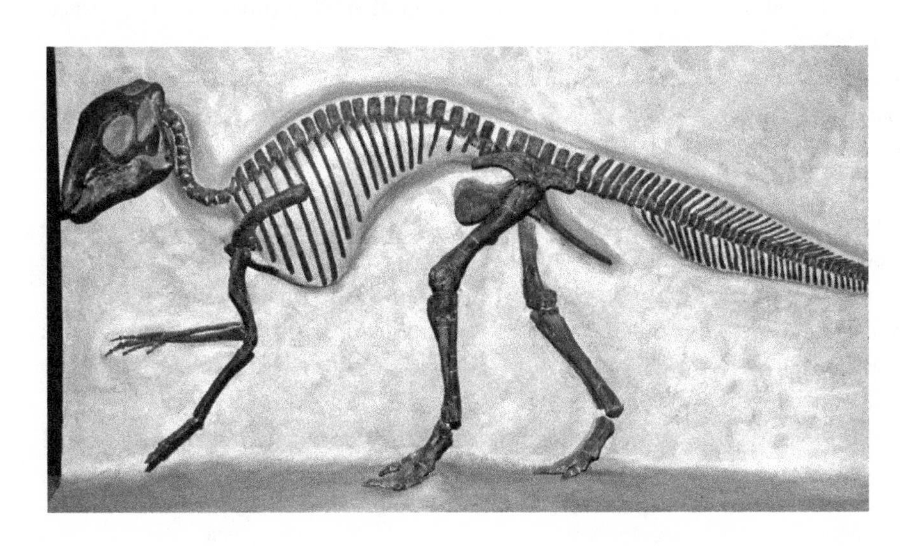

**FIGURE 13.12**
This skeleton of a very young *Maiasaura* is about 1 meter long and provides a remarkable record of the early growth of a hadrosaurid. (Courtesy Robert M. Sullivan)

suggest not just parental care, but even "family groups" of dinosaurs living together (see box 13.1). Thus, there appears to have been a spectrum of parental care among dinosaurs, ranging from no care at all to the feeding and protection of hatchlings to living in "family groups."

## Attack and Defense

How predatory dinosaurs hunted and how they and other dinosaurs defended themselves are subjects for which there is some good, hard evidence and much unfounded speculation.

How did theropods hunt? Here, we enter the realm of speculation, although some interesting suggestions have been made. One is that some theropods, such as *Allosaurus*, when hunting large prey, such as sauropods, hunted in packs, like wild dogs do today. An *Allosaurus* attack on a sauropod is well documented by a skeleton of *Apatosaurus* in which some of the caudal vertebrae were bitten off. The bite marks match the tooth spacing of *Allosaurus*, and broken teeth of *Allosaurus* were associated with the sauropod fossil. Pack hunting by smaller theropods when attacking large prey items seems necessary if such comparatively small predators were to take much larger prey.

A different, ambush style of predation has been suggested for some of the larger theropods. This style is analogous to that of the great white shark, which attacks its victim from behind and below, inflicting a terrible bite and then retreating and waiting until its victim either goes into shock or bleeds to death. Another, slightly different, view sees large theropods as the "big cats" (lions and tigers) of the Mesozoic. This view suggests that large carnosaurs either ran down or ambushed their victims, delivering a single (when possible) killing blow with their huge, tooth-filled jaws.

It is important to emphasize that these different ideas about how theropods hunted are speculation. In general, most paleontologists view the very large theropods as solitary hunters and many of the smaller theropods as pack hunters when taking big game, such as sauropods. However, many of the truly small theropods (less than three meters long) probably were solitary hunters of small animals, such as insects and lizards.

Defensive behavior varied greatly in dinosaurs. Among the predatory dinosaurs, perhaps the most that can be said is that they followed the maxim "the best defense is a good offense." Their speed and agility, and their slashing teeth and claws, must have been used to defend against enemies, as with many predators today.

Among the plant-eating dinosaurs, three different defensive strategies appear to have evolved (figure 13.13). In the sauropods, huge body size and powerful whip-like tails provided defense against enemies. Some sauropods, such as the titanosaurs, evolved body **armor**. But defensive armor was the hallmark of most ornithischians. This armor ranged from the impervious plating of ankylosaurs to the spiked tails of stegosaurs.

Ornithopods were the exception. They lacked body armor and must have defended themselves from attackers in other ways—by speed, camouflage, fleeing to the water,

A

B

**FIGURE 13.13**
At least three defensive strategies evolved in plant-eating dinosaurs. Ceratopsians (*A*) and ankylosaurs (*B*) had body armor. Sauropods (*not shown*) were gigantic. The defensive strategy of hadrosaurs is not easily determined. (© Mark Hallett. Reproduced with permission of Mark Hallett Paleoart)

or some sort of group defensive behavior. But it is not clear which one (or more) of these strategies ornithopods employed. Indeed, defensive behavior probably varied greatly among the ornithopods, from the small and speedy *Hypsilophodon* to the large, spike-thumbed *Iguanodon*.

## Group Behavior

The evidence for **group behavior** (gregariousness or sociality) among dinosaurs can be listed as follows:

1. Display structures—crests on allosaur skulls, tubes and crests on hadrosaur skulls, and so forth—suggest sociality among some groups of dinosaurs. These display structures presumably would have enabled the recognition of potential mates or opponents in a social group.

2. Sexual dimorphism (differences between males and females of the same species) of these display features and other structures (for example, tusks) in some dinosaurs also supports sociality. Often (though not always), sexual dimorphism among living social animals allows them to distinguish males from females and provides males with display and defensive structures that aid in the defense of territory and the acquisition of mates.

3. The change in shape during the growth of some dinosaur display structures could indicate the need to distinguish juveniles from adults in a social group.

4. Multiple dinosaur fossils (mass death assemblages) might also indicate group behavior. Dinosaurs that lived in groups would, occasionally, die in groups. However, many dinosaur mass death assemblages, such as the Late Jurassic fossils at Dinosaur National Monument in Utah, represent river-transported accumulations of carcasses. Not all dinosaur mass death assemblages actually represent groups of animals that lived and died together.

5. The evidence for parental care and nesting behavior discussed earlier also suggests some sort of group behavior among some dinosaurs.

6. Finally, and regarded by many paleontologists as the strongest evidence, multiple trackways of dinosaurs that walked in the same direction suggest social behavior. Indeed, many dinosaur track sites preserve more than one trackway of the same type of dinosaur, all heading the same way and, in many cases, at the same estimated speed (figure 13.14). Because so many tracksites document this pattern, an alternative interpretation—that at each site individual dinosaurs walked, at different times, to a common goal—seems unlikely.

These six points of evidence suggest some form of group behavior among dinosaurs, especially theropods, ornithopods, and sauropods. But they don't allow us to infer the exact types of **social structures** of these dinosaurs. Some books talk of dinosaurs living in "herds." But a **herd** is a complicated and specific kind of social group in which a dominant animal (usually a male) leads other animals. No unequivocal evidence for herd behavior exists among dinosaurs, although much evidence for social groupings does exist for many different types of dinosaurs.

## Summary

1. Fossilized skin impressions suggest many kinds of dinosaurs had reptilian scales covering their bodies. Many (possibly all) theropods had feathers or feather-like body coverings.
2. Dinosaur coloration does not fossilize and is usually arrived at in renderings by analogy to the coloration of living reptiles or birds.
3. Restorations of the external appearance of dinosaurs have evolved with changing ideas about the biology and behavior of dinosaurs.
4. Dinosaur weight estimates are based mostly on scale models and are only as accurate as the models themselves.
5. Living reptiles experience indeterminate growth, whereas mammals and birds experience determinate growth.

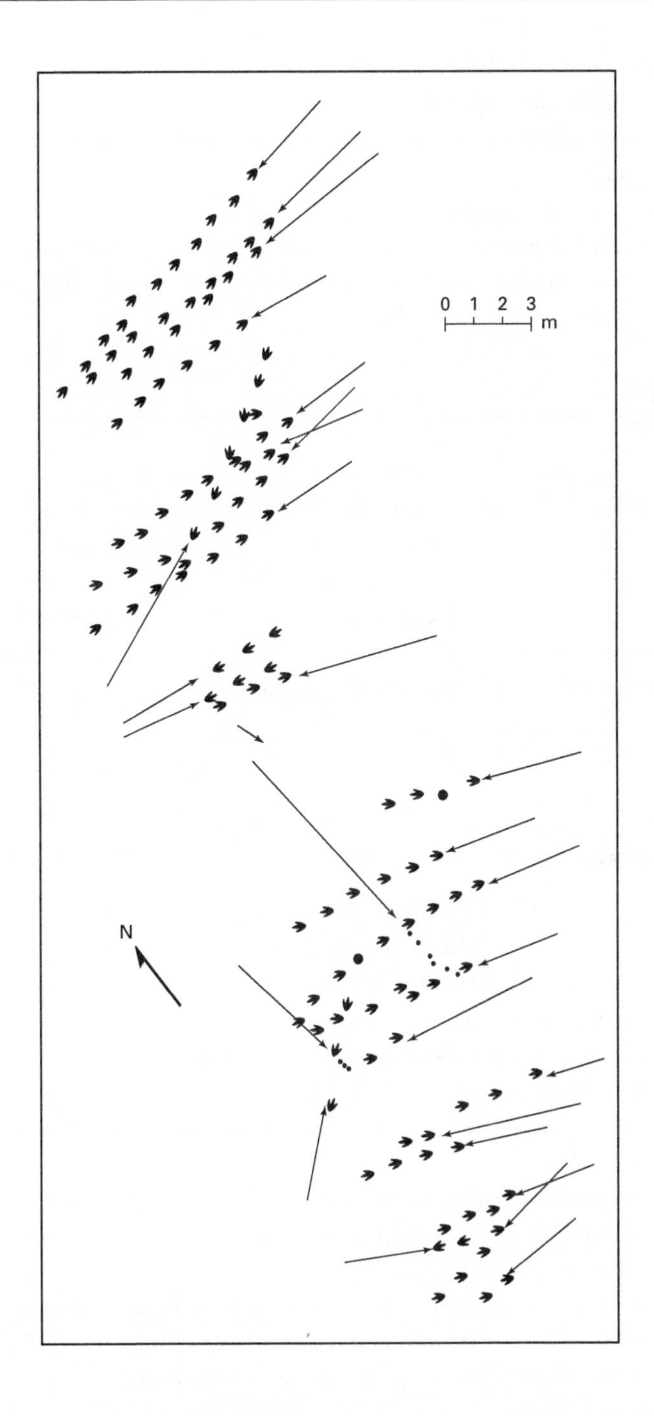


**FIGURE 13.14**
Most of these Early Jurassic dinosaur trackways (from Massachusetts) are heading in the same direction and suggest the possibility of group behavior. Note that 20 trackways of the same kind of dinosaur are heading west. (Drawing by Network Graphics)

6. Dinosaurs grew quickly and experienced determinate growth. They achieved gigantic size by an acceleration of growth rate.
7. Although jaw and tooth structures allow paleontologists to distinguish meat-eating from plant-eating dinosaurs, it is seldom possible to determine exactly what food a dinosaur ate.
8. The skeletons of almost all dinosaurs identify them as ground-dwelling walkers and runners. Only a few lived in the trees or in water.

9. All dinosaurs probably laid eggs, some in nests that may have been protected and, after hatching, tended by adult dinosaurs.
10. Dinosaurs defended themselves in a variety of ways, from speedy escape to impervious body armor.
11. Several lines of evidence suggest group behavior, especially in some theropods, ornithopods, and sauropods. But the exact kinds of social structures of these dinosaurs cannot be determined, and there is no evidence of herd behavior among dinosaurs.

## Key Terms

armor
cannibalism
clutch
*Coelophysis*
coloration
cross-sectional area
determinate growth
feeding range
group behavior
herd
indeterminate growth
locomotion
*Maiasaura*
mass death assemblage
nest
scale model
skin impression
social structure
soft-tissue anatomy
weight estimate

## Review Questions

1. When a restoration of a dinosaur is examined, what features are inferred from sound skeletal evidence, and what features are based on speculation?
2. How have changing ideas about dinosaurs influenced dinosaur restorations?
3. How are dinosaur weights estimated?
4. What do we now know about dinosaur growth rates?
5. How long did dinosaurs live?
6. How do paleontologists determine what dinosaurs ate, and what are the limitations of these determinations?
7. What does the locomotion of dinosaurs tell us about where they lived?
8. What sort of parental care did hadrosaurids confer on their young? What evidence supports this?
9. List the defensive strategies employed by different types of dinosaurs. What is the evidence to indicate which dinosaurs employed a particular strategy?
10. What is the evidence for group behavior among dinosaurs?

## Further Reading

Barrett, P. M. 2014. Paleobiology of herbivorous dinosaurs. *Annual Review of Earth and Planetary Science* 42:207–230. (Reviews the evolution of dinosaur herbivory)

Bell, P. R. 2012. Standardized terminology and potential taxonomic utility for hadrosaurid skin impressions: A case study for *Saurolophus* from Canada and Mongolia. *PLoS ONE* 7:e31295. (Detailed study of the skin impressions of the hadrosaur *Saurolophus*)

Brusatte, S. L. 2012. *Dinosaur Paleobiology*. Chichester: Wiley-Blackwell. (Book-length review of all aspects of dinosaur biology)

Case, T. J. 1978. Speculations on the growth rate and reproduction of some dinosaurs. *Paleobiology* 4:320–328. (A classic technical analysis of dinosaur growth rates based on the growth rates of living reptiles)

Christian, A. 2002. Neck posture and overall body design in sauropods. *Mitteilungen Museum für Naturkunde der Humboldt-Universität zu Berlin, Geowissenschaftliche Reihe* 5:271–281. (Analyzes the neck postures and feeding ranges of sauropods)

Colbert, E. H. 1962. The weights of dinosaurs. *American Museum Novitates* 2076:1–16. (Weight estimates of dinosaurs using scale models)

Erickson, G. M. 2014. On dinosaur growth. *Annual Review of Earth and Planetary Science* 42:675–697. (Current knowledge of growth in dinosaurs)

Erickson, G. M., P. J. Makovicky, P. J. Currie, M. A. Norell, S. A. Yerby, and C. A. Brochu. 2004. Gigantism and comparative life-history parameters of tyrannosaurid dinosaurs. *Nature* 430:772–775. (Posits rapid growth rates for tyrannosaurids based on growth lines in bones)

García-Ortiz, E., and F. Pérez-Lorente. 2014. Palaeoecological inferences about dinosaur gregarious behavior based on the study of tracksites from La Rioja area in the Cameros basin (Lower Cretaceous, Spain). *Journal of Iberian Geology* 40:113–127. (Reviews the evidence for gregarious behavior in dinosaurs based on trackways)

Horner, J. R. 1984. The nesting behavior of dinosaurs. *Scientific American* 250:130–137. (A popular review of the dinosaur nests from Montana)

Lingham-Soliar, T., and G. Plodowski. 2010. The integument of *Psittacosaurus* from Liaoning Province, China: Taphonomy, epidermal patterns and color of a ceratopsian dinosaur. *Naturwissenschaften* 97:479–486. (Scales and color pattern of an exceptionally preserved fossil of *Psittacosaurus*)

Ostrom, J. H. 1972. Were some dinosaurs gregarious? *Palaeogeography, Palaeoclimatology, Palaeoecology* 11:287–301. (A now classic, technical review of the trackway evidence for social behavior in dinosaurs)

Sander, P. M., C. T. Gee, J. Hummel, and M. Clauss. 2010. Mesozoic plants and dinosaur herbivory, pp. 331–359, in C. T. Gee, ed., *Plants in Mesozoic Time: Morphological Innovations, Phylogeny, Ecosystems*. Bloomington: Indiana University Press. (Reviews dinosaur herbivory)

Schweizer, M. H. 2011. Soft tissue preservation in terrestrial Mesozoic vertebrates. *Annual Review of Earth and Planetary Science* 39:187–216. (Reviews soft tissues in Mesozoic vertebrates, including dinosaurs, and how they are identified)

# 14

# HOT-BLOODED DINOSAURS?

PALEONTOLOGISTS long considered dinosaurs to have had the reptilian metabolism popularly referred to as cold-blooded. But, in 1970, paleontologist John Ostrom of Yale University suggested that dinosaur metabolisms may have been more mammal- or bird-like than reptilian. There soon followed an article in *Scientific American* by paleontologist Robert Bakker titled "Dinosaur Renaissance" in which he presented evidence and analysis to support warm-bloodedness in all dinosaurs. Bakker thereby initiated debate, still ongoing, about the nature of dinosaur metabolism. Much progress has been made in this debate, but some key issues remain unresolved. In this chapter, I review the evidence for dinosaur metabolism and evaluate the debate over this complex and fascinating subject.

## SOME TERMS AND CONCEPTS

Before studying dinosaur metabolism, we need to become conversant with some basic terminology and concepts. **Metabolism** is best defined as the chemical processes that provide energy to and repair the cells of an organism. In popular terms, vertebrate metabolisms either are **cold-blooded** (fishes, amphibians, and reptiles) or **warm-blooded** (mammals and birds). The equivalent technical terms are **ectothermic** (cold-blooded) and **endothermic** (warm-blooded or its exaggerated synonym, **hot-blooded**), based on the Greek roots *ecto* (outside), *endo* (inside), and *thermos* (temperature).

Ectotherms receive most, or all, of their body heat from an external source, usually directly from the sun (figure 14.1). In contrast, endotherms generate most, or all, of their body heat internally. Endotherms characteristically have high rates of heat production. Ectotherms, however, typically have slower metabolisms. Vertebrates that maintain a nearly constant body temperature (usually +/− 2 degrees Celsius) are called **homeotherms**, whereas those in which body temperature varies daily, seasonally, or

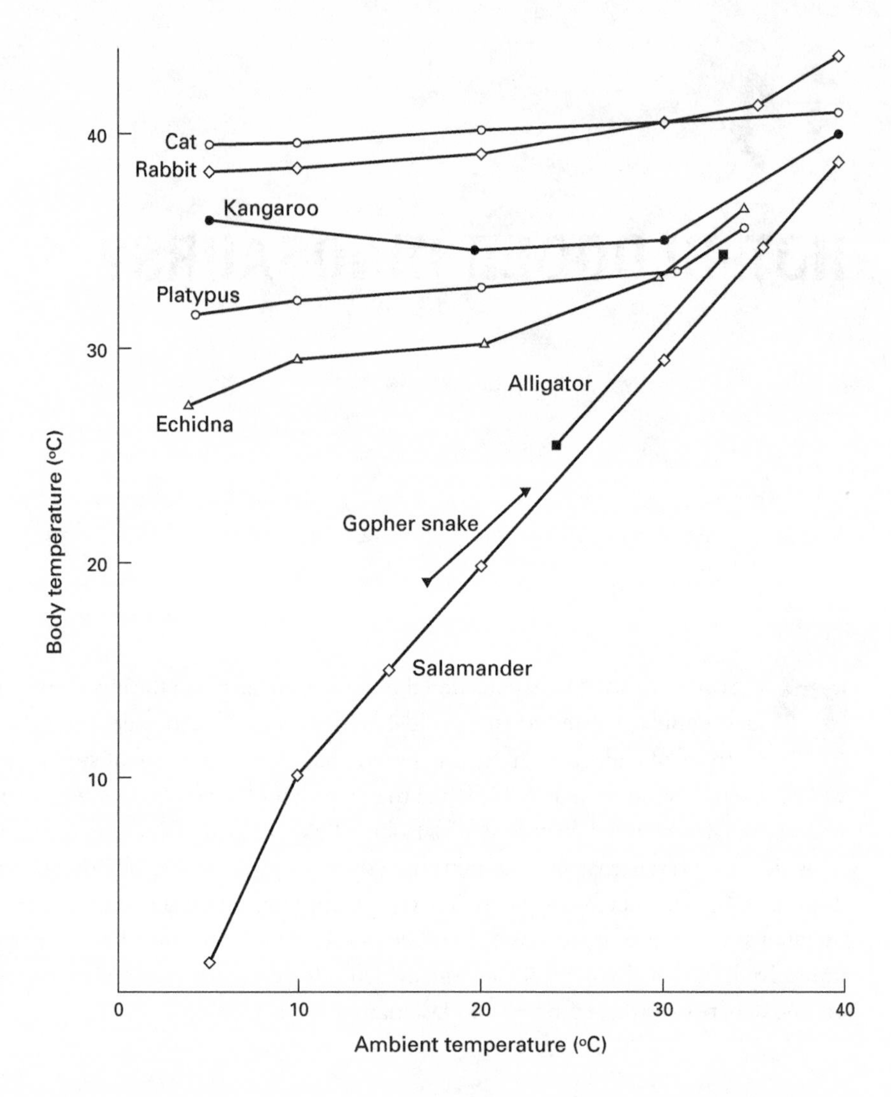


**FIGURE 14.1**
Endotherms maintain a relatively constant body temperature, whereas the body temperature of ectotherms varies with the ambient (outside) temperature. (Drawing by Network Graphics)

throughout the life cycle are called **heterotherms** (from the Greek *homeos* [similar] and *heteros* [different]).

Today, different metabolisms are found in different kinds of vertebrates. The very diverse and dominant land vertebrates, mammals and birds, are endotherms, whereas the less diverse reptiles and amphibians are ectotherms. The current success of endotherms relative to ectotherms has incorrectly fueled the notion that ectothermy is an inferior type of metabolism (box 14.1). Also, the great difference between living endotherms and ectotherms suggests a marked dichotomy in vertebrate metabolism, with endotherms on one side and ectotherms on the other. But, endothermy and ectothermy describe only the extremes of a broad spectrum of vertebrate metabolisms (see figure 14.1). Some ectotherms, such as tuna, maintain high body temperatures, whereas some endotherms, such as shrew tenrecs, are hard pressed to maintain a high, constant body temperature. There is great complexity and variety in existing vertebrate metabolism, and there is no reason why such variety should not have been present during the past, even in the age of dinosaurs.

## Box 14.1

### Misconceptions About Metabolism

The fact that today vertebrate endotherms are more successful than ectotherms has helped create the misconception that ectothermy is an inferior metabolism (box figure 14.1). Furthermore, because we are mammals, most of us see mammals as superior to cold-blooded amphibians and reptiles. Indeed, many people find reptiles repulsive, for the reasons so well stated by Carolus Linnaeus in 1797: "Reptiles are abhorrent because of their cold body, pale color, cartilaginous skeleton, filthy skin, fierce aspect, calculating eye, offensive smell, harsh voice, squalid habitation, and terrible venom." Small wonder that most of our pets are mammals and birds, not reptiles.

**BOX FIGURE 14.1**
Ectothermic reptiles are typically seen as sluggish and inferior to endothermic mammals.

But it is simply a mistake to view ectotherms as inferior to endotherms. Instead, we should realize that both represent equally viable types of vertebrate metabolism, each with advantages and disadvantages. Consider some of the advantages of the relatively slow ectothermic metabolism of an animal receiving most of its body heat from the sun. Such an animal needs to eat much less than a comparably sized endotherm, thus reducing the problems associated with obtaining food. Most ectotherms can endure much greater temperature extremes than endotherms can because their body temperature can vary widely and does not need to be held within a narrow range. These advantages explain, for example, why in some of the hottest deserts on Earth, in western Australia, ectothermic lizards abound, and there are few mammals. The endothermic mammals simply cannot find enough food or endure the extreme temperatures of these deserts.

Although ectothermy may not be inferior to endothermy, the question of superior metabolism has often been behind efforts to demonstrate dinosaurs were endotherms. Some paleontologists simply cannot believe that dinosaurs could have been as successful as they were if they had been ectotherms. No doubt, few living lizards would agree!

## THE EVIDENCE

Metabolism is determined by chemical reactions in the enzymes, blood, and internal organs. None of these structures in dinosaurs are known to have fossilized, so it might seem there is very little direct evidence of dinosaur metabolism. Indeed, wouldn't we need a time machine and a thermometer in order to determine the body temperatures and metabolisms of dinosaurs?

Fortunately, fossils preserve quite a variety of evidence of dinosaur metabolism, although much of it is indirect. This evidence can be organized into several categories, and most of it, as we shall see, has been interpreted in different ways.

### Posture and Gait

Living ectotherms have sprawling postures and gaits, whereas living endotherms, with a few exceptions, have upright postures and gaits (figure 14.2). The exceptions are mostly aquatic mammals, such as seals and walruses, which have sprawling limbs admirably adapted to propelling them through water. Living vertebrates show a nearly perfect correlation between posture and metabolism. This correlation may indicate that extinct animals with an upright posture, the dinosaurs, were endotherms.

**FIGURE 14.2**
Upright posture characterizes living endotherms, whereas living ectotherms have a sprawling posture. (© Scott Hartman)

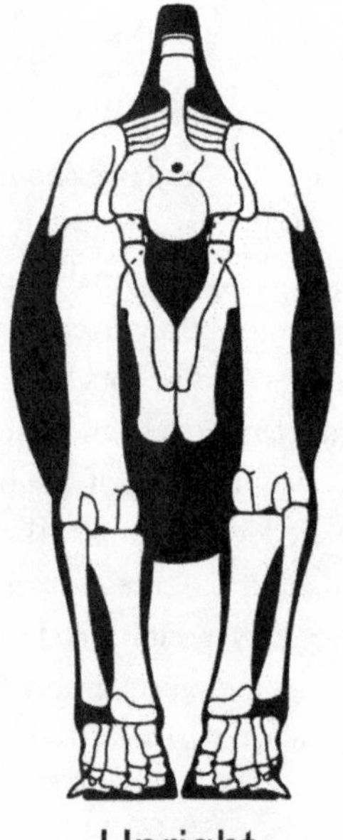
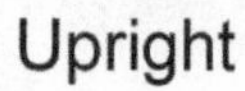

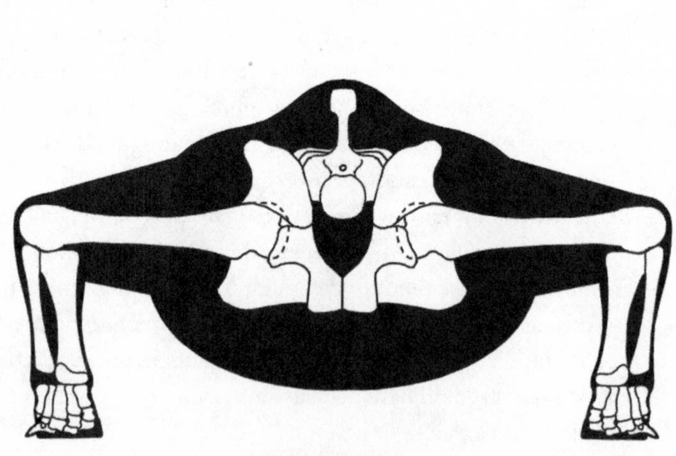

Two criticisms can be leveled at this conclusion. First, no cause-and-effect relationship has been established between posture and metabolism. Maybe ectotherms can have an upright posture, maybe some (the dinosaurs?) did in the past, and it is only a coincidence that all living ectotherms happen to be sprawlers.

A second criticism is that not all dinosaurs may have had an upright posture. Stegosaurs, ankylosaurs, and ceratopsians have been reconstructed with upright hind limbs and sprawling or semi-sprawling forelimb postures. What does this indicate about their metabolisms? Does it mean their posterior halves were endothermic and anterior halves ectothermic? Of course not. It probably indicates that factors other than metabolism, such as the heavy skulls and armor of stegosaurs, ankylosaurs, and ceratopsians, which forced a slow and powerful sprawling forelimb posture, also determined a dinosaur's posture.

Despite these criticisms, the correlation between posture and metabolism in living vertebrates is striking. A subset of this correlation, that all living bipeds are endotherms, also suggests that bipedal dinosaurs were endotherms. But, to believe that these correlations indicate that dinosaurs were endothermic, we must agree that posture is controlled by metabolism, a conclusion still open to some debate.

## Speed, Activity Level, and Agility

High levels of activity characterize living endotherms, whereas today's ectotherms generally are slower, more sluggish, and less agile. The qualifier "generally" needs to be used here because some living ectotherms, for example sea turtles, are capable of great speed and/or agility, if only for short periods of time. Nevertheless, evidence for speed, sustained activity, and agility in dinosaurs would be more consistent with their having had an endothermic, rather than ectothermic, metabolism.

Speed, sustained activity, and agility in dinosaurs is evident in their skeletal structures, brain size, brain complexity, and in the speeds estimated from dinosaur trackways. As we saw in chapters 4 through 9, the skeletons of some dinosaurs, especially small theropods and ornithopods, in many ways resemble those of fast-running birds and mammals. Key features of this resemblance include elongate, slender limbs and limb joints, indicating an ability to flex acutely at the elbow, wrist, knee, and ankle joints. Hollow bones, large claws, limb proportions indicative of habitual bipedality, and long, rigid tails for precision balance while running contribute to the impression of speed and agility in some dinosaurs (figure 14.3). Some theropod and ornithopod dinosaurs thus seem to have been every bit as quick and agile as living birds and mammals, suggesting that these dinosaurs may have been endotherms.

Some paleontologists assert that the larger dinosaurs, such as ceratopsians, galloped like living endothermic rhinoceroses. But this argument is inconsistent with the limb structures and trackways of the large, quadrupedal plant-eating dinosaurs, which indicate they were slow, powerful walkers. The skeletal evidence for speed, high levels of activity, and agility in dinosaurs, and the inference of endothermy from this evidence, is mostly confined to theropods and ornithopods.

**Relative brain size** and complexity provide a second line of evidence for speed, high activity levels, and agility in some dinosaurs. These characteristics of living endotherms

**FIGURE 14.3**
The theropod *Deinonychus* well displays many skeletal features suggestive of speed, sustained high activity, and agility. (© Scott Hartman)

require great motor and sensory control by larger and more complex brains than those of living ectotherms. If we compare the relative brain sizes of dinosaurs to that of a living ectothermic lizard, theropods and most ornithopods appear to have been relatively "brainy" (figure 14.4). But sauropod, ankylosaur, stegosaur, and ceratopsian brain sizes fall well below the line set by the lizard.

**Brain complexity** in dinosaurs can be inferred only from endocasts of empty fossil skulls (see box 8.2). An empty skull acts as a mold that reproduce the brain's overall

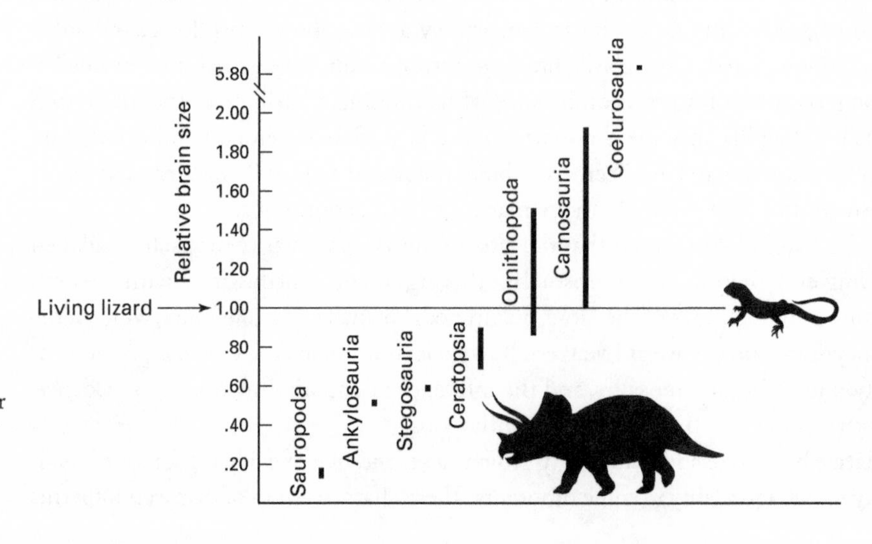


**FIGURE 14.4**
Dinosaurs had a range of relative brain sizes, some larger and some smaller than that of a living lizard. (Drawing by Network Graphics)

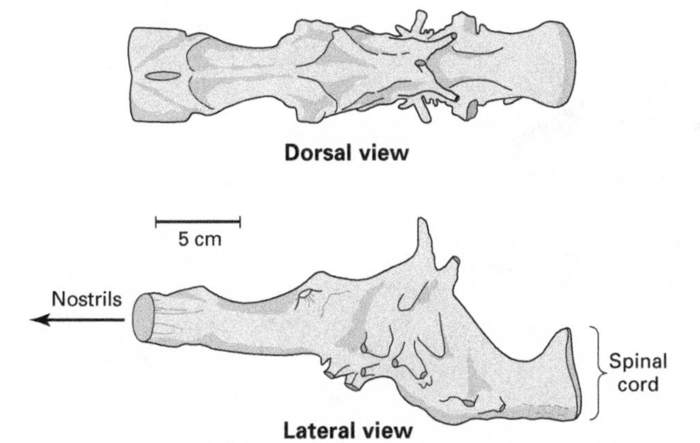


FIGURE 14.5
The brain of *Tyrannosaurus rex* is typically reptilian in its level of complexity. (Drawing by Network Graphics)

configuration and location and the number of associated blood vessels and nerves. Dinosaur brain casts, however, reveal typically reptilian levels of brain complexity (figure 14.5). These casts do not preserve special structures, such as extremely large cerebral hemispheres, that might be linked to speed, high activity levels, and agility.

A third line of evidence for dinosaur speed, activity, and agility comes from their trackways. In chapter 12, we saw that dinosaur speeds can be estimated from trackways. Most dinosaur trackways indicate slow walking, which is the normal speed of all living vertebrates, ectotherms and endotherms alike. But, a few trackways document small theropods running as fast as a living antelope.

So, the evidence for speed, high activity levels, and agility among dinosaurs seems to support endothermy only in small theropods and ornithopods. But some paleontologists point out, as I did at the beginning of this discussion, that some living ectotherms can be very fast, active, and agile. It might be that speedy, active, and agile theropods and ornithopods were ectotherms.

## Feeding Adaptations

Living endotherms maintain a constant, high body temperature and a fast metabolism by consuming and processing large amounts of food (energy). This means they eat more relative to body weight and process that food more quickly than do ectotherms. Key to rapid processing are the teeth, jaws, and skulls of many endotherms, which allow the food to be broken down rapidly into small pieces, thereby increasing the food's surface area and releasing important nutrients so that the endothermic digestive system can rapidly assimilate the nutrients. If dinosaurs had tooth, jaw, and skull structures that indicated extensive processing of the food in the mouth, this might be evidence that they had an endothermic metabolism.

When we look at the teeth, jaws, and skulls of dinosaurs, however, only among ornithopods do we find structures similar to the "**food processors**" of living endotherms. The powerful jaws and extensive dental batteries of ornithopods, such as hadrosaurs,

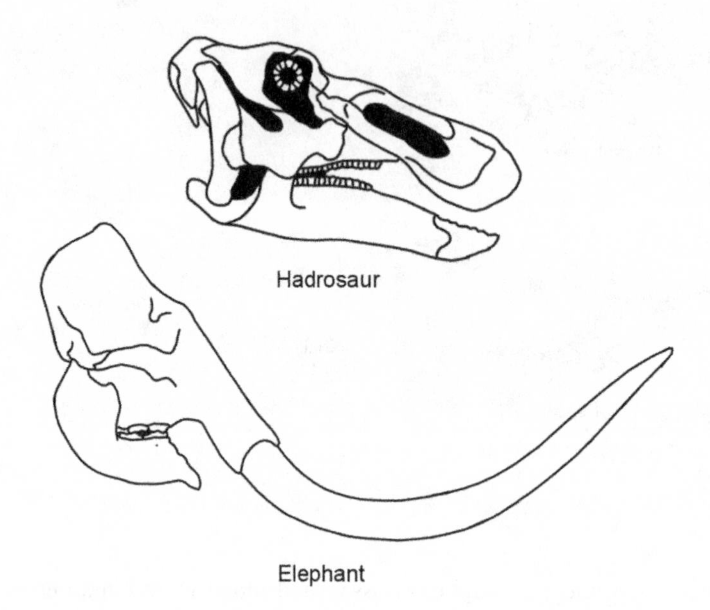


**FIGURE 14.6** The dental batteries and powerful jaws of hadrosaurs resemble those of some living plant-eating mammals, such as the elephant. (© Scott Hartman)

are similar to the jaws and teeth of living endothermic horses and elephants that extensively grind the vegetation they consume (figure 14.6).

This might be taken to indicate that only the ornithopods possessed "food processors" suggestive of endothermy. But, the teeth, jaws, and gastric mills of some theropods, prosauropods, and some of the armored ornithischians don't indicate food processing inferior to that of living plant-eating birds. Predatory theropod food processing doesn't look much different from the slicing of meat undertaken by living meat-eating mammals, such as wild dogs and cats. We also need to remember that the large size of most dinosaurs would have forced them to process and consume large amounts of food, whether they were ectotherms or endotherms. Furthermore, the ornithopod grinding mechanism may simply reflect their dietary preference for tough, hard-to-process plant foods.

We simply cannot be certain what the "food processors" of dinosaurs indicate about their metabolism. Many needed to consume large amounts of food to support huge body masses. Specialization in tough food items, or the need to feed a rapid, endothermic metabolism, could explain the tooth, jaw, and skull structures of some dinosaurs, especially the ornithopods.

## Bone Microstructure

The compact (external) layer of bone of many living ectotherms contains few channels for blood vessels. In contrast, the **compact bone** of many living endotherms is full of large numbers of blood vessels (figure 14.7). This is thought to reflect the quick metabolism of endotherms, which requires the rapid exchange of elements stored in

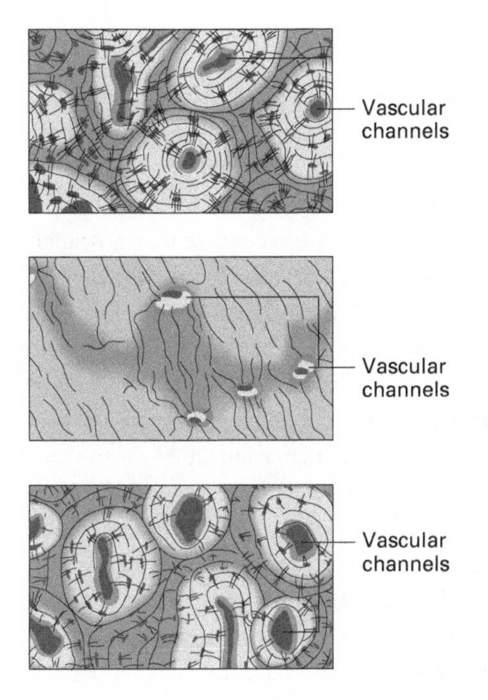


FIGURE 14.7
Dinosaur compact bone (*below*) has many channels for blood vessels, like that of living endotherms (*above*), and unlike that of living ectotherms (*center*). (Drawing by Network Graphics)

bone, such as calcium and phosphorus. It stands to reason that if dinosaur compact bone had many channels for blood vessels, then this would indicate endothermy.

Well-preserved dinosaur bone can be cut into thin wafers, and the microscopic structure of the bone can be determined. This has been done for all major groups of dinosaurs and reveals compact bone with numerous channels similar to that of many living endotherms (see figure 14.7).

This might seem conclusive evidence for endothermy in all dinosaurs, but the correlation between metabolism and compact-bone microstructure is not a perfect one in living vertebrates. Bone from different parts of the skeleton (for example, limbs versus vertebrae) has different microstructure, in part because of the different stresses placed on different parts of the body. Also, some large ectotherms (turtles and crocodiles) have compact bone with many channels, whereas some small living endotherms (certain mammals and birds) lack numerous blood channels in their compact bone. This suggests that bone microstructure may be more related to size than to metabolism, making the **bone microstructure** of dinosaurs inconclusive evidence of endothermy.

Another aspect of dinosaur bone microstructure is the presence of growth rings in the bone. These growth rings have been found in the bone of many kinds of dinosaurs, including ornithopods, theropods, and ceratopsians. The rings are characteristic of the bone of living ectothermic reptiles. Periodic pauses in bone growth due to seasonal (or annual) temperature fluctuation cause these rings to form. However, the bone of some dinosaurs, such as the ornithopod *Dryosaurus*, lacks growth rings and more resembles endothermic bone. Furthermore, the microstructure of some dinosaur bone closely resembles the microstructure of bird bone, which indicates rapid bone growth in dinosaurs, as in endothermic birds.

## Blood Pressure

An endothermic metabolism requires high **blood pressure** and rapid blood circulation to move energy quickly through the body. Thus, living endotherms have consistently higher blood pressures than do ectotherms (figure 14.8). If we could estimate dinosaur blood pressures, this might indicate whether they were endotherms.

But, how can we do this? It turns out quite easily if we recognize that a primary function of the vertebrate heart is to pump blood to the brain, which is usually elevated above the level of the heart. In other words, the vertical distance between the heart and the brain should be related, in some way, to blood pressure, because sufficient pressure must be maintained to move blood to the brain, or the animal will die.

It is possible to estimate blood pressures based on the vertical distance between the dinosaur's heart and its brain (figure 14.9). Such estimates are based on the relationship between blood pressure and the distance between the heart and the brain of living vertebrates and also require certainty of the posture of the dinosaur. These estimates suggest high endothermic blood pressures for most dinosaurs (*Triceratops* is a notable exception) and incredibly high blood pressures (more than 400 millimeters of mercury!) for sauropods. It is questionable whether blood pressures as high as those estimated for sauropods could be maintained by any vertebrate circulatory system without an explosion. Thus, it seems likely that sauropods may have used arterial valves or muscular contractions in the neck, as do living giraffes, to help bring blood to the brain and thus would have had much lower blood pressures than estimated. Estimates of blood pressures, however, are consistent with endothermy in most dinosaurs, although critics point out that large vertical heart–brain distances in dinosaurs could be the determining factor here, not metabolism.

A second, and speculative, aspect of dinosaur blood pressure concerns the structure of the dinosaurian heart. Living endotherms have a fully divided **four-chambered heart** that separates aerated from nonaerated blood and thus acts as a double pump,

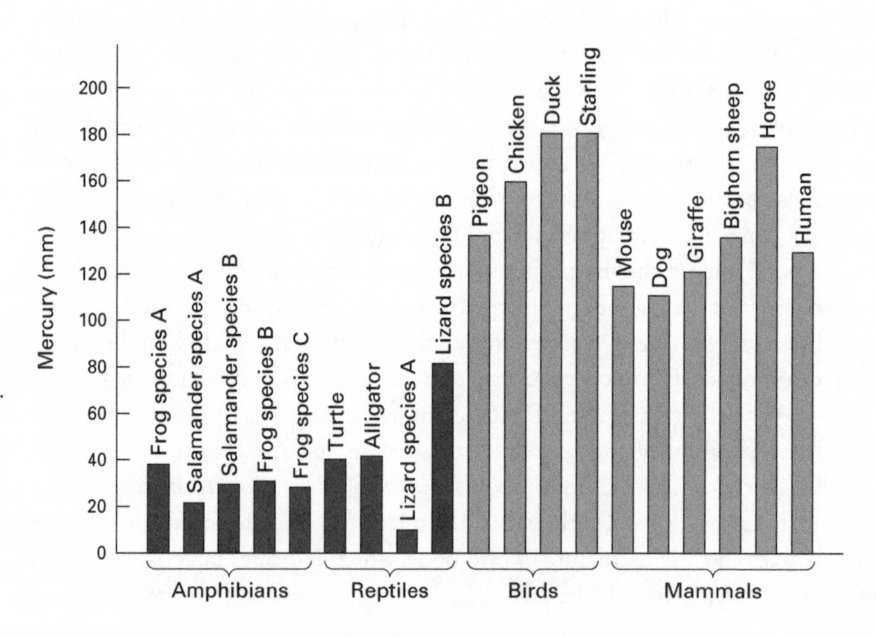


**FIGURE 14.8** Living endotherms have higher blood pressures than do ectotherms. (Drawing by Network Graphics; modified from J. H. Ostrom. 1980. The evidence for endothermy in dinosaurs, pp. 82–105, in R. D. K. Thomas and E. C. Olson, eds., *A Cold Look at the Warm-Blooded Dinosaurs*. Copyright © 1980 American Association for the Advancement of Science)

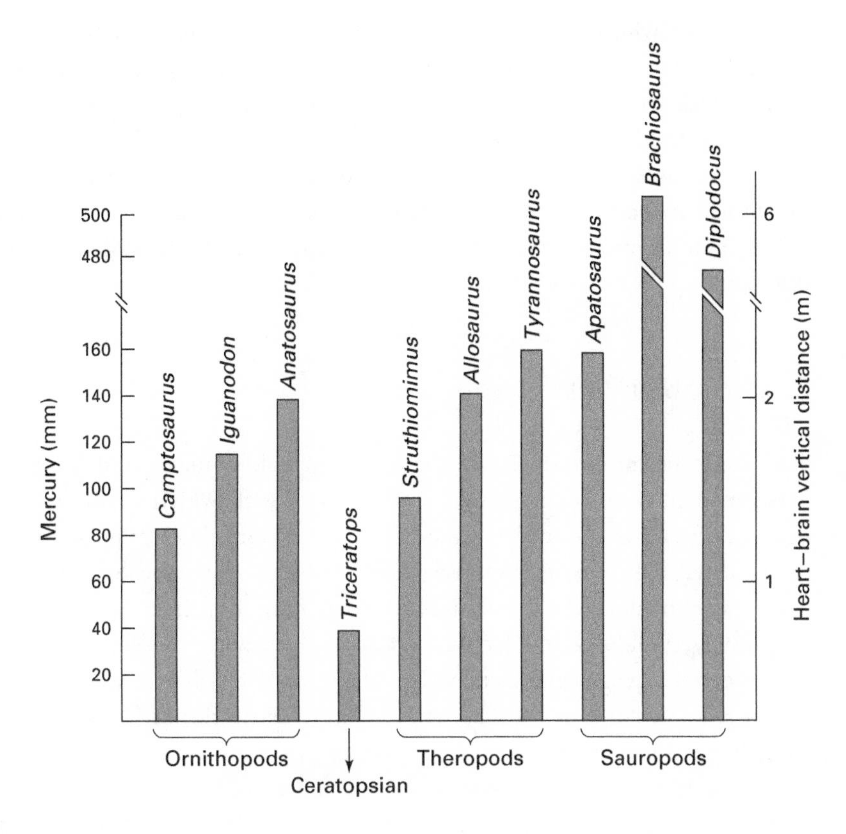


**FIGURE 14.9** Dinosaur blood pressures can be estimated from their heart–brain vertical distances. (Drawing by Network Graphics; modified from J. H. Ostrom. 1980. The evidence for endothermy in dinosaurs, pp. 82–105, in R. D. K. Thomas and E. C. Olson, eds., *A Cold Look at the Warm-Blooded Dinosaurs*. Copyright © 1980 American Association for the Advancement of Science)

producing high pressure for the aerated blood and lower pressure for the nonaerated blood. This mechanism is especially significant because lowering the pressure of the nonaerated blood prevents the rupture of tiny blood vessels characteristic of the lungs of endotherms. In contrast, ectothermic hearts have only two chambers that do not efficiently separate the aerated from the nonaerated blood. An exception is living crocodiles, which have an imperfectly divided four-chambered heart (figure 14.10).

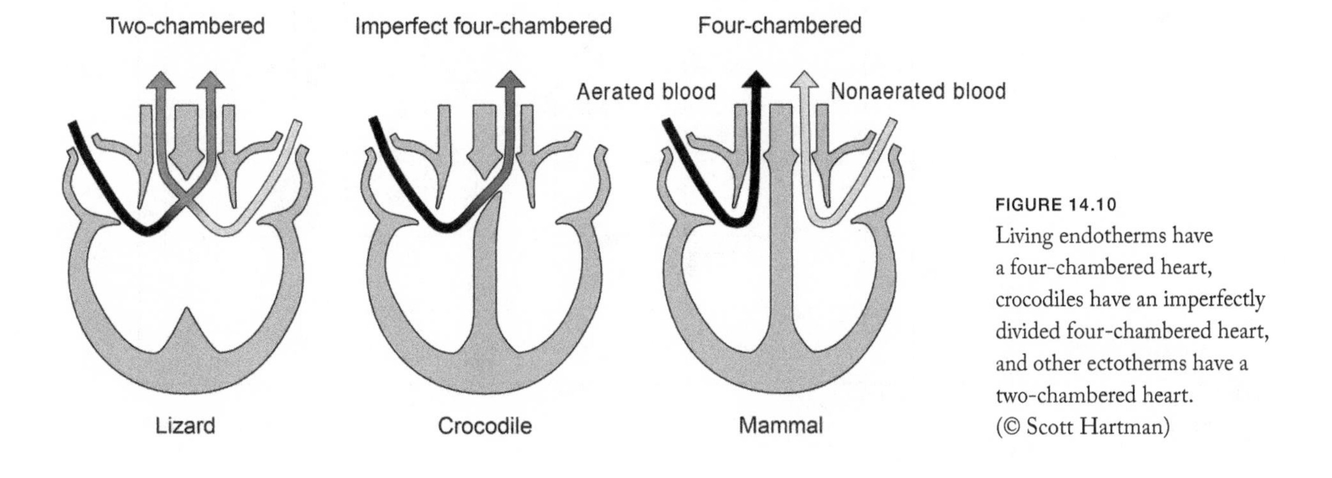


**FIGURE 14.10** Living endotherms have a four-chambered heart, crocodiles have an imperfectly divided four-chambered heart, and other ectotherms have a two-chambered heart. (© Scott Hartman)

The closest living relatives of dinosaurs—birds and crocodiles—have four-chambered hearts. So, many paleontologists believe that dinosaurs had four-chambered hearts as well. Dinosaur hearts have not fossilized, so this remains speculation. Indeed, a supposed "fossilized heart" of a dinosaur, with four chambers, has been demonstrated to be nothing more than a piece of ironstone. But, if dinosaurs did have four-chambered hearts, and if their high estimated blood pressures don't simply reflect large vertical heart–brain distances, then these lines of evidence suggest endothermy among most dinosaurs.

## Geographic Distribution

Today, ectotherms cannot live in the extremely cold climates near the poles simply because there is not enough solar energy there with which to warm their bodies (figure 14.11). However, endothermic mammals and birds, such as polar bears and penguins, are capable of living in those colder regions of the globe. So, if the geographic distribution of dinosaur fossils indicates that they lived in cold climates, this also might suggest endothermy in dinosaurs.

The current geographic distribution of dinosaur fossils encompasses Cretaceous localities as far north as Alaska, the Northwest Territories of Canada, and Svalbard (Spitzbergen) and as far south as Antarctica. These are places where ectotherms do not live today. But, the Cretaceous world was not as cold toward the poles as is today's world, and because of continental drift, these Cretaceous dinosaur localities were also not as poleward as they are today. Despite this, these locations would still have been far enough poleward during the Cretaceous to have experienced the "**polar night**"—winter months of virtually continuous darkness. This darkness would have excluded ectotherms from living in poleward regions, even if there were no polar ice caps and temperatures were warmer during the Cretaceous.

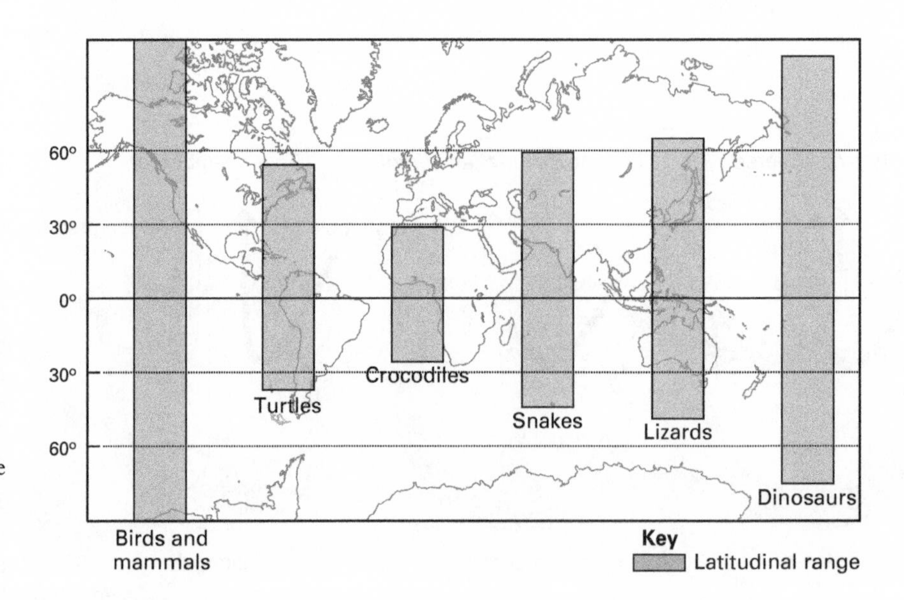


**FIGURE 14.11**
Today, ectotherms do not live in cold, poleward regions. (Drawing by Network Graphics)

It also has been suggested that the poleward Cretaceous dinosaur fossils do not represent dinosaurs that lived in these regions year round. Instead, they may represent dinosaurs that migrated over large areas and were only living poleward during the warmer, brighter portion of the year. So, a hadrosaurid might have migrated the 3,000 kilometers from Alaska to Alberta in a given year, a trek of 60 days at a speed of 50 kilometers per day. To have migrated so far, a dinosaur would have to be endothermic—or would it? Given the possibility of migration, it is difficult to argue that the geographical distribution of dinosaurs tells us something conclusive about their metabolism.

## Bird Ancestry

Dinosaurs are the ancestors of birds (see chapter 15). Living birds are endotherms, and it may be that all extinct birds, including the Late Jurassic *Archaeopteryx*, the first bird, were endotherms. Indeed, some have argued that the small size, skeletal structure, insulating feathers and powered flight of *Archaeopteryx* strongly suggest an endothermic metabolism (figure 14.12). The question thus arises, did endothermy in birds first evolve in *Archaeopteryx*, or did the theropod ancestors and close relatives of birds have an endothermic metabolism? Or, how many times did endothermy arise in the origin and evolution of birds?

**FIGURE 14.12**
Living birds are endotherms.

This question can't be answered definitively. As already discussed, many features of the small theropods are consistent with endothermy. Indeed, some small theropods from the Lower Cretaceous of China have feather-like body coverings. These and many other theropods are strikingly bird-like. That the theropods gave rise to the birds and are very bird-like in many ways is also consistent with endothermy. Although the endothermy of birds may suggest endothermy in at least some theropods, it provides no clues to the metabolism of the other dinosaurs.

## Social Behavior

Today, complex social behaviors are characteristic of many endotherms and uncommon among ectotherms. The evidence reviewed in chapter 13 suggests that some dinosaurs may have lived in groups and had some form of sociality based on their trackways, visual display, and inferred parental care of young dinosaurs. On face value, this is consistent with endothermic dinosaurs.

But, some living reptiles form social groups to hunt and administer minimal care to their young, and not all living endotherms form social groups; some live solitary lives. Indeed, social behavior is not normally viewed as being caused by metabolism, but instead is related to other factors, such as the distribution of food resources in an animal's habitat. So, the inferred social behavior of some dinosaurs at best is consistent with, but not strong evidence for, their endothermy.

## Predator–Prey Ratios

Living endotherms need to consume more energy than do comparably sized ectotherms. This means that a 150-kilogram lion eats more food, and more frequently, than a 150-kilogram crocodile. Therefore, in the wild, a lion must have access to more food items (prey) than a crocodile (figure 14.13). The **predator–prey ratio,** the ratio of

**FIGURE 14.13**
A 150-kilogram lion (endotherm) eats more food than a 150-kilogram crocodile (ectotherm) in a given period of time. (Drawing by Network Graphics)

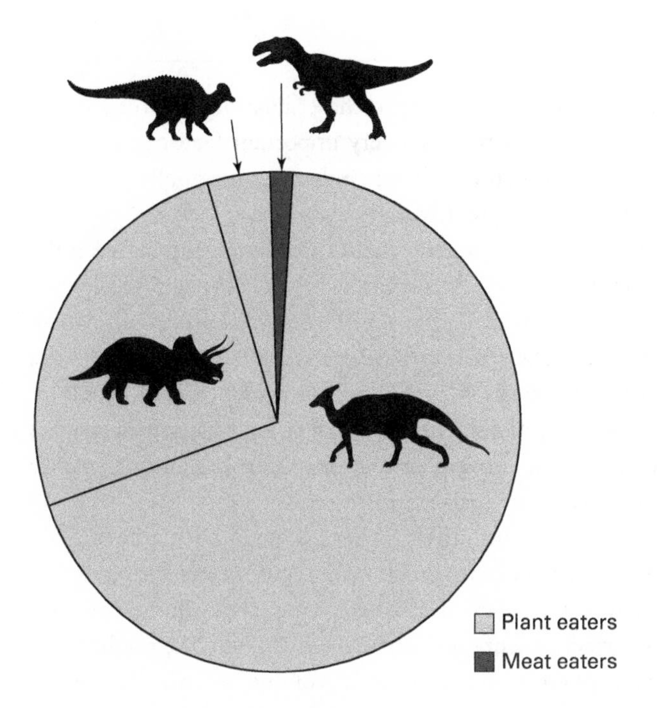


FIGURE 14.14
Predatory dinosaurs are extremely rare in the dinosaur collections from the Upper Cretaceous of Dinosaur Provincial Park in Alberta, Canada. (Drawing by Network Graphics)

the body mass of predators to that of their potential prey, thus should differ for endothermic and ectothermic predators. Thus, endothermic predator–prey ratios should be lower (less predator mass per prey) than the ratio for ectotherms. Extrapolating this to dinosaurs, one can predict that predatory dinosaurs should be rare relative to their potential prey dinosaurs if the predatory dinosaurs were endotherms and more common if they were ectotherms.

A survey of dinosaur fossil collections reveals that predatory dinosaurs are relatively rare. For example, the most famous dinosaurian predator of all time, *Tyrannosaurus rex*, is known from a few skeletons. One of the most extensive dinosaur collections ever made, from the Upper Cretaceous badlands in Dinosaur Provincial Park, in Alberta, Canada, contains only 3 to 5 percent predatory dinosaurs (figure 14.14). On face value, the scarcity of predatory dinosaurs suggests that they (and only they, because this fact tells us nothing about the metabolism of their prey) were endotherms.

There are, however, two insurmountable problems with using predator–prey ratios to infer dinosaur metabolism. First, it is not clear that the ratio of predators to prey today is determined simply by the food requirements of the predators. Other factors, such as food availability for the prey, are important as well. The second problem, and a very large one, is that we cannot be certain that collections of dinosaur fossils reflect the actual predator–prey ratios of dinosaurs. Taphonomic processes destroy many potential fossils and may have biased the dinosaur fossil record against predators or toward prey (see box 3.1). If dinosaur predators and prey did not always inhabit the same environments, then fossil collections would not necessarily reflect their actual abundances. In light of these problems, the predator–prey ratios of dinosaurs seem an unsatisfactory way to infer the metabolism of predatory dinosaurs.

## Body Size

Body size has already been mentioned as a factor in the evaluation of several of the lines of evidence for dinosaur metabolism. It is a very important factor in any consideration of vertebrate metabolism because body size influences metabolism, and the metabolism of a vertebrate must be consistent with its body size.

To understand why this is so, we need to understand the relationship between the body size of an animal and its surface area. This relationship is best portrayed by two spheres, one large and one small (figure 14.15). The volume of a sphere is $4/3\pi r^3$, where $r$ is the radius of the sphere. The surface area of a sphere is $2/3\pi r^2$, where, again, $r$ is the radius. As a sphere becomes larger, its volume increases as the cube of its radius, whereas its surface area increases only as a square of its radius. The guaranteed result is that a larger sphere has a smaller surface area relative to its volume than does a smaller sphere, which has much more surface area relative to volume.

Although this may not strike you as intuitively correct, a quick calculation should convince you. Consider two spheres, one with a radius of 5 centimeters and the other with a radius of 10 centimeters. Calculate the surface areas and volumes of both spheres, and then divide the surface area of each sphere by its volume. The result is that the surface area of the smaller sphere is 59 percent of its volume, whereas the surface area of the larger sphere is only 30 percent of its volume. Smaller spheres have larger surface areas relative to larger spheres.

If we transfer this basic geometry to animals, we realize that small animals have much more surface area relative to their volume than do large animals. Much of metabolism is generating body heat, and the surface area of an animal determines how readily it can acquire heat from an external source, or how rapidly it loses its body heat. Small animals, such as mice and hamsters, have such large surface areas relative to their volume (or mass) that they lose heat very fast. As endotherms, they have an insulating coat of fur to help retard heat loss and use their very rapid metabolism to generate more heat to replace that being lost rapidly. They also shiver frequently to generate more body heat and hide in burrows or under vegetation to slow heat loss.

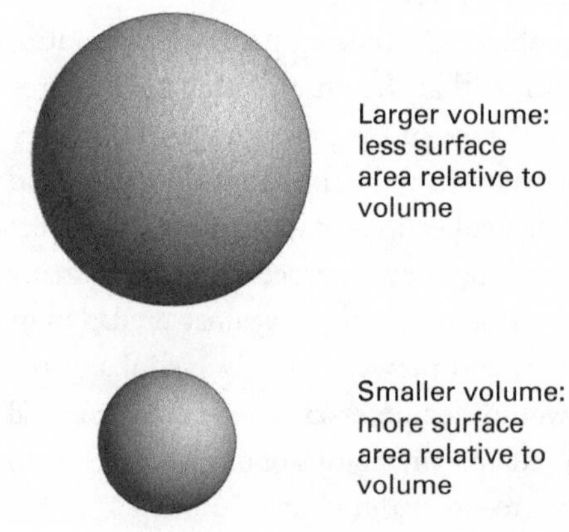


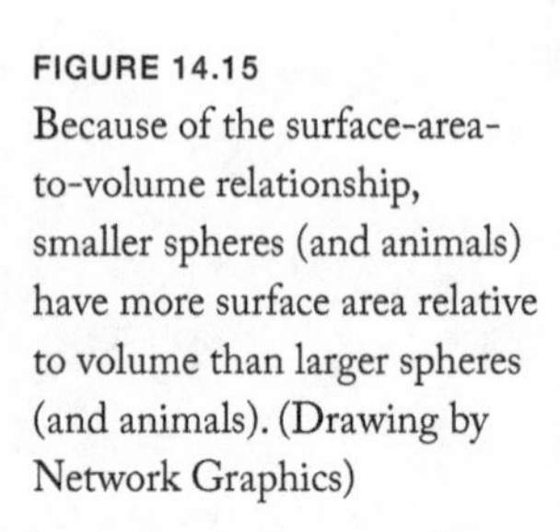

**FIGURE 14.15**
Because of the surface-area-to-volume relationship, smaller spheres (and animals) have more surface area relative to volume than larger spheres (and animals). (Drawing by Network Graphics)

In contrast, an elephant has a much lower surface-area-to-volume ratio than a mouse. Its problem, as an endotherm, lies in overheating because an elephant loses heat very slowly. For this reason, elephants have little in the way of insulating body hair and use their large ears as heat radiators by pumping blood into large vessels in the ears, thereby cooling it by increasing the blood's surface area. Elephants also employ behavioral mechanisms to avoid overheating, such as bathing in rivers several times a day.

Ectotherms that receive most of their body heat from the sun also are affected by the surface-area-to-volume relationship. Because of their relatively large surface areas, small lizards can heat and cool themselves rapidly. But, because of their large size, crocodiles are slower to warm up and cool down.

Dinosaurs also must have been affected by the **surface-area-to-volume relationship**, and this must constrain our interpretation of dinosaur metabolism. Large dinosaurs, those that weighed about 1,000 kilograms or more, had very low surface areas relative to their volumes. This has inspired calculations that suggest that very large dinosaurs, especially full-grown sauropods, would have overheated if they had had an endothermic metabolism. Indeed, these dinosaurs would have been inertial homeotherms ("**gigantotherms**"), huge animals with a nearly constant body temperature (homeotherms) due to their large size (which, via the surface-area-to-volume relationship, produces thermal inertia) even though they might have had relatively slow, ectothermic metabolisms (box 14.2). Small dinosaurs, including baby and juvenile sauropods, based on their surface-area-to-volume relationship, could have been either ectotherms or endotherms.

Dinosaur body size considered in light of the surface-area-to-volume relationship makes endothermy in adult sauropods and some of the other larger dinosaurs (those with a body weight well above 1,000 kilograms) seem improbable. It suggests that the largest dinosaurs were gigantotherms, but leaves open the question of the metabolism of small dinosaurs. Indeed, small juvenile dinosaurs may have had a metabolism different from their metabolism later in life at large adult sizes, so many dinosaurs could have been heterotherms.

## Bone Chemistry

The most common form (what chemists call an isotope) of the oxygen atom in nature has an atomic weight of 16 (abbreviated $^{16}O$), whereas the next most common form has an atomic weight of 18 ($^{18}O$). The bone of any vertebrate incorporates both forms of oxygen into its mineral matrix. The relative amount (ratio) of $^{18}O$ to $^{16}O$ in the bone depends on temperature. Some scientists have argued that in an endotherm, the ratio of $^{18}O$ and $^{16}O$ should vary little between the limb bones and the bones in the core of the body (such as vertebrae) because the body temperature is nearly the same at both locations. Conversely, they argue, the temperatures of the extremities and body core are very different in an ectotherm, so the oxygen ratios should be very different at each location.

A consistency of oxygen ratios has been demonstrated in the extremities and core of the skeleton of *Tyrannosaurus rex*, suggestive of endothermy. However, there may be a problem with the basic argument underlying this conclusion. Various studies of living

Box 14.2

## Gigantothermy

Chapter 6 discussed the metabolism of sauropod dinosaurs, identifying them as animals that maintained a nearly constant body temperature by virtue of their great bulk; such creatures are referred to as inertial homeotherms. The term "inertial homeotherm" can be replaced by the more colorful word **gigantotherm**, referring to an animal that maintains a constant, high body temperature by virtue of its large size.

How large does an animal have to be to be a gigantotherm? Recent studies suggest that ectotherms at least as large as 1,000 kilograms are gigantotherms. The most studied example of a living ectothermic gigantotherm is the leatherback turtle. This large reptile lives in the sea, migrating from the tropics to the Arctic oceans on a regular basis. The 1,000-kilogram adult leatherbacks have metabolic rates well below those of comparably sized mammals. Yet gigantothermy enables leatherbacks to stay warm in frigid waters. If anything, the leatherbacks have trouble dumping heat and staying cool in tropical waters and on the warm beaches where they lay their eggs.

BOX FIGURE 14.2
Very large dinosaurs, such as sauropods, were gigantotherms.
(© Mark Hallett. Reproduced with permission of Mark Hallett Paleoart)

The example of the leatherback shows that large size can be conducive to maintaining a constant body temperature. All large vertebrates, especially the large dinosaurs (box figure 14.2), must experience some degree of gigantothermy.

mammals and birds show that temperature varies considerably between their cores and extremities, at least as much as in living alligators. Also, factors other than body temperature, including diet and water consumption, can affect the oxygen isotope ratios in bone. The reasoning behind using oxygen ratios to determine metabolism thus appears to be flawed.

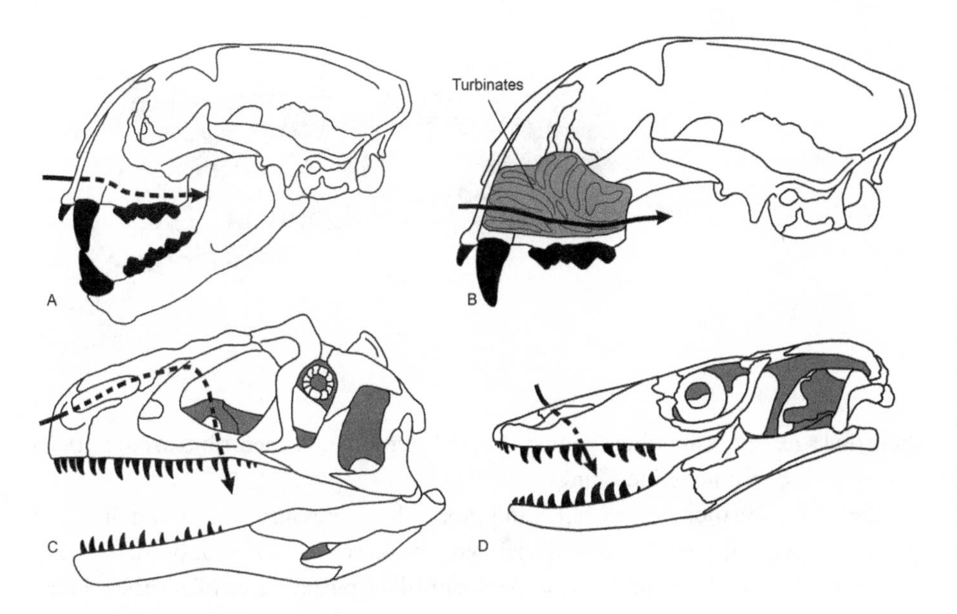


**FIGURE 14.16** Living endotherms (*A*) have respiratory turbinates (*B*) that warm and moisten the air that is inhaled. Theropod dinosaurs (*C*) lacked respiratory turbinates, so air would have flown directly into the mouth, as in a living ectothermic lizard (*D*). (© Scott Hartman)

## Respiratory Turbinates

Living mammals and birds have small bones in their nasal passages called **respiratory turbinates** (figure 14.16). These bones increase the surface area over which blood and moist tissues are exposed to air. By conserving water, respiratory turbinates can thus play a role in the rapid breathing and high rate of oxygen consumption characteristic of the endotherms.

Most dinosaur skulls appear to lack respiratory turbinates, so air would have flowed directly from their nostrils into their mouths without first being warmed or moistened. Some argue that this provides strong evidence that dinosaurs were not endotherms like living birds and mammals. Nevertheless, recent studies show that in living birds, respiratory turbinates are not important for water conservation. These endotherms conserve water in other ways, and endothermic dinosaurs may also have conserved water without using respiratory turbinates. Thus, the case for dinosaur ectothermy based on their lack of respiratory turbinates is not as strong as has been argued by some paleontologists.

## Lungs

Living tetrapods have two kinds of lungs. Mammals have **alveolar lungs** made of millions of tiny, very vascularized, spherical alveoli. Other tetrapods, including birds, have **septate lungs** in which the entire lung is like one huge alveolus and has vascularized septa that penetrate it from its perimeter. Airflow (ventilation) is bidirectional in both kinds of lung, but there are two very different ways the lungs are ventilated. Mammals and crocodiles use a diaphragm muscle between the lung and the pelvis to act like a piston to change lung volume (figure 14.17). In contrast, birds expand and contract

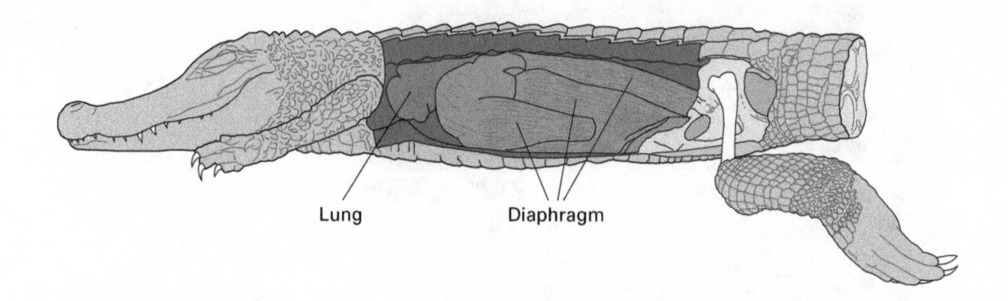


**FIGURE 14.17**
Dinosaurs likely had septate lungs ventilated by a diaphragm muscle, like living crocodiles. (Drawing by Network Graphics)

their rib cage, thus using their jointed ribs and sternum and their rib (costal) muscles to change the volume of their lungs.

Given their relationship to birds and crocodiles, dinosaurs almost certainly had septate lungs. But, they do not have jointed ribs or sternal bones. Indeed, their hip structure suggests dinosaurs used a diaphragm-like piston to ventilate their lungs. *Archaeopteryx* is similar to the dinosaurs in this regard. Thus, the question is, did dinosaurs ventilate their lungs rapidly enough to maintain an endothermic metabolism? In other words, would a diaphragm muscle ventilate a septate lung rapidly enough to provide enough oxygen to an endotherm? This is a tough question to answer, given that no living endotherm ventilates a septate lung with a diaphragm. Lung structure thus seems to tell us little about dinosaur metabolism.

## WHAT TYPE(S) OF METABOLISM DID DINOSAURS HAVE?

Having reviewed many lines of information bearing on hypotheses of dinosaur metabolism, you should see how difficult it is to present a simple answer to the question about what type(s) of metabolism dinosaurs had. The following answers, however, are offered by different paleontologists:

1. All dinosaurs were ectotherms.
2. All dinosaurs were endotherms.
3. Dinosaurs were a diverse group of animals metabolically; some (at least some theropods and ornithopods) were endotherms, and others were ectotherms.
4. Large dinosaurs, the sauropods and big ornithischians, were gigantotherms as adults. Juveniles of these large dinosaurs, and the smaller dinosaurs, may have been either endotherms or ectotherms.

The information presented in this chapter suggests that neither of the first two extreme views of dinosaur metabolism is correct. The third and fourth views, some combination of them or some variant, appear to best explain the evidence. There is, indeed, no simple answer to the question of what type of metabolism the dinosaurs had.

## Summary

1. Warm-blooded vertebrate metabolisms are endothermic and fast.
2. Cold-blooded vertebrate metabolisms are ectothermic and slow.
3. Numerous lines of evidence have been brought to bear on the question of dinosaur metabolism, including posture and gait; speed, activity level, and agility; feeding adaptations; bone microstructure; blood pressure; geographic distribution; bird ancestry; social behavior; predator–prey ratios; body size; the ratio of oxygen isotopes in bone; the presence versus absence of respiratory turbinates; and lung function.
4. Many of these lines of evidence are consistent with endothermy in at least some theropods and ornithopods, but most evidence does not support endothermy in the other dinosaurs.
5. Very large dinosaurs, such as the sauropods, were gigantotherms.
6. The evidence does not support extreme views of dinosaur metabolism; that is, that all dinosaurs were ectothermic or all were endothermic.
7. The evidence suggests a probable variety of metabolisms in dinosaurs, including endotherms and ectotherms, some of which may also have been heterotherms or gigantotherms.

## Key Terms

alveolar lung
blood pressure
bone microstructure
brain complexity
cold-blooded
compact bone
ectothermic
endothermic
"food processor"
four-chambered heart
gigantotherm
heterotherm
homeotherm
hot-blooded
metabolism
"polar night"
predator–prey ratio
relative brain size
respiratory turbinate
septate lung
surface-area-to-volume relationship
warm-blooded

## Review Questions

1. Define the following terms: endotherm, ectotherm, homeotherm, heterotherm, gigantotherm.
2. What are some common misconceptions about ectotherms, and why are they wrong?
3. Which of the 13 lines of evidence presents the most convincing evidence for endothermic dinosaurs? Which presents the weakest?
4. Explain the surface-area-to-volume relationship and its bearing on dinosaur metabolism.
5. What type(s) of metabolism did dinosaurs have? Defend your answer.

## Further Reading

Bakker, R. T. 1975. Dinosaur renaissance. *Scientific American* 232:48–78. (The now classic, original article that argues all dinosaurs were endotherms)

Bakker, R. T. 1986. *The Dinosaur Heresies*. New York: Morrow. (Much of this book argues the case for endothermic dinosaurs)

Cleland, T. P., M. K. Stoskopf, and M. H. Schweitzer. 2011. Histological, chemical, and morphological reexamination of the "heart" of a small Late Cretaceous *Thescelosaurus*. *Naturwissenschaften* 98:203–211. (Re–evaluates the supposed fossil dinosaur heart and concludes that it is inorganic)

Paul, G. 2012. Evidence for avian-mammalian aerobic capacity and thermoregulation in Mesozoic dinosaurs, pp. 819–872, in M. K. Brett-Surman, T. R. Holtz, Jr., and J. O. Farlow, eds., *The Complete Dinosaur*. 2nd ed. Bloomington: Indiana University Press. (Argues the case for dinosaur endothermy)

Reid, R. E. H. 2012. "Intermediate" dinosaurs: The case updated, pp. 873–921, in M. K. Brett-Surman, T. R. Holtz, Jr., and J. O. Farlow, eds., *The Complete Dinosaur*. 2nd ed. Bloomington: Indiana University Press. (Argues the case for unique dinosaur metabolism, somewhere between ectothermy and endothermy)

Ruben, J. A., et al. 2012. Metabolic physiology of dinosaurs and early birds, pp. 785–817, in M. K. Brett-Surman, T. R. Holtz, Jr., and J. O. Farlow, eds., *The Complete Dinosaur*. 2nd ed. Bloomington: Indiana University Press. (Argues the case for dinosaur ectothermy)

Spotila, J. R., M. P. O'Connor, P. Dodson, and F. V. Paladino. 1991. Hot and cold running dinosaurs: Body size, metabolism and migration. *Modern Geology* 16:203–227. (Presents the evidence for dinosaur gigantothermy)

# 15

# DINOSAURS AND THE ORIGIN OF BIRDS

SEVERAL times in this book, dinosaurs are identified as the ancestors of birds. This should have a major effect on our view of the dinosaurs. Instead of thinking of them as huge lizard- or elephant-like animals, we should see dinosaurs as bird-like behaviorally (see chapter 13). Dinosaurs are no longer considered just a long-extinct group of animals because their descendants, the birds, live today as one of the world's most successful animal groups. Indeed, some paleontologists do not make a distinction and simply regard birds as dinosaurs. In this chapter, I develop the evidence for a dinosaur ancestry of birds and discuss some aspects of the evolution of the "feathered dinosaurs."

## WHAT IS A BIRD?

There are about 10,000 species of living **birds**, making them the most diverse group of vertebrates except for the bony fishes. All birds belong to a single class of vertebrates, the **Aves** (from the Latin for "bird"). It is difficult to imagine a person who has not seen a bird and cannot readily distinguish it from other animals. This distinction, of course, is based primarily on the powered flight and feathers that are the hallmarks of birds. The **feathers** of birds are intricate structures that not only form flight surfaces (flight feathers) but also insulate the bird (downy feathers) (figure 15.1).

Like their flight feathers, skeletal features characteristic of birds are associated with flight. The limb bones are hollow and are very lightly constructed (thin walls). Braces and struts within bird long bones provide structural support for their thin walls (figure 15.2), and the long bones are **pneumatic**. This means that air passes through small openings (ducts) in the bones when the bird respires. These **pneumatic ducts** are connected to **air sacs** and the bird's lungs (figure 15.3). Avian air sacs not only lighten a bird, but they also supplement its lungs by acting both as a supercharger that increases

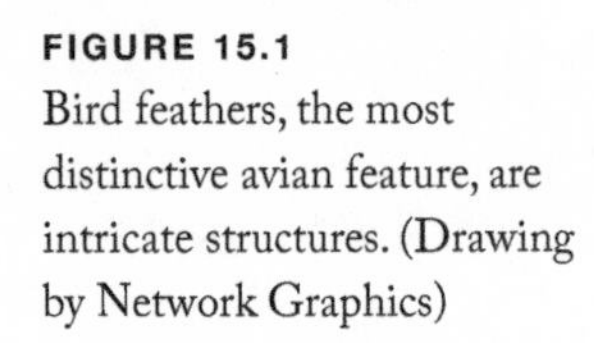

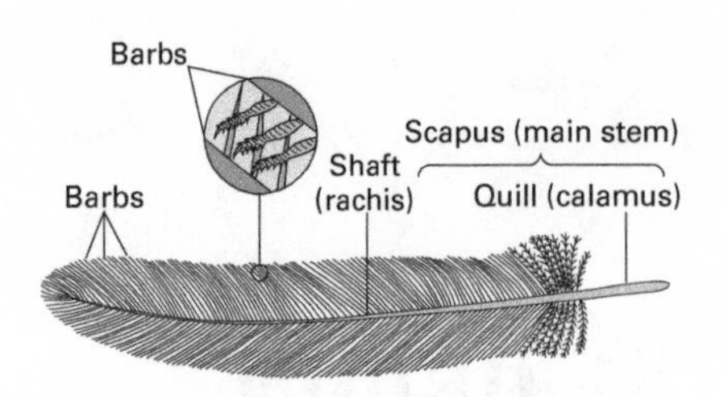


**FIGURE 15.1**
Bird feathers, the most distinctive avian feature, are intricate structures. (Drawing by Network Graphics)

the efficiency of respiration and as a cooling system for the fast avian metabolism. Because birds lack sweat glands, they are strictly air cooled by their lungs and air sacs.

A second unique aspect of the avian skeleton is the fusion of many bones to form more rigid structures, which is particularly evident in the hind limb, pelvis, sacrum, and skull (figure 15.4). For example, the collarbones (clavicles) are fused to form a single bone, the **furcula** (the "wishbone"). Such fused structures contain far fewer separate bones than the same structures in most other vertebrates, including theropod dinosaurs.

Other distinctive features of all but the most primitive birds include modifications of the forelimb skeleton to form wings (figure 15.5). The sternum (breastbone) is very large and keeled. The coracoid bone is very long and forms a brace with the sternum, and a large groove for the passage of a large tendon is present in the shoulder girdle. These structures are designed to anchor the huge and powerful chest muscles needed to flap the wings for flying. The limb itself is modified to form a wing. The carpals are fused so that the wrist is a simple hinge joint. The digits and metacarpals also are fused, and the joints at the wrist and elbow are modified to allow folding of the wing. Bumps on the ulna anchor flight feathers.

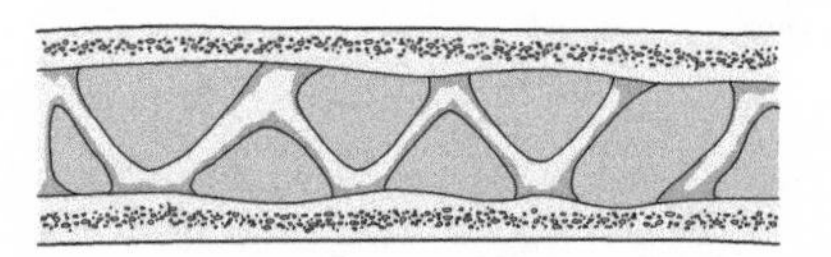

**FIGURE 15.2**
Avian long bones are hollow, have thin walls, and are braced by struts. (Drawing by Network Graphics)

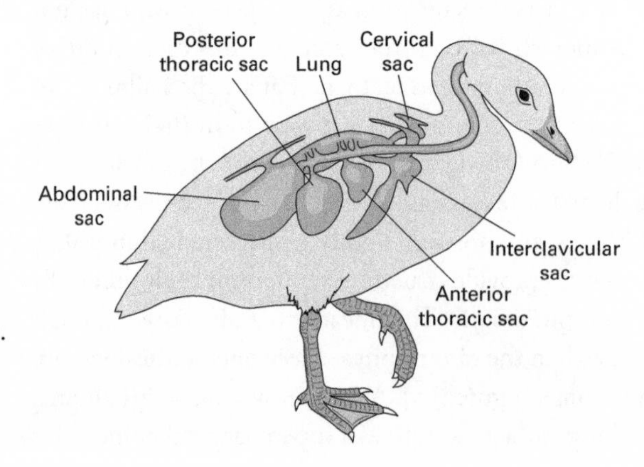


**FIGURE 15.3**
The respiratory system of a bird is more than just its lungs. It includes air sacs connected to hollow bones. (Drawing by Network Graphics)

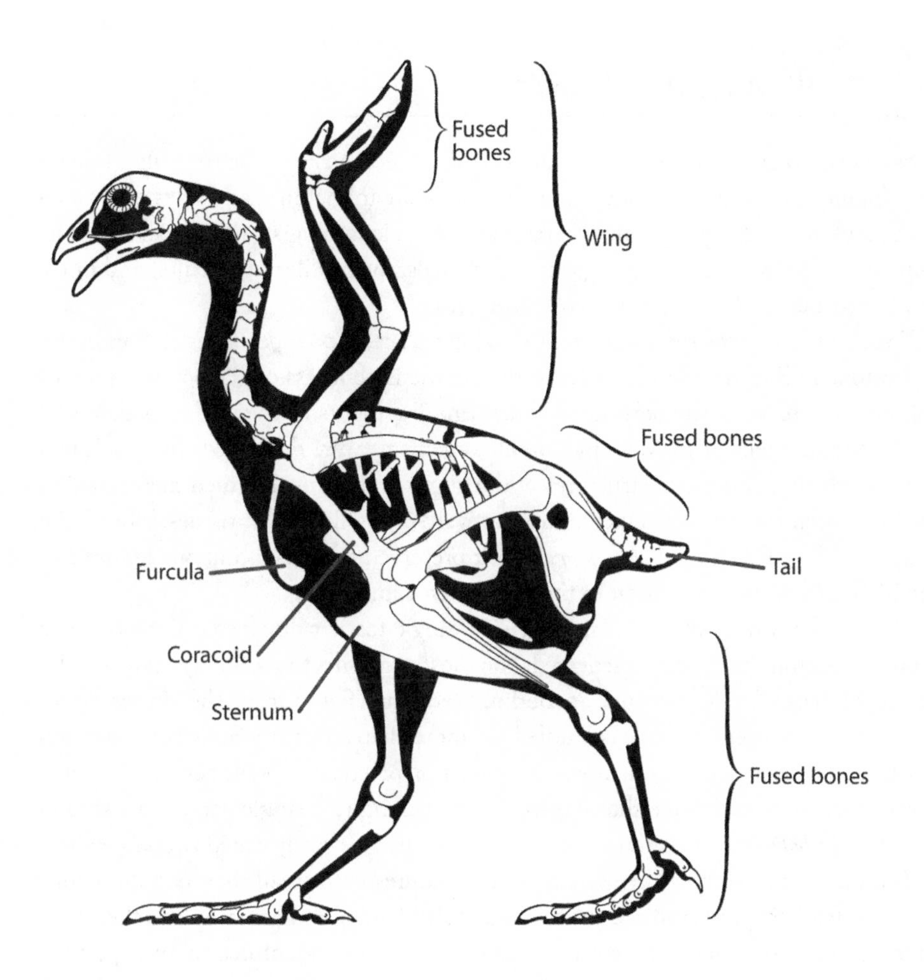


**FIGURE 15.4**
This skeleton of a chicken shows the fused bones in the limbs and hip that are characteristic of birds. (© Scott Hartman)

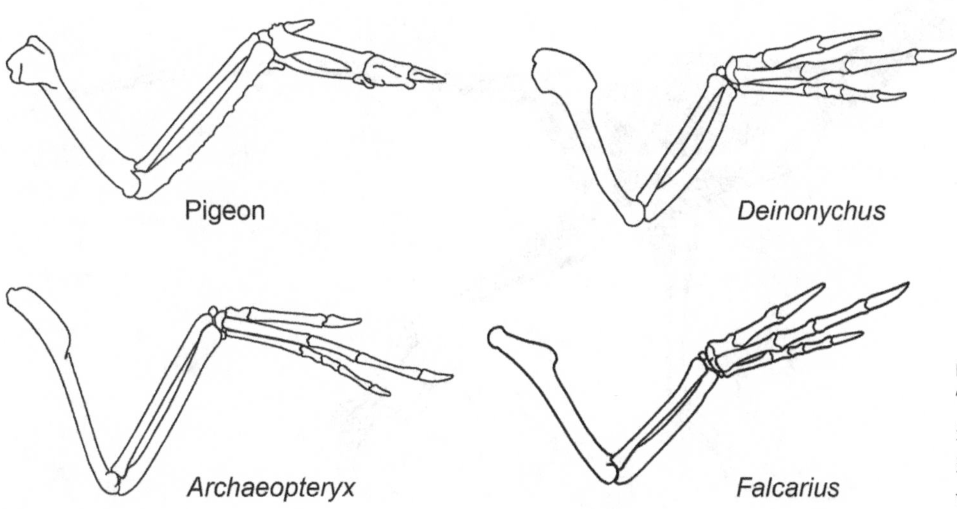


**FIGURE 15.5**
The shoulder girdle and forelimb skeleton of a bird is highly modified to form a wing. (© Scott Hartman)

## THE GENUS *ARCHAEOPTERYX*

Most paleontologists consider the Late Jurassic *Archaeopteryx* to be the oldest known bird (figure 15.6). It provides a unique glimpse of an animal in many ways part dinosaur, part bird. *Archaeopteryx* thus represents what paleontologists would call a "transitional form" between two major groups of animals, the reptiles (dinosaurs) and birds. In popular terms, *Archaeopteryx* is a "**missing link.**"

Fossils of *Archaeopteryx* come from only one place, the Upper Jurassic **Solnhofen Limestone** in Bavaria, Germany. There, during the Late Jurassic, a lagoon was present behind coral reefs, at the bottom of which tiny fragments of calcium carbonate and the carbonate shells of microscopic marine organisms were deposited. The result was the preservation of delicate structures of the larger organisms that died there, making the Solnhofen Limestone an incredible graveyard of fossilized soft tissues. Indeed, the hallmark of the fossils of *Archaeopteryx*, their preserved feathers, is but one example of the high quality of fossilization in the Solnhofen Limestone.

The first known fossil of *Archaeopteryx* is a single feather discovered in 1860. Since then, 11 skeletons have been discovered. The most exquisite, found in 1877, now resides in the Museum für Naturkunde in Berlin, Germany (figure 15.7). The *Archaeopteryx* specimens have been intensively studied for more than a century, and more has been written about them than about any other collection of a dozen fossils pertaining to one genus. Indeed, not all agree that the specimens belong to a single genus, so a second genus, *Wellnhoferia*, has been named for some fossils previously called *Archaeopteryx*.

The skeleton of *Archaeopteryx* shares many features with small theropod dinosaurs. Without feathers preserved, we might reasonably identify *Archaeopteryx* as a theropod. Indeed, some specimens of *Archaeopteryx* were originally identified as theropod (or pterosaur) before their feather impressions were recognized.

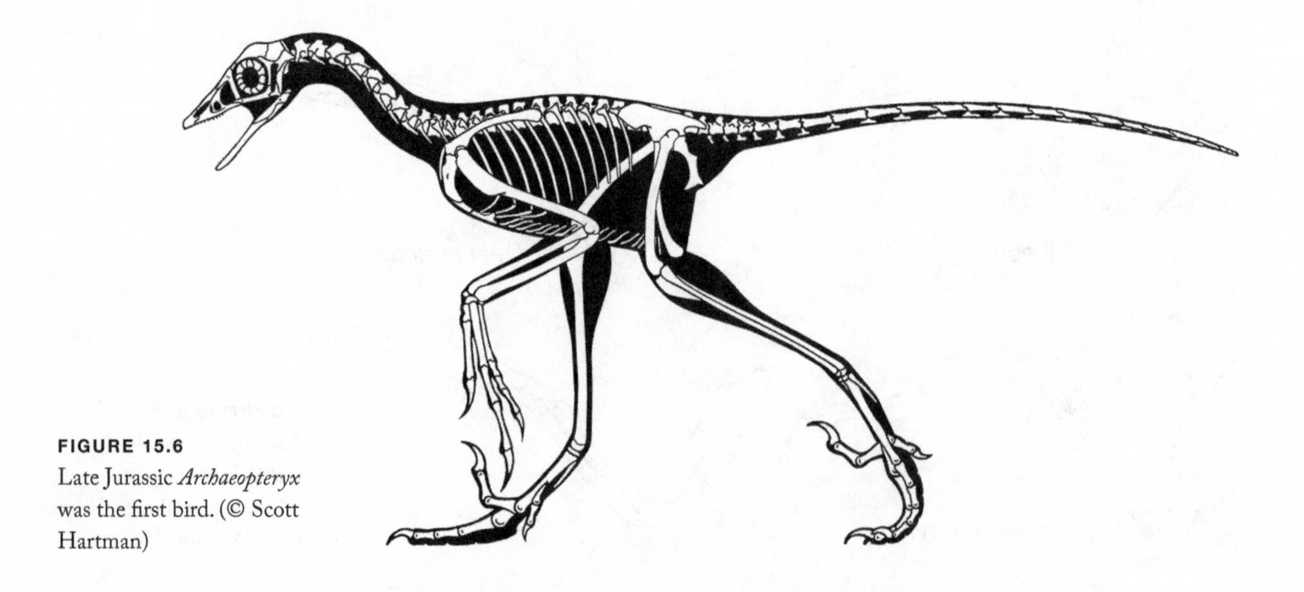

**FIGURE 15.6**
Late Jurassic *Archaeopteryx* was the first bird. (© Scott Hartman)

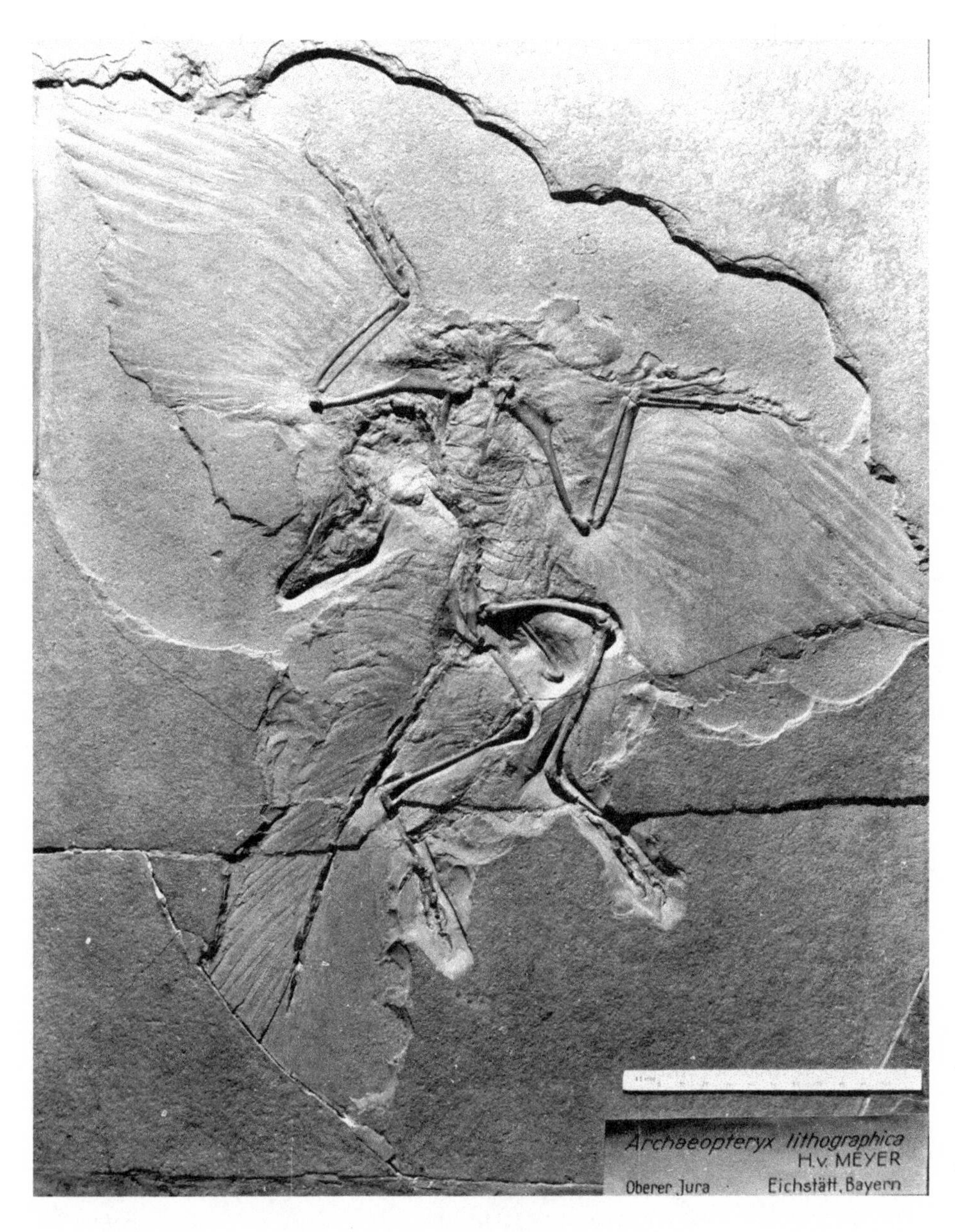


**FIGURE 15.7**
The Berlin specimen of *Archaeopteryx* is the most exquisitely preserved fossil of the first bird. (Courtesy John Ostrom, Yale Peabody Museum of Natural History)

*Archaeopteryx* was about the size of a crow or pigeon. It had a wingspan of about half a meter and an estimated weight of 300 to 500 grams. Its skull had teeth, and very few of its bones were fused to each other—dinosaurian, not avian features. The hollow limb bones of *Archaeopteryx* had thick walls without pneumatic openings, unlike the thin-walled, pneumatic limb bones of later birds.

No sternum (breastbone) is known for *Archaeopteryx*, so it apparently lacked the main anchor for flight muscles seen in other birds. Although *Archaeopteryx* had a furcula (distally fused clavicles) like other birds, this feature was also present in many theropods (see chapter 5). The shape of the pelvis of *Archaeopteryx*—the pubis apparently pointed posteriorly—was very avian and also resembled that of some theropods. The long, bony tail of *Archaeopteryx* also sets it apart from other birds.

The forelimb of *Archaeopteryx* was theropod-like in some features. It had three fully developed digits and a flexible and unmodified wrist, radius, and ulna. Yet, the forelimb of *Archaepteryx* was also very avian. It had claws adapted to tree climbing and bore an elliptical wing made of asymmetrical flight feathers.

The hind limb of *Archaeopteryx* was also very bird-like. The head of the femur was turned medially, and the knee and ankle joints acted as hinges. The fibula was reduced, the proximal tarsals were fused to the tibia, and the distal tarsals were fused to the metatarsals. The long metatarsals were partly fused to each other, and there were three slender, forward-facing digits, as well as a fourth digit that faced backward. *Archaeopteryx* evidently used its hind feet for climbing and perching. Many of these avian features of the hind limb of *Archaeopteryx* were also present in some small theropods. Compare them, for example, in *Archaeopteryx* and *Compsognathus*, a theropod of similar size contemporaneous with *Archaeopteryx* (see figure 5.9).

There was actually only one feature of *Archaeopteryx* that long identified it as a bird. This was its feathers, which in structure and arrangement clearly ally *Archaeopteryx* with birds. To call *Archaeopteryx* a "feathered dinosaur," or, in other words, to recognize it as an animal with an essentially theropod skeleton and avian feathers, is a reasonable conclusion.

## NONDINOSAURIAN ANCESTORS OF BIRDS

Although most paleontologists agree that dinosaurs were the ancestors of birds, a vocal group of dissenters exists. Some of these paleontologists argue that birds evolved directly from Triassic archosaurs more primitive than dinosaurs. This old idea is difficult to disprove because Triassic archosaurs certainly were sufficiently primitive to have been the ancestors of birds. But, to argue for a Triassic archosaur as the ancestor of birds is to ignore the many evolutionary novelties shared by dinosaurs and birds that indicate they are more closely related to each other than either is to a Triassic archosaur. Indeed, those who so argue also claim that the similarities of dinosaurs and birds are largely the result of convergent evolution (box 15.1).

### Box 15.1

#### *Scansoriopteryx*

Although first suggested in the nineteenth century, the dinosaurian ancestry of birds did not gain much credence until the 1970s. At that time, paleontologist John Ostrom drew attention to the many similarities shared by the skeletons of *Archaeopyteryx* and theropod dinosaurs. He thus demonstrated that *Archaeopteryx* had a basically theropod skeleton, with a few specializations seen in birds, and the asymmetrical flight feathers distinctive of birds. Subsequent cladistic analyses and discoveries of "feathered dinosaurs" in the Lower

Cretaceous of northeastern China, including *Caudipteryx* and *Microraptor*, decided the dinosaurian origin of birds in the mind of most paleontologists. Yet, there remains a group of scientific skeptics who argue that the dinosaur–bird similarities are the result of evolutionary convergence inherited from very separate ancestors. To them, bird ancestry must be sought in archosaurs other than the dinosaurs.

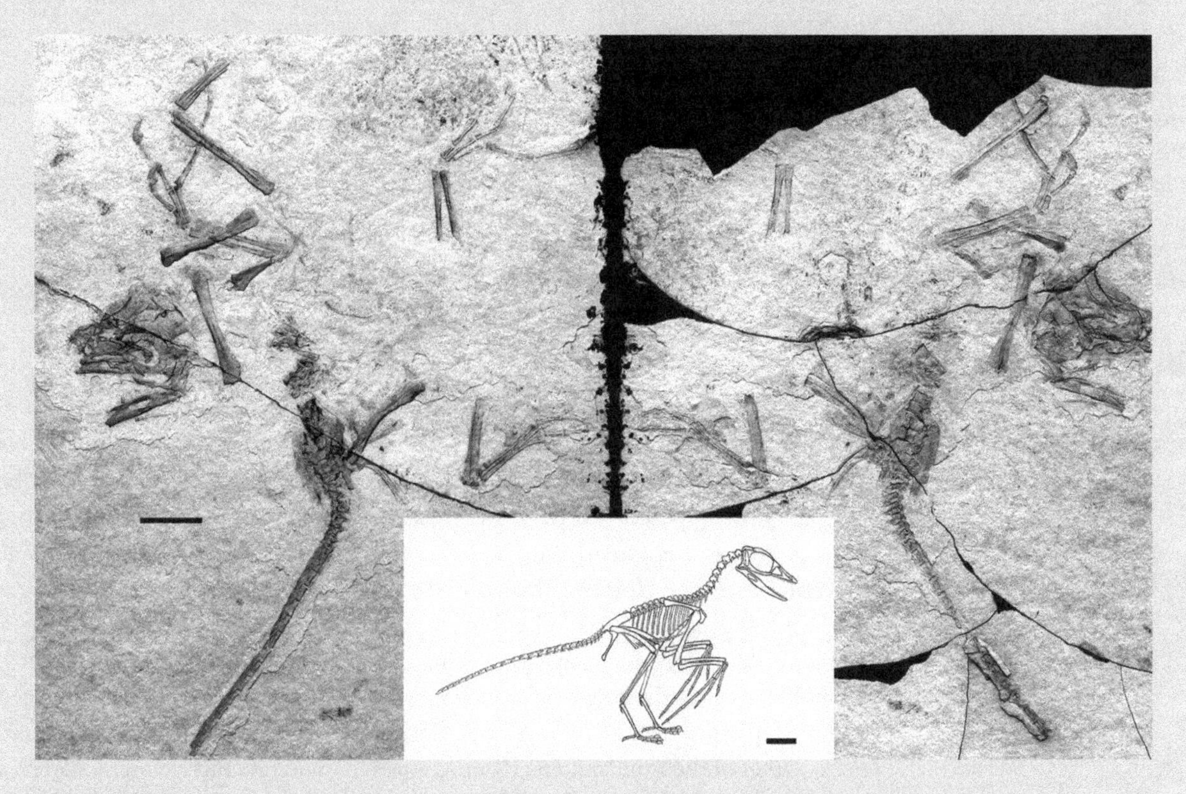

**BOX FIGURE 15.1**
*Scansoriopteryx*, from the Middle Jurassic of China, has been identified as a maniraptoran dinosaur or an early bird. (Photograph and illustration © Stephen Czerkas)

The latest addition to this discussion is ***Scansoriopteryx***, a Middle Jurassic animal first described as a maniraptoran dinosaur (box figure 15.1). In a recent restudy of this animal, paleontologist Stephen Czerkas and ornithologist Alan Feduccia argue that *Scansoriopteryx* is a bird older than *Archaeopteryx*. They assign it to Aves based on its perching foot, gliding wing, and asymmetrical wing feathers. Furthermore, Czerkas and Feduccia draw attention to the nondinosaurian features of *Scansoriopteryx*, including the relative lengths of its arm and hand, the digit proportions of its hand, and its hip socket structure, which they point out is very different from the hip socket of upright walking dinosaurs. *Scansoriopteryx* not only has feathered wings and arms, but a feathered hind wing. Czerkas and Feduccia argue that *Scansoriopteryx* was a small arboreal bird not derived from dinosaurs. They see it as a four-winged arboreal glider that represents an early stage of avian evolution prior to *Archaeopteryx*. As Czerkas and Feduccia conclude, "The origin of Aves is all the more complex because convergence within Dinosauria has obscured the true avian ancestry that, instead of being from within Dinosauria, is actually from nondinosaurian archosaurs." The debate goes on!

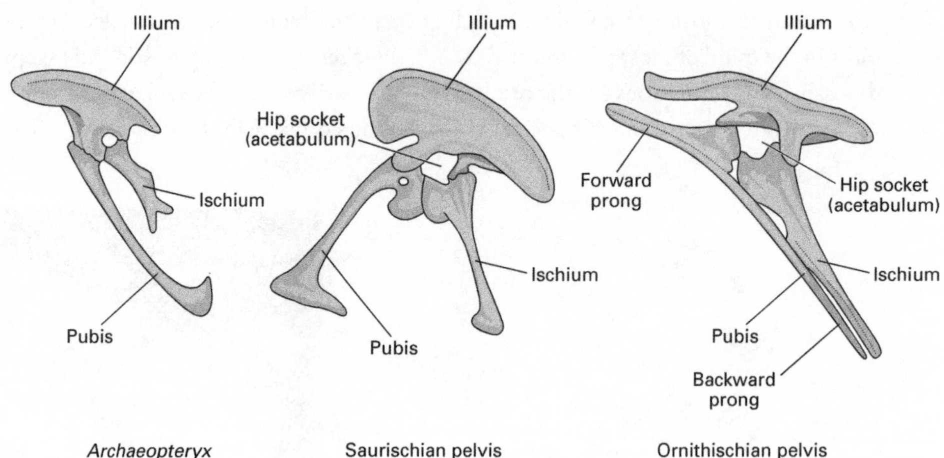


**FIGURE 15.8**
The avian pelvis is a modified saurischian pelvis that superficially resembles the ornithischian pelvis. (© Scott Hartman)

One challenge to the dinosaurian origin of birds comes from the claim that the three digits of the theropod hand are not the same digits as those in the hand of a bird. The dinosaur digits are supposedly I-II-III (the "thumb" is digit I), whereas the bird wing digits are supposedly II-III-IV. This means some evolutionary convergence must have occurred to produce strikingly different forelimbs in theropod dinosaurs and birds (see figure 15.5). However, this claim has been countered by analysis that suggests that such a shift in digits could be achieved by a relatively minor gene change in the embryonic development of the earliest birds.

Two ideas about the origin of birds to avoid are that the ornithischian dinosaurs and the pterosaurs were avian ancestors. Certainly, the pelvis of birds does somewhat resemble that of the ornithischian ("bird-hipped") dinosaurs. But a closer look reveals that the avian pelvis is actually a highly modified saurischian pelvis (figure 15.8). Thus, the pubis of the saurischian pelvis was shifted backward, the ischium was shortened, and the posterior portion of the ilium was lost to produce an avian pelvis.

Pterosaurs were flying archosaurs that first appeared during the Late Triassic and became extinct at the end of the Cretaceous. Pterosaurs lacked feathers and the evolutionary novelties shared by dinosaurs and birds. As noted in chapter 4, pterosaurs were archosaurs closely related to dinosaurs, but that's as close as their relationship was to birds. Flight must have evolved twice, quite separately, among the archosaurs—once in pterosaurs and once in birds.

## ORIGIN AND EVOLUTION OF AVIAN FLIGHT

Because the hallmarks of birds, including *Archaeopteryx*, are their feathers and other skeletal modifications for flight, it is natural to ask how avian flight originated. Two major hypotheses have been advanced and remain the subject of discussion.

The oldest idea is that the immediate ancestors of *Archaeopteryx* lived in the trees; they were arboreal. These arboreal bird ancestors evolved feathers to glide from branch to branch and to the ground. Flight powered by flapping the wings evolved later. Indeed, Archaeopteryx itself has claws that indicate it was capable of climbing trees. This hypothesis of the origin of avian flight can be termed the **arboreal hypothesis**.

An alternative hypothesis is that *Archaeopteryx* and its immediate ancestors were ground-living, fast runners (cursorial). According to this hypothesis, feathers evolved to insulate the dinosaur and, on the arms, may have provided additional surface area to aid in catching insects (figure 15.9). Enlargement of these feathers followed to provide support and thrust when the newly evolved wings were flapped. This hypothesis can be termed the **cursorial hypothesis**.

A criticism of the arboreal hypothesis points out that the wings of gliders, such as "flying" squirrels, are very different from those of birds. For example, in gliders, the wing membrane is attached to the body and extends between the fore and hind limbs, very different than the wing membrane of birds. Indeed, *Archaeopteryx* was not an arboreal glider, so no evidence is available for this gliding stage of evolution required by the arboreal hypothesis.

The arboreal hypothesis argues that birds took advantage of gravity, evolving at small size to take advantage of the "cheap energy" it provides when jumping out of a tree. Indeed, flight in arboreal gliders (lizards, snakes, mammals), pterosaurs, and bats has long been agreed to have started in high places. The relatively recent discovery of small, feathered theropods, such as *Microraptor*, that look like they could climb trees, also lends support to the arboreal origin of avian flight.

The principal criticism of the cursorial hypothesis is that it is challenged by gravity. Furthermore, the development of wings and feathers on the ground apparently indicates they first evolved for some function other than flying. This led to the idea that catching insects with feathered hands may have originally driven the evolution of the feathered wing. But, some paleontologists have difficult imagining what adaptive advantage the early stages of flapping the arms would have conferred on theropods. If anything, this might have slowed a theropod trying to catch insects or running away from a predator.

As is often the case with two competing ideas in science, the correct answer may blend portions of both. The currently known array of maniraptoran theropods and other possible bird ancestors includes animals capable not only of ground running but also of running up into trees, climbing in trees, and gliding from branches

**FIGURE 15.9**
Two hypothetical stages in the cursorial origin of avian flight show a proto-bird catching insects with feathered hands (*above*) and a later, winged proto-bird like *Archaeopteryx* (*below*). (Drawing by Network Graphics)

(box 15.2). Feathers may have first evolved as insulation and camouflage and later developed further into aerofoils used in flying. In other words, the early steps of bird origins may have moved from the ground and then into the trees, thus combining elements of both hypotheses.

Box 15.2

## Flightless Descendants of Birds?

The discovery of very bird-like theropods in the Lower Cretaceous lake beds of northeastern China was at first heralded as evidence establishing dinosaurs as the ancestors of birds. Indeed, dromaeosaurs, such as *Microraptor*, had asymmetrical flight feathers, the hallmark of birds. If such feathers diagnose birds, how could such dromaeosaurs be distinguished from birds? Isn't *Microraptor* (and similar feathered dromaeosaurs) a bird?

Some paleontologist say yes! They identify many of the bird-like features of dromaeosaurs, such as *Microraptor*, as evidence that dromaeosaurs are descendants of birds, not dinosaurs. Indeed, some paleontologists have suggested that many of the maniraptorans are secondarily flightless descendants of birds. But, not all paleontologists agree with this.

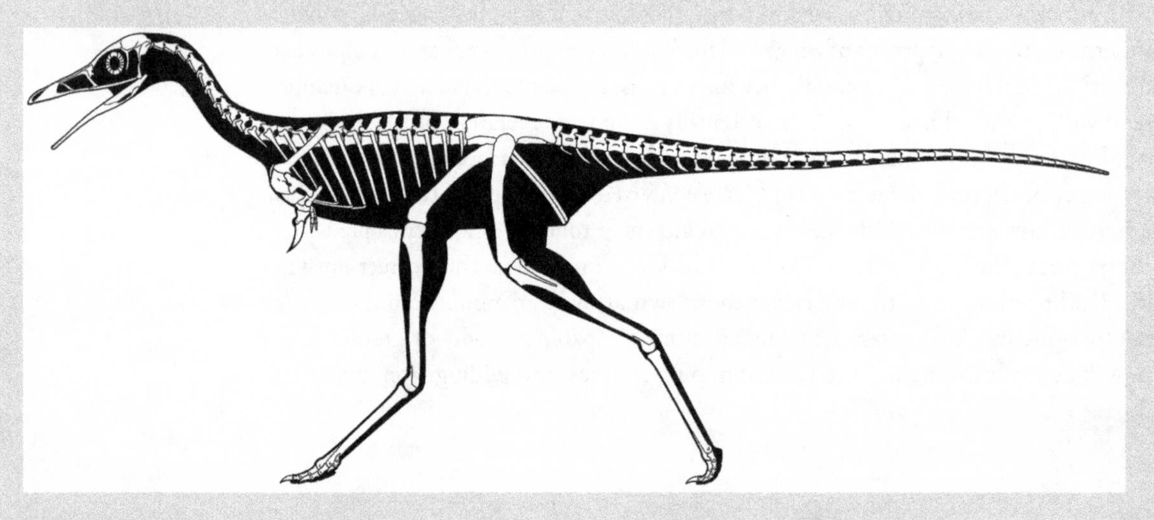

**BOX FIGURE 15.2**
*Mononykus*, from the Upper Cretaceous of Asia, has been called a flightless bird or a theropod dinosaur. (© Scott Hartman)

One of the most bizarre of the maniraptorans is ***Mononykus***, from the Upper Cretaceous of Mongolia, with its very peculiar, spike-like forelimbs (box figure 15.2). Although originally described as a flightless bird, some paleontologists argue that *Mononykus* is a bizarre theropod dinosaur. When it comes to deciding whether some maniraptorans were birds or dinosaurs—the choice can truly be difficult! Stay tuned for more on this subject.

## EVOLUTION OF BIRDS

A book on dinosaurs would be remiss if it did not include a brief overview of the evolution of the descendants of dinosaurs, the birds.

Despite claims of older birds, most paleontologists regard Late Jurassic *Archaeopteryx* as the oldest known bird. Cretaceous birds were much more advanced than *Archaeopteryx*, although some still had teeth (figure 15.10). The past few decades have seen the discovery of many Early Cretaceous birds from China (box 15.3). These include long-tailed, clumsy fliers like ***Confuciusornis***, perhaps the best known Cretaceous bird represented by more than 1,000 fossil specimens (figure 15.11).

Indeed, a diverse group of early birds appeared during the Cretaceous, the **enantiornithines.** These are the so-called "opposite birds" because the articulation of their shoulder bones, the coracoid and scapula, appears to be opposite to the anatomy seen in modern birds. Enantiornithine bird fossils are best known from the Cretaceous of

**FIGURE 15.10**
Wingless diving birds of the Early Cretaceous, such as *Baptornis*, still had teeth. (Drawing by Network Graphics)

### Box 15.3

### Chinese Cretaceous Birds

During the Early Cretaceous, northeastern China was a land of towering volcanoes and deep lake basins. Paper-thin layers of shale and limestone deposited in these lakes preserved a remarkable fossil record, including a surprising diversity of very modern-looking early birds. The first of these birds to be described, the sparrow-size *Sinornis*, was a fully powered flier capable of perching in trees (box figure 15.3). This primitive bird had a very archaic skull and skeleton that included teeth, a clawed hand, and limited fusion of the limb bones. Yet *Sinornis* flew and perched very much like a modem bird. There is a huge evolutionary distance between *Sinornis* and the oldest bird, *Archaeopteryx*; evidently, during the 25 million years that separates them, much evolution took place in the early history of birds.

(*continued*)

**BOX FIGURE 15.3**
*Sinornis*, from the Lower Cretaceous of China, was a surprisingly advanced bird. (© Scott Hartman)

*Sinornis* was only one of many new birds recently described from the Lower Cretaceous of northeastern China. One of the most remarkable of these is *Confuciusornis*, known from hundreds of specimens preserved with feather impressions. *Confuciusornis* is about the same size as *Archaeopteryx*. But, even though *Confuciusornis* is the oldest known toothless bird, it still has a clawed hand and some other primitive features. *Confuciusornis* is thus a classic example of what has been called mosaic evolution; it has a mixture of primitive (dinosaur-like) and advanced (modern bird–like) features. *Confuciusornis, Sinornis*, and other Early Cretaceous birds from China, as well as other Early Cretaceous birds from Spain and Mongolia, attest to an explosive early evolution of birds after their first appearance during the Late Jurassic. Birds were clearly very successful from the start.

China, Argentina, and Spain. ***Sinornis***, from the Lower Cretaceous of China, is a representative enantiornithine.

During the Early Cretaceous, highly specialized, flightless diving birds also evolved. This early diversity among birds may indicate an ancestry older than *Archaeopteryx* or that other Late Jurassic birds remain to be discovered. A major change took place by the end of the Cretaceous when the enantiornithines totally disappeared, and

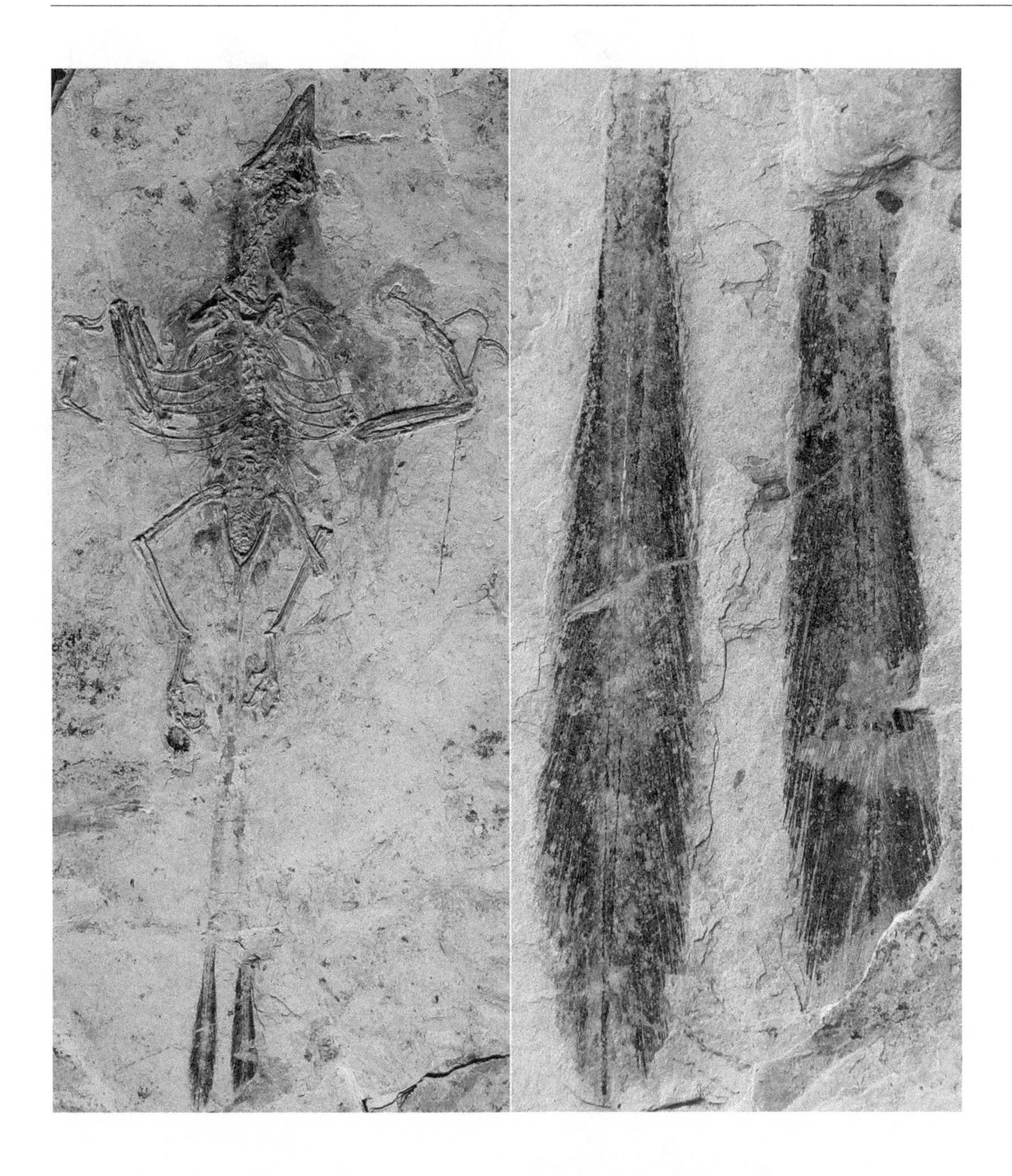

**FIGURE 15.11**
A superbly preserved *Confuciusornis* from the Lower Cretaceous of northeastern China (*left*) with detail of its tail feathers (*right*). (Courtesy Zhang Zihui)

primitive, toothed birds also became extinct. Indeed, one paleontologist has referred to the Cretaceous as the "age of archaic birds." Truly modern, toothless birds appeared by the beginning of the Cenozoic. During the past 66 million years, birds have been one of the most successful groups of vertebrates (figure 15.12).

Birds have delicate, hollow bones; most are small animals; and many live in dry or forested environments. For these reasons, the fossil record of birds is less extensive than that of reptiles and mammals. Yet, despite this, we have fossils of water birds, land birds, and giant, flightless birds (**ratites**) at least as far back as the Paleocene. The giant, flightless birds are particularly interesting because they evolved from flying ancestors and so represent a return to the flightless habits of the theropod ancestors of birds (figure 15.13).

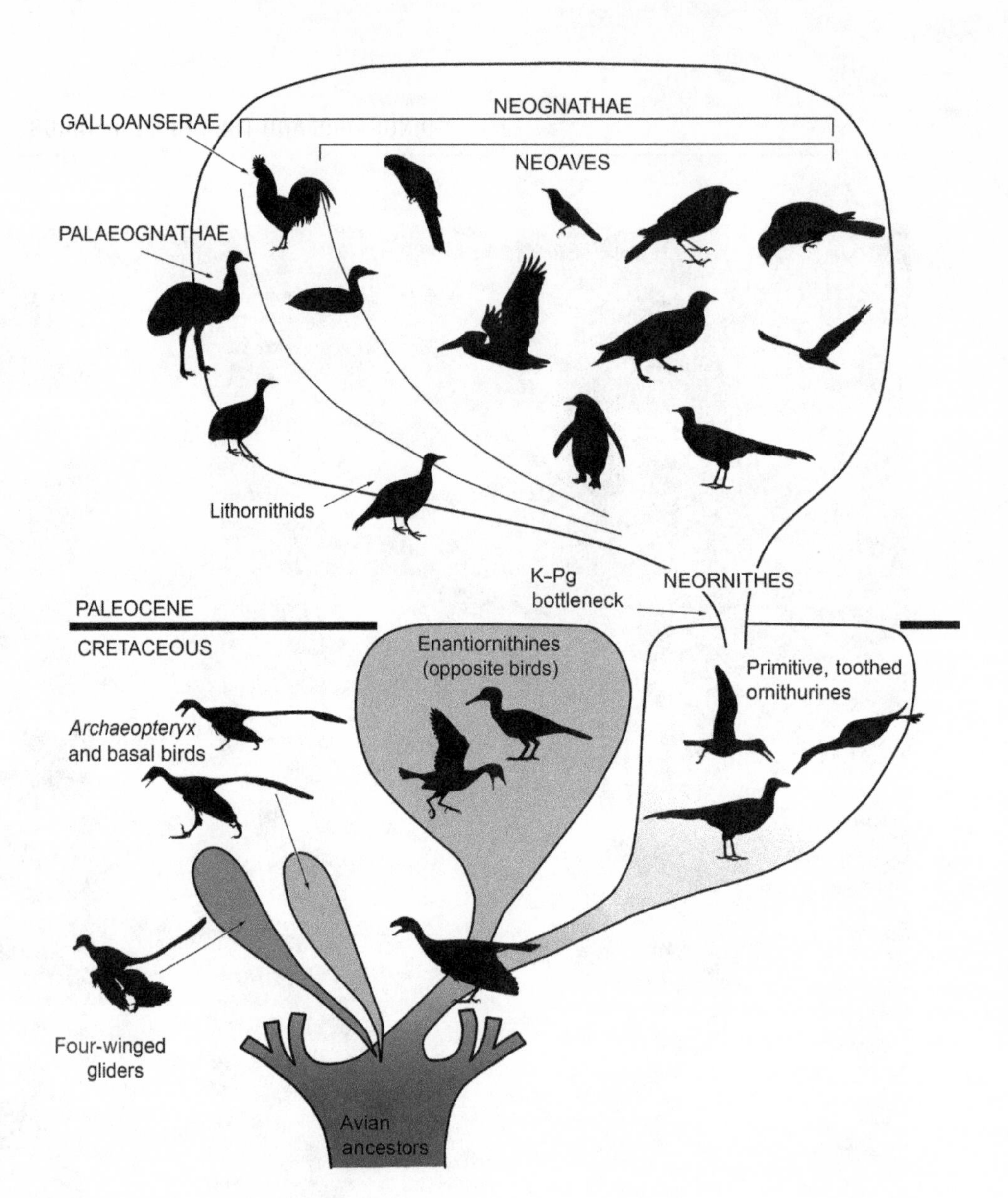


**FIGURE 15.12**
Birds have been very diverse and successful and underwent a Cretaceous diversification of archaic birds, followed by a second, Cenozoic, diversification of modern birds. (© Scott Hartman)

**FIGURE 15.13**
Giant, flightless birds, such as *Diatryma*, appeared as early as 58 million years ago. (Drawing by Network Graphics)

## SIGNIFICANCE OF DINOSAURS AS BIRD ANCESTORS

As stated at the beginning of this chapter, identifying theropod dinosaurs as the ancestors of birds should have a major impact on how we view the dinosaurs. We should no longer think of them as reptile-like in many aspects of their biology. Instead, much about dinosaur biology and behavior was very avian, as explained elsewhere in this book.

If dinosaurs were so bird-like, should we classify them with reptiles? Perhaps not, although various answers to this question have been proposed (box 15.4). Regardless of the classification used, the close relationship of birds to dinosaurs is well established.

### Box 15.4

### Birds as Dinosaurs

The cladogram of theropod dinosaurs presented earlier in this book (see figure 5.1) can be modified slightly to show birds where most paleontologists believe they belong, as close relatives of the maniraptoran theropods (box figure 15.4). If this cladogram is turned directly into a classification, that classification would simply identify birds as a particular type of tetanuran. In this classification, birds are just another type of dinosaur.

Some paleontologists have taken such a classification to heart. They cast doubt on the significance of "dinosaur extinction," because, after all, the avian dinosaurs (= feathered dinosaurs, = birds) are still with us, some 10,000 species strong (see chapter 16). These paleontologists speak flippantly of "carving the dinosaur" at a Thanksgiving dinner. In a more serious vein, they argue that putting the birds in a class distinct from that which contains the dinosaurs obscures their close relationship.

This may be the case, but putting birds into the same class as dinosaurs also obscures the key evolutionary novelties of birds: their feathers, skeletal features, and physiological mechanisms for sustained, powered flight. Despite the many similarities of theropod dinosaurs to birds, and their shared evolutionary novelties, birds are unique vertebrates. It is this uniqueness, and the great diversity of birds, that long ago led to their recognition as a separate class of vertebrates called Aves. For these reasons, some paleontologists prefer to distinguish this class from dinosaurs. But, others place dinosaurs (or at least theropod dinosaurs) in the Aves, and, as mentioned earlier, still others abandon Aves as a separate class and include birds in the Theropoda. Another solution was to introduce the term "Avialae" to encompass all the early birds, including *Archaeopteryx*, as distinct from birds as we know them, the Aves.

(*continued*)

These differing classifications of birds reflect different philosophies of how a phylogeny should be turned into a classification; they do not reflect different ideas on the phylogenetic relationships of birds. All but a few paleontologists agree that the birds, or the nondinosaurian Aves, or the avian theropods, or whatever you choose to call them, were descended from theropod dinosaurs.

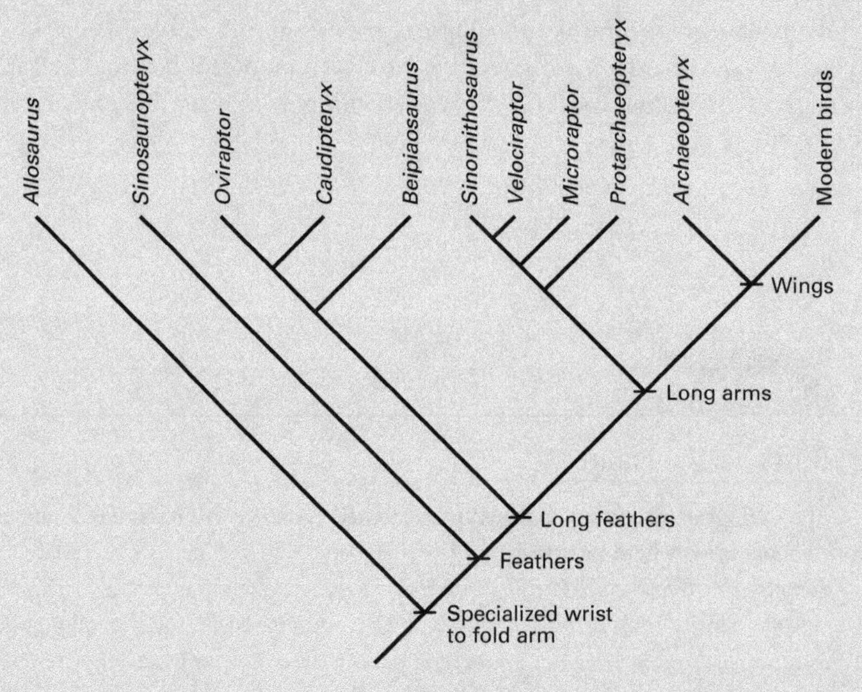


**BOX FIGURE 15.4**
This cladogram indicates that birds are closely related to dromaeosaurs. (Drawing by Network Graphics)

## Summary

1. Birds are feathered vertebrates with skeletons highly modified to be lightweight and rigid to enable sustained powered flight.
2. *Archaeopteryx*, from the Upper Jurassic of Germany, is the oldest known bird.
3. Many skeletal features of *Archaeopteryx* are very similar to those of small theropod dinosaurs.
4. Evolutionary novelties of the skeleton shared by theropod dinosaurs and birds provide strong evidence that birds descended from dinosaurs.
5. Some paleontologists argue that the dinosaur–bird similarities are the result of evolutionary convergence and that the ancestry of birds lies among more primitive archosaurs.

6. There are two hypotheses of the origin of avian flight: the arboreal and the cursorial hypotheses.
7. The correct hypothesis of the origin of avian flight may blend portions of both the arboreal and cursorial hypotheses by recognizing possible bird ancestors capable not only of ground running but also of running up into trees, climbing in trees, and gliding from branches.
8. Modern birds evolved by the beginning of the Cenozoic.
9. Recognition of dinosaurs as the ancestors of birds forces us to rethink much of dinosaur anatomy and behavior and stress their bird-like aspects.

## Key Terms

air sac
arboreal hypothesis
*Archaeopteryx*
Aves
bird
*Confuciusornis*
cursorial hypothesis
enantiornithine
feather
furcula
"missing link"
*Mononykus*
pneumatic
pneumatic duct
ratite
*Scansoriopteryx*
*Sinornis*
Solnhofen Limestone

## Review Questions

1. How is a bird's skeleton designed to make powered flight possible?
2. What evolutionary novelties were shared by theropod dinosaurs and birds?
3. Why do most paleontologists consider *Archaeopteryx* to be a bird? What similarities does it show to theropod dinosaurs?
4. What other possible ancestors of birds have been identified, and what might they suggest about the similarities of dinosaurs and birds?
5. Compare and contrast the arboreal and cursorial hypotheses of the origin of avian flight.
6. What do Cretaceous birds suggest about the origin and early diversification of the Aves?
7. What impact does their recognition as the ancestors of birds have on our ideas about dinosaurs?

## Further Reading

Chiappe, L. M. 2007. *Glorified Dinosaurs: The Origin and Early Evolution of Birds.* Hoboken, N.J.: Wiley. (Book-length treatment of the origin of birds from dinosaurs)

Currie, P. J., et al., eds. 2004. *Feathered Dragons: Studies on the Transition from Dinosaurs to Birds.* Bloomington: Indiana University Press. (A collection of 14 articles on theropods and birds)

Czerkas, S. A., and A. Feduccia. 2014. Jurassic archosaur is a non-dinosaurian bird. *Journal of Ornithology* 155:841–851. (Argues that *Scansoriopoteryx* is a bird not descended from dinsoaurs)

Feduccia, A. 2014. Avian extinction at the end of the Cretaceous: Assessing the magnitude and subsequent explosive radiation. *Cretaceous Research* 50:1–15. (Argues for a major extinction of birds at the end of the Cretaceous followed by the explosive diversification of modern birds)

Feduccia, J. A. 1996. *The Origin and Evolution of Birds*. Cambridge, Mass.: Harvard University Press. (A comprehensive review of the fossil record of birds, with a particularly detailed treatment of their Cenozoic evolution)

Feduccia, J. A. 2014. *Riddle of the Feathered Dragons: Hidden Birds of China*. New Haven, Conn.: Yale University Press. (Book-length treatment of the origin of birds, but from the perspective that they did not evolve from dinosaurs)

Naish, D. 2012. Birds, pp. 379–423, in M. K. Brett-Surman, T. R. Holtz, Jr., and J. O. Farlow, eds., *The Complete Dinosaur*. 2nd ed. Bloomington: Indiana University Press. (Concise review of bird origins and evolution)

Ostrom, J. H. 1976. *Archaeopteryx* and the origin of birds. *Biological Journal of the Linnaean Society* 8:91–182. (A comprehensive and now classic technical article on the origin of birds)

Paul, G. S. 2002. *Dinosaurs of the Air: The Evolution and Loss of Flight in Dinosaurs and Birds*. Baltimore: Johns Hopkins University Press. (Argues that many bird-like theropods are the flightless descendants of birds)

Pickrell, J. 2014. *Flying Dinosaurs: How Fearsome Reptiles Became Birds*. New York: Columbia University Press. (Very readable account of the discoveries, debates, and debacles of the science of the dinosaur–bird connection)

Xu, X., et al. 2003. Four-winged dinosaurs from China. *Nature* 421:335–340. (Provides a description of *Microraptor*)

## Find a Dinosaur!

To find the oldest fossil bird, *Archaeopteryx*, you must look hard, because *Archaeopteryx* fossils are rare. The pigeon-size *Archaeopteryx* remains one of the rarest of fossils, known only from about a dozen specimens collected from Upper Jurassic rocks in southern Germany. The famous Berlin specimen can be seen at the Museum für Naturkunde in Berlin, Germany. At the Natural History Museum in London, England, the London Specimen is on display. Other *Archaeopteryx* fossils on display in Europe are in smaller museums in the Netherlands and Germany. Closer to home, what has been called the Thermopolis specimen is on display at the Wyoming Dinosaur Center in Thermopolis. Excellent replicas (casts) of *Archaeopteryx* abound, particularly of the Berlin and London specimens. They are on display at many museums worldwide.

# 16

# THE EXTINCTION OF DINOSAURS

THE most frequently asked question about dinosaurs is, why (or how) did they become extinct? To attempt to answer that question, we need to examine one of the most highly charged scientific debates about dinosaurs. This debate is ongoing, and it has produced two radically different explanations of dinosaur extinction. One explanation links the extinction of dinosaurs to a single cause: the impact of an asteroid that exploded upon striking the earth about 66 million years ago. The other explanation sees dinosaur extinction as due to multiple causes, including the outgrowth of changes in topography, climate, vegetation, and/or animal life because of falling sea levels and massive volcanism culminating in an asteroid impact at the end of the Cretaceous. Other explanations of dinosaur extinction, some quite well known, are based on much less (or no) evidence than these two possibilities (box 16.1). Therefore, this chapter focuses on the two current hypotheses about the cause of dinosaur extinction.

## THE TERMINAL CRETACEOUS EXTINCTION

Dinosaur extinction was part of a more pervasive extinction that occurred at the end of the Cretaceous, usually called the **terminal Cretaceous extinction**, or the extinction at the Cretaceous–Tertiary boundary (or K–T boundary for short, K standing for "Cretaceous" and T for "Tertiary"). The terminal Cretaceous extinction was only one of several major extinctions in the history of life (figure 16.1). But it was not the largest; that "honor" is held by the extinction that occurred at the end of the Permian, 251 million years ago.

The terminal Cretaceous extinction was not just the extinction of the dinosaurs. It included the extinction of several types of organisms, both in the sea and on land. Paleontologists estimate that at least 15 percent of the families (or approximately 100 families) of shelled marine invertebrates became extinct in the sea during the

## Box 16.1

### Extinction Explanations Without Convincing Evidence

Many explanations of dinosaur extinction lack supporting evidence, but, ironically, these are some of the most widely known explanations. Perhaps the best known explanation is that dinosaurs became extinct because Late Cretaceous mammals ate their eggs (box figure 16.1). Today, very few mammals eat eggs, humans and mongooses notwithstanding. There is no evidence of egg eating by Late Cretaceous mammals. Furthermore, mammals coexisted with dinosaurs throughout the Mesozoic reign of the dinosaurs. Why, then, would **egg-eating mammals** have had an adverse effect on dinosaurs only at the end of the Cretaceous? Egg eating by mammals thus can be rejected as an explanation of dinosaur extinction.

**BOX FIGURE 16.1**
Mammals eating dinosaur eggs is a popular idea about the cause of dinosaur extinction for which there is no evidence. (© Mark Hallett. Reproduced with permission of Mark Hallett Paleoart)

Some explanations of dinosaur extinction rely on extreme changes in climate at the end of the Cretaceous. One argument is that dinosaurs were so large they could not hibernate and were thus unable to cope with extremely cold climates at the end of the Cretaceous. A contrasting argument is that temperatures at the end of the Cretaceous became so hot that dinosaurs literally "roasted" to death. Both of these explanations point to the small size and/or hibernating ability of the mammals and small reptiles that survived the Cretaceous as reasons why they were able to cope with one or the other temperature extreme. The problem is that there is no evidence for either an extremely hot or cold climate at the end of the Cretaceous, although the impact of an asteroid may have produced a geologically short period of intense dark and cold.

Today, some plants, called converters, absorb and concentrate radioactive elements like **selenium**, which are poisonous in large amounts. Large amounts of selenium are present in some volcanic ashes. Thus arose the argument that increased volcanic activity at the end of the Cretaceous produced increased selenium in plants eaten by herbivorous dinosaurs that was then passed on to the meat-eating dinosaurs when they preyed upon the plant-eaters. The theory states that the selenium poisoned the dinosaurs, bringing about their extinction. Of course, a fossil record of the selenium-converting plants does not exist, and no explanation can be offered as to why nondinosaurian plant-eaters, some mammals and nondinosaurian reptiles, were not poisoned. Selenium poisoning is an inadequate explanation of dinosaur extinction.

The longest-running extraterrestrial cause of dinosaur extinction is a **supernova**. It has been estimated that a nearby supernova could generate a tremendous rise in the radiation in the atmosphere, immediately killing or causing massive chromosomal mutations in the dinosaurs that would have made them sterile or at least incapable of producing viable offspring. At present, however, no remnant of a nearby 66-million-year-old supernova has been detected by astronomers. Also, the supernova explanation is hard pressed to explain why many organisms survived the Cretaceous.

Thinning of and trace elements in dinosaur eggshells at the end of the Cretaceous as a cause of dinosaur extinction was discussed in box 12.4. There, we saw that high levels of trace elements in dinosaur eggshells may have been introduced during fossilization. Also, it has not been demonstrated that the dinosaur eggs with relatively thin shells were not viable eggs.

A similarly problematic explanation is that disease extinguished the dinosaurs. Not only is there no evidence for this, but a worldwide epidemic or plague among the dinosaurs exceeds the scope of any known disease and seems implausible.

This review of some popular, but unsupported, explanations of dinosaur extinction still leaves a residuum of fantastic ideas beyond scientific inquiry. The most popular of these is that extraterrestrial beings either hunted the dinosaurs into extinction or trapped them all and carried them off into space. Clearly, these ideas are the stuff of science fiction stories, not of the scientific inquiry into dinosaur extinction.

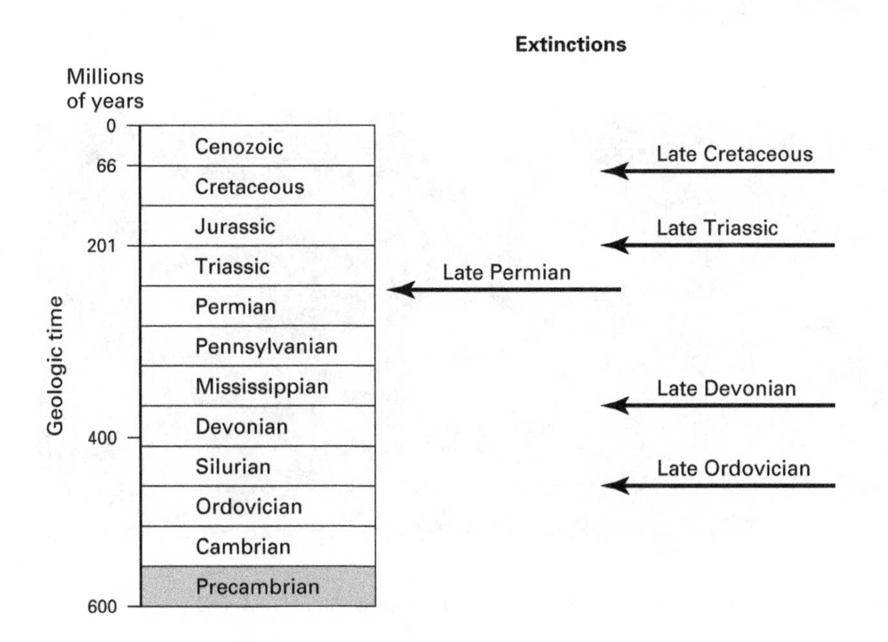


**FIGURE 16.1**
The terminal Cretaceous extinction was one of several major extinctions in the history of life. (Drawing by Network Graphics)

terminal Cretaceous extinction. Particularly hard-hit groups were the **ammonoid** cephalopods, relatives of living octopuses and squids, which suffered total extinction; clams and snails, which suffered significant losses including the total extinction of the **rudists**, reef-building clams, and the **inoceramids**, thin-shelled, plate-like clams; and the marine reptiles, the **mosasaurs** (marine lizards) and the long-necked **plesiosaurs** (figure 16.2). Major changes also occurred in the marine plankton, and the microscopic shelled protozoans known as **Foraminifera** also suffered heavy losses (figure 16.3). On land, the pterosaurs and dinosaurs became extinct, many types of **marsupial mammals** disappeared, and a few types of plants died out (figure 16.4). However, most mammals, reptiles (lizards, snakes, turtles, and crocodiles), and birds did not suffer a major extinction.

**FIGURE 16.2**
Several types of marine organisms became extinct at the end of the Cretaceous. Those shown here are the ammonoid cephalopods (*left*), the ichthyosaurs (*center*), and the plesiosaurs (*right*). (Drawing by Network Graphics)

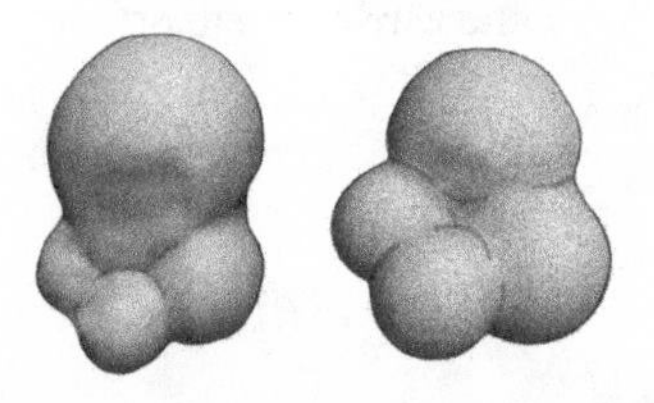

**FIGURE 16.3**
Foraminifera are microscopic shelled protozoans; they suffered heavy losses during the terminal Cretaceous extinction. (Drawing by Network Graphics)

**FIGURE 16.4**
Dinosaurs became extinct at the end of the Cretaceous. (© Mark Hallett. Reproduced with permission of Mark Hallett Paleoart)

It is important to recognize that more than just dinosaurs became extinct at the end of the Cretaceous. The broad nature of the extinctions calls for an explanation beyond one which would account for just the disappearance of the dinosaurs. Furthermore, the selectivity of the extinctions (many types of organisms survived the Cretaceous) needs to be explained. Hence, we are looking for a cause of dinosaur extinction that is consistent with what we know about both the breadth and the selectivity of the terminal Cretaceous extinction.

## NATURE OF THE EVIDENCE

We must not only consider the breadth and selectivity of the terminal Cretaceous extinction in the search for a cause, but the evidence of the extinction—the rocks and fossils and their distribution in geologic time and space—must also be considered. Indeed, the evidence takes center stage in the search for the cause of the extinction because most debate about the cause of the terminal Cretaceous extinction focuses on disagreements over the evidence and how it should be interpreted.

This issue is best understood if we make a clear, conceptual distinction between the evidence itself, or the pattern of the terminal Cretaceous extinction, and the processes that produced it. This distinction between **pattern** and **process** is a fundamental one in paleontology. It involves distinguishing the fossils themselves and the rocks that contain them as patterns to be described from the processes (for example, evolution and sedimentation) that produced the patterns. Processes can only be inferred from patterns, not observed, much as a detective might infer the process of a crime from the patterns represented by the evidence of the crime.

The pattern of the terminal Cretaceous extinction, and particularly the extinction of dinosaurs, is a subject of great disagreement among paleontologists. This disagreement lies at the core of debate over the cause of the extinction. Different patterns, especially in terms of the distributions of dinosaur fossils through geologic time, suggest different causes of extinction. One possible pattern suggests a sudden and simultaneous cause of dinosaur extinction, whereas a different pattern suggests a more gradual and long-term cause of the extinction (figure 16.5). Which pattern best describes the fossil record of the last dinosaurs?

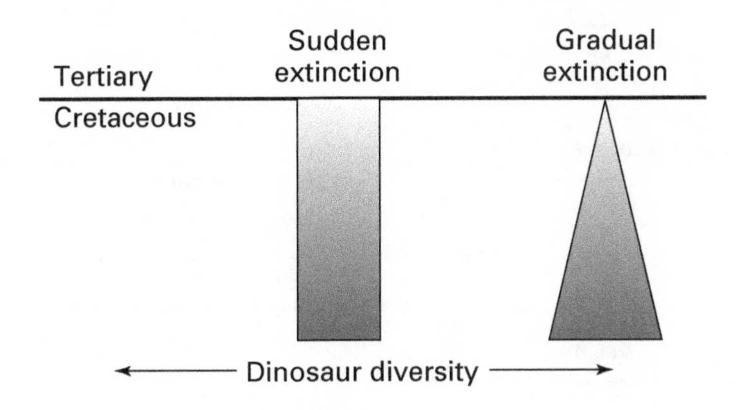


**FIGURE 16.5**
Two patterns of dinosaur extinction, one sudden and the other gradual, have been described. A sudden extinction is suggested by samples of dinosaur fossils that indicate unchanged dinosaur diversity until the very end of the Cretaceous. Diminishing diversity throughout the latest Cretaceous seems to be indicated by other samples (or analyses), and this is suggestive of a more gradual extinction. (Drawing by Network Graphics)

Before answering this question, let us consider another aspect of the nature of the evidence of dinosaur extinction: its completeness. The last interval of Cretaceous time is termed the **Maastrichtian**, after the town of Maastricht in Belgium, and it lasted about 4 million years, from 70 to 66 million years ago. Maastrichtian dinosaurs were the last dinosaurs and are known from a variety of locations worldwide (figure 16.6). But most of these Maastrichtian dinosaur localities are 1 million years or more older than the end of the Cretaceous and thus do not include the last dinosaurs, which are those of latest Cretaceous age. Furthermore, many of these localities yield only a few dinosaur fossils or dinosaur eggs or their fossils have not yet been collected and studied extensively.

Only late Maastrichtian dinosaur localities in western North America yield enough dinosaur fossils that have been collected and studied sufficiently to produce a detailed pattern of the distribution of the last dinosaurs in geologic time. These localities extend in an arc from the high plains of southern Alberta to the Big Bend of Texas (figure 16.7). Although some other localities have been analyzed to explain dinosaur extinction, almost all localities that can produce a reliable pattern are restricted to a small portion of the world in western North America. This, of course, should make us cautious about the pattern of dinosaur extinction, because the pattern can be drawn from only a limited area and then assumed to represent a global pattern.

**FIGURE 16.6**
Maastrichtian dinosaur localities in western North America provide the only extensive record of dinosaur extinction. (Drawing by Network Graphics)

**FIGURE 16.7**
Maastrichtian badlands in North America, like these in northwestern New Mexico, yield the best fossil record of the last dinosaurs.

Bearing this in mind, two patterns of dinosaur extinction have been constructed from the late Maastrichtian localities in western North America. One pattern is of decreasing dinosaur abundance and diversity during the Maastrichtian; the other is of an undiminished or increasing dinosaur presence throughout the Maastrichtian (see figure 16.5). The first pattern, of course, suggests a gradual disappearance of dinosaurs over a period of a few million years, whereas the second suggests a sudden, catastrophic extinction of the dinosaurs.

Many paleontologists believe that the first pattern accurately reflects what happened to dinosaurs at the end of the Cretaceous. But others believe that strong evidence supports the second pattern. Debate over the two patterns is ongoing, with new evidence being accumulated and described to support one pattern or the other. This makes it difficult to choose one pattern over the other at present. Instead, we can best focus our attention on how the two patterns have been arrived at and why they differ.

Documentation of how dinosaur diversity (the number of different kinds of dinosaurs) changed during the Maastrichtian is only as good as the sampling and identifications on which it is based. We need extensive collections of dinosaur fossils from as many Maastrichtian fossil layers as possible and identifications of the dinosaur fossils that are as accurate as possible (ideally including the genus and species). This might sound straightforward, but extensive collections of dinosaur fossils are expensive and time consuming. Also, many of the fossils are not complete dinosaur skeletons or skulls and thus are difficult to identify precisely (figure 16.8).

**FIGURE 16.8**
Fragmentary dinosaur fossils of Maastrichtian age, such as this vertebra of a ceratopsian dinosaur from New Mexico, are difficult to identify precisely.

Different amounts of collecting, different degrees of geologic time resolution (in other words, many versus few fossil layers), and differing identifications account for the two conflicting patterns seen in the Maastrichtian dinosaurs of western North America. Also, collections from different areas, even from two nearby locations in Montana, suggest different patterns, partly because of different degrees of preservation at the different locations.

Two contrasting patterns of dinosaur extinction have thus arisen. Even a third one has been proposed that claims dinosaurs survived the Cretaceous (box 16.2). Choosing among these patterns is critical, because each suggests a very different cause of dinosaur extinction, one gradual, the other sudden. But, as of this writing, the jury is still

Box 16.2

## Paleocene Dinosaurs?

For about 150 years, paleontologists have marked the end of the Cretaceous with the disappearance of the dinosaurs. This means that the youngest dinosaur fossils are of Late Cretaceous age, and, conversely, that the discovery of a dinosaur fossil is taken to indicate the rocks at the discovery site are no younger than Late Cretaceous. Despite this, some paleontologists claim to have discovered younger dinosaur fossils, of Paleocene age, thus demonstrating that not all of the dinosaurs became extinct at the end of the Cretaceous. These claims have come from such far-flung locales as China, India, Peru, and Bolivia, as well as in from the United States, in New Mexico and Montana. If substantiated, these claims seriously challenge well-accepted ideas about the timing of dinosaur extinction.

But, most of these claims cannot be substantiated simply because the Paleocene age of the dinosaur fossils cannot be verified. This is because few of the supposed Paleocene dinosaur fossils have been found associated with other types of fossils of unquestioned Paleocene age, or directly above such fossils in the strata. A good example of this is provided by supposed Paleocene dinosaur footprints from Bolivia. These footprints are found in an exposure of a rock formation that contains no Paleocene fossils. But, many kilometers away, the same formation contains fossils of mammals and fishes of Paleocene age, so it has been suggested that the entire formation, including the dinosaur footprints, is of Paleocene age. This, despite the fact that many kilometers of forest intervene between the two exposures, so it cannot be demonstrated that the dinosaur footprints are the same age or younger than the Paleocene fishes and mammals. They might just as well be older!

A recent report of Paleocene dinosaurs from New Mexico also fails to convince because a detailed evaluation of that work shows that the claimed association of Paleocene pollen with dinosaur fossils cannot be replicated (box figure 16.2). A lack of direct association with Paleocene fossils eliminates all but one of the current claims of Paleocene dinosaurs. This remaining claim comes from Montana, where dinosaur fossils are found in direct association, mixed in the same rock layer, with fossil mammals and pollen of Paleocene age. These dinosaur fossils are isolated teeth—mostly of hadrosaurids, ceratopsians, and small theropods—and fragments of bone. This might seem convincing proof of Paleocene dinosaurs were it not for the possibility that these dinosaurs have been reworked from older Cretaceous strata.

A fossil is said to be **reworked** if, after initial fossilization in the rock, it was exhumed and redeposited in an overlying, younger layer of rock. Reworking of small fossils may be common, especially in fluvial environments. Imagine, for example, isolated dinosaur teeth and parts of dinosaur bones buried and fossilized on a Late Cretaceous river floodplain. Subsequently, during the Paleocene, large river channels carve out and scour into the floodplain. In so doing, older Upper Cretaceous floodplain deposits, containing the dinosaur fossils, are eroded and carried off by the running water. This mixes the Cretaceous fossils with younger sediments and fossils, and subsequently deposits them on channel bottoms and margins, producing a mixture of Cretaceous and Paleocene fossils.

Most paleontologists believe this is the explanation of how dinosaur fossils and Paleocene mammal and pollen fossils came to be associated in Montana. They point to the small size of the dinosaur fossils and the fact that the association is in river-channel deposits as evidence that the dinosaur fossils have been reworked. But, proponents of the Paleocene age of these dinosaur fossils point to the lack of damage, abrasion, and weathering of the dinosaur teeth as evidence against their having been reworked. They argue that we should

(*continued*)

**BOX FIGURE 16.2**
This hadrosaur femur from New Mexico was found stratigraphically above fossils of Paleocene pollen. It is almost certainly a reworked fossil from underlying Cretaceous strata.

see some mechanical damage to the dinosaur teeth by the process of erosion, exhumation, transport, and reburial that reworking entails.

But is this really the case? Cretaceous shark teeth found reworked in rocks as young as Eocene show no evidence of mechanical damage. Therefore, we might expect reworking of dinosaur teeth to occur without appreciable alteration of the fossils.

There thus seems to be no convincing evidence for Paleocene dinosaurs. The search for them, however, continues, and we can expect claims of Paleocene dinosaurs to continue for as long as dinosaur extinction is studied.

out on which pattern is the accurate one. Until this matter is decided, debate about dinosaur extinction will continue.

## SINGLE CAUSE: ASTEROID IMPACT

Having revealed the impasse over the pattern of dinosaur extinction, it might seem anticlimactic to discuss possible causes. But, the two current explanations of dinosaur extinction present possible explanations of the alternative patterns, and one or the other, or a combination of both, stands a good chance of being correct.

In 1979, Nobel physics laureate Luis Alvarez, his geologist son, Walter Alvarez, and two nuclear chemists, Frank Asaro and Helen Michel, proposed that an asteroid the size of a mountain (10 kilometers in diameter) collided with Earth 66 million years ago and caused the terminal Cretaceous extinction (figure 16.9). They based this proposition on the chemical analysis of a thin, 66-million-year-old clay layer at **Gubbio** in northern Italy. This clay layer was deposited by wind-blown dust that settled at the bottom of the sea at the end of the Cretaceous, and the chemical analysis revealed it contains an unusually high concentration of the platinum-group metal **iridium** (figure 16.10). Iridium is very rare on the earth's surface except in some rock deposits mined for gold and platinum. The rocks at Gubbio are clearly not such a deposit. Most of the iridium on Earth comes from meteorites and other matter from space, so the Alvarez team decided that the high concentration in the 66-million-year-old clay at Gubbio must have settled in the dust produced by a huge asteroid impact.

This idea elicited both criticism and enthusiasm in the scientific community, and much effort was expended in the 1980s to either refute or support the theory of the asteroid impact and its inferred results. Further chemical analysis of the Gubbio clay layer revealed it contains other elements (such as osmium) not common on Earth except as the result of meteoritic infall. The high concentration of iridium was also identified in 66-million-year-old rock layers at dozens of other localities worldwide (figure 16.11). Another peculiarity of these layers is that they contain microscopic droplets of "**shocked quartz**," a type of quartz grain with laminar deformations known to be generated only by high-speed shock in laboratories, nuclear test sites, and near impact craters.

The search for the terminal Cretaceous impact crater also energized research into the detection of ancient impact craters on the earth's surface (box 16.3). This research identified about 16 impact craters that could be the right age to have been the site of the terminal Cretaceous asteroid impact. Most of these craters are in and around the

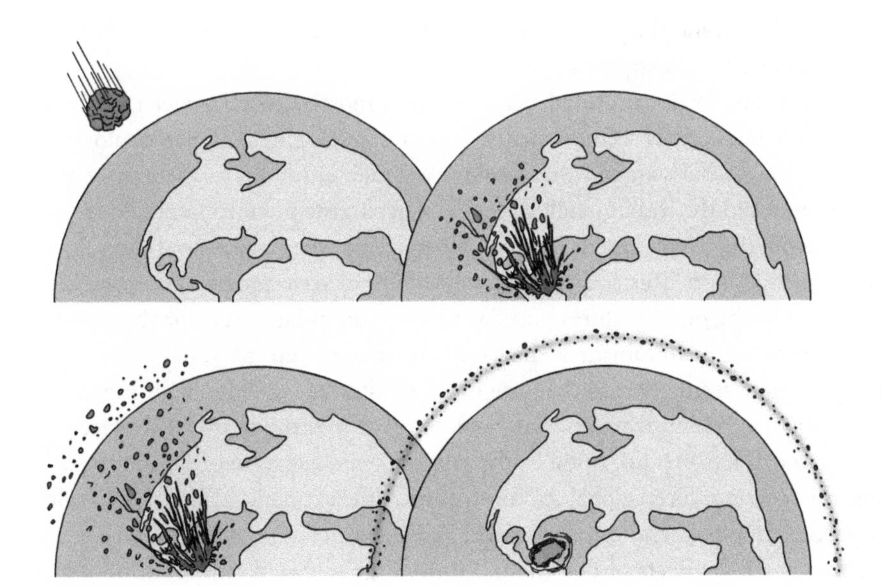

**FIGURE 16.9**
An asteroid impact at the end of the Cretaceous may have looked like this. (Drawing by Network Graphics)

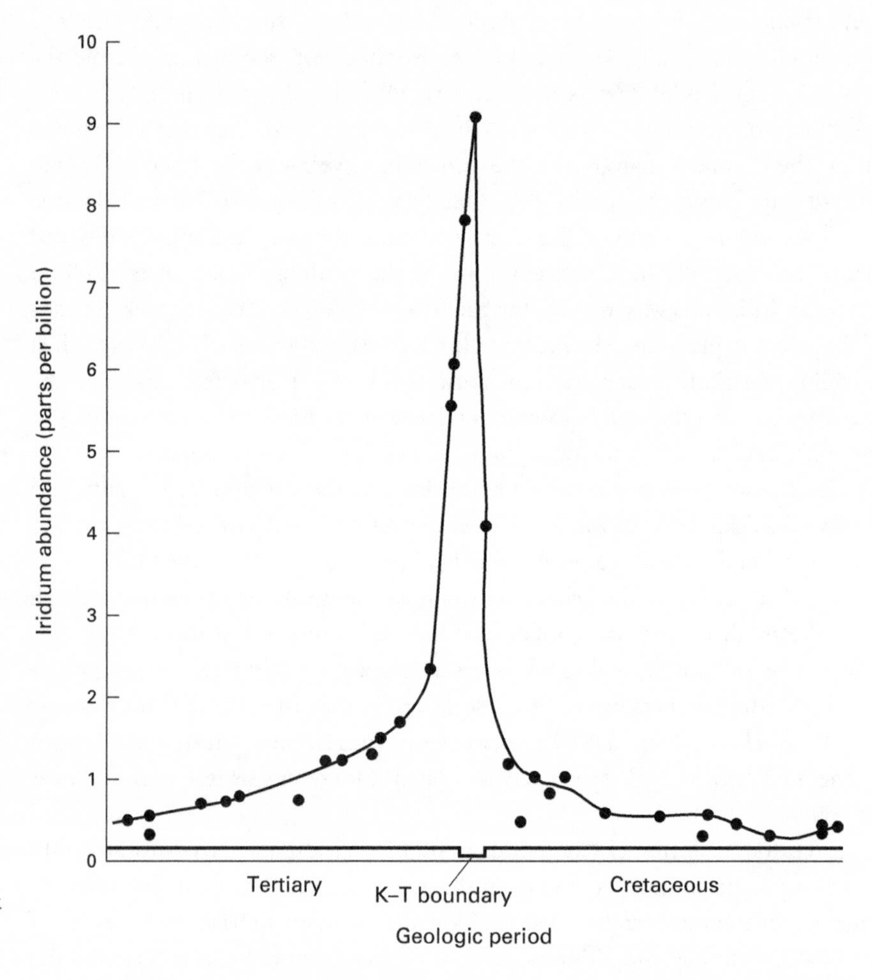


**FIGURE 16.10**
In northern Italy, the clay layer that marks the end of the Cretaceous contains an unusually high abundance of iridium. (Drawing by Network Graphics)

Atlantic Ocean basin, and the possibility of multiple asteroid impacts at the end of the Cretaceous has also been suggested.

At the time of this writing, the evidence that one or more asteroids struck the Earth at the end of the Cretaceous seems conclusive to most geologists and paleontologists. But, some geologists have argued that massive volcanic eruptions at the end of the Cretaceous produced the iridium-rich layer. These geologists point to huge Late Cretaceous volcanoes that were present in India, Siberia, and other locations as the source of the iridium-rich layer. This source, however, could not have provided the shocked quartz or some of the other features peculiar to the iridium layer, and for this reason most scientists discount volcanism as the likely source of the layer.

Although it now seems certain that one or more asteroids collided with Earth 66 million years ago, how and whether that would have caused the terminal Cretaceous extinction is much less certain. Initially, the Alvarez team suggested that the impact generated dust that produced global darkness, causing the extinctions. This dust would have stopped plant photosynthesis and cooled global temperatures, resulting in a deep freeze. Other possible effects of the asteroid impact include acid rain resulting from

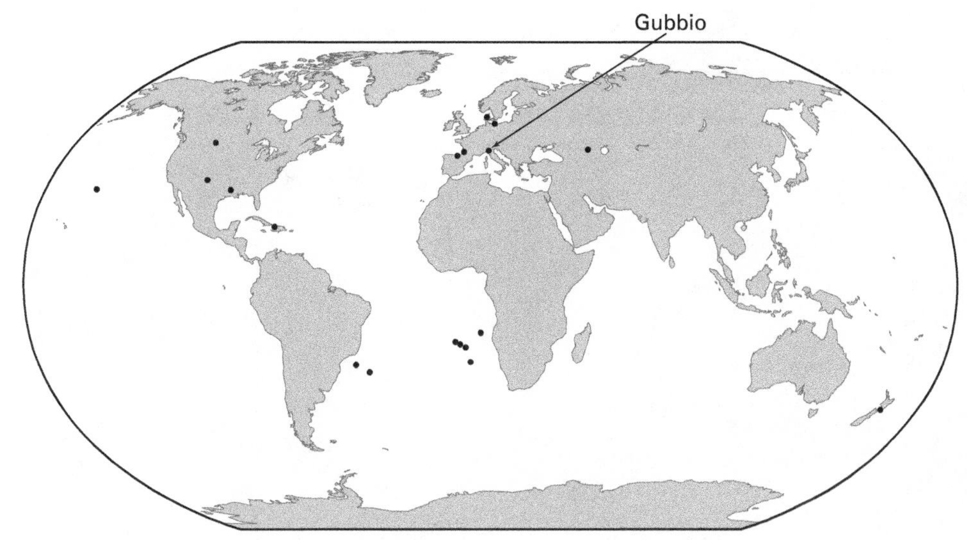


**FIGURE 16.11**
An iridium-rich clay layer has been identified at the Cretaceous–Tertiary boundary at many sites worldwide. (Drawing by Network Graphics)

Box 16.3

## The Chicxulub Impact Structure

In 1990, a group of American geologists announced the discovery of a huge impact structure of terminal Cretaceous age, and this structure has since been proclaimed the "smoking gun" behind the terminal Cretaceous extinctions. Located on the northwestern coast of Mexico's Yucatán Peninsula, the **Chicxulub** (pronounced CHICKS-uh-loob) structure is a circular collapse feature with an estimated diameter of about 180 kilometers (box figure 16.3). Diverse and impressive evidence makes Chicxulub look like the place where a huge asteroid crashed into the earth at the end of the Cretaceous. This evidence includes the following points:

1. The Chicxulub structure itself has many features diagnostic of an impact crater, including the concentrically collapsed crust under it.
2. A layer of melted rock underlies the structure. The numerical age of this melt layer has been calculated at 66 million years, exactly the Cretaceous–Tertiary boundary. And the melted rocks are unusually rich in iridium.
3. Above the melted rocks are well-sorted debris that is most easily understood as rock fragments that fell back into the crater after the impact.
4. Tidal wave deposits of the same age as the structure have been identified in the southern Gulf of Mexico and the western Atlantic.
5. Deposits of rock debris thrown out of the crater (called **ejecta**) have been found in Haiti and northeastern Mexico. These deposits are full of shocked mineral grains and melted rock droplets.

*(continued)*

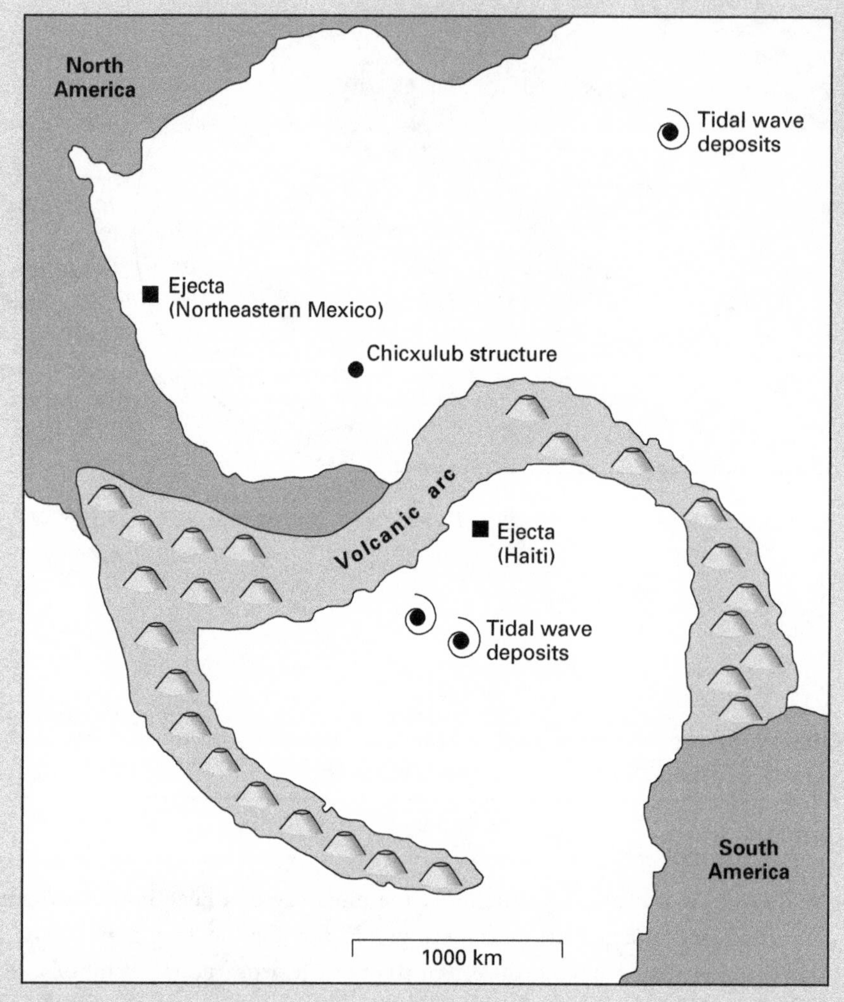


**BOX FIGURE 16.3**
This paleogeographic map of the Caribbean at the end of the Cretaceous shows the Chicxulub impact structure (now under the seafloor) and some related rock features. (Drawing by Network Graphics)

The case for Chicxulub being the site where the asteroid plowed into Earth at the end of the Cretaceous seems incontrovertible, though some critics still argue that it predates the terminal Cretaceous extinctions by as much as 300,000 years.

At the end of the Cretaceous, much of the rock present at the site of the Chicxulub structure consisted of anhydrite, a mineral rich in sulfur. Some scientists suggest that the impact released huge amounts of sulfur into the atmosphere. The sulfur would have combined with water, producing sulfuric acid and a devastating acid rain. Acidic dust would then have filled the upper atmosphere, blocking out sunlight and causing a decade of freezing or near-freezing temperatures worldwide.

heating of the atmosphere by the entering asteroid or a worldwide greenhouse effect (after the initial deep freeze following impact) produced by water vapor that would have been lofted into the atmosphere if the asteroid landed in the ocean.

Currently, there is no clear agreement on the precise effect of the asteroid impact, but there is general agreement that it produced a **global catastrophe**. Whether this catastrophe caused the terminal Cretaceous extinctions, or whether it was a violent perturbation weathered by most organisms, is still debated.

We might best think of this by concluding that the iridium-rich layer is a pattern indicative of a specific process: an asteroid impact. Now, the question becomes, does the pattern of dinosaur and other Late Cretaceous fossil distribution support the asteroid impact and the resulting catastrophe as the cause of the extinction?

One advantage of the asteroid-impact explanation is that it predicts a sudden and simultaneous pattern of terminal Cretaceous extinction. Many fossil patterns, such as those constructed by some paleontologists for Maastrichtian dinosaurs, don't indicate a sudden and simultaneous end. But, proponents of the asteroid-impact explanation argue that these patterns are based on incomplete sampling of fossils or incomplete preservation of latest Maastrichtian fossils. There is some evidence to support their contention. For example, dinosaur collecting by paleontologist Peter Sheehan and colleagues of the Milwaukee Public Museum in eastern Montana and western North Dakota has been interpreted to reveal no evidence of a gradual decline of dinosaurs at the end of the Cretaceous. In contrast, other intensively sampled locations suggest gradual decline and an **ecological collapse** of dinosaurs during the Maastrichtian. One of these studies represents the most serious challenge to the sudden, catastrophic extinction of dinosaurs posited by the asteroid-impact explanation.

## MULTIPLE CAUSES

The badlands developed in the upper Maastrichtian and Paleocene strata in eastern **Montana** represent the most extensively sampled record of dinosaur extinction on Earth. The upper Maastrichtian rocks are the **Hell Creek Formation** overlain by lower Paleocene strata of the Tullock Formation. The end of the Cretaceous, and the boundary of the two formations, has generally been placed at a layer of coal called the Z-coal bed. An iridium-rich layer is found just below this coal bed.

Beginning in the 1950s, paleontologists William Clemens of the University of California at Berkeley and Robert Sloan of the University of Minnesota, as well as numerous colleagues and students, collected fossils from these rocks. This collecting has produced an unparalleled record of the evolution of terrestrial plants and animals during the transition from the Cretaceous to the Tertiary.

Much of this record does not suggest a sudden, catastrophic extinction at the end of the Cretaceous (figure 16.12). For example, there is evidence that placental mammals increased in diversity throughout the Maastrichtian. Nondinosaurian reptiles, including lizards, snakes, turtles, and crocodiles, show very little extinction at the end of the Cretaceous, with only about 30 percent of their genera disappearing. Dinosaurs do go extinct at the end of the Cretaceous, but there was a marked decline in diversity in the

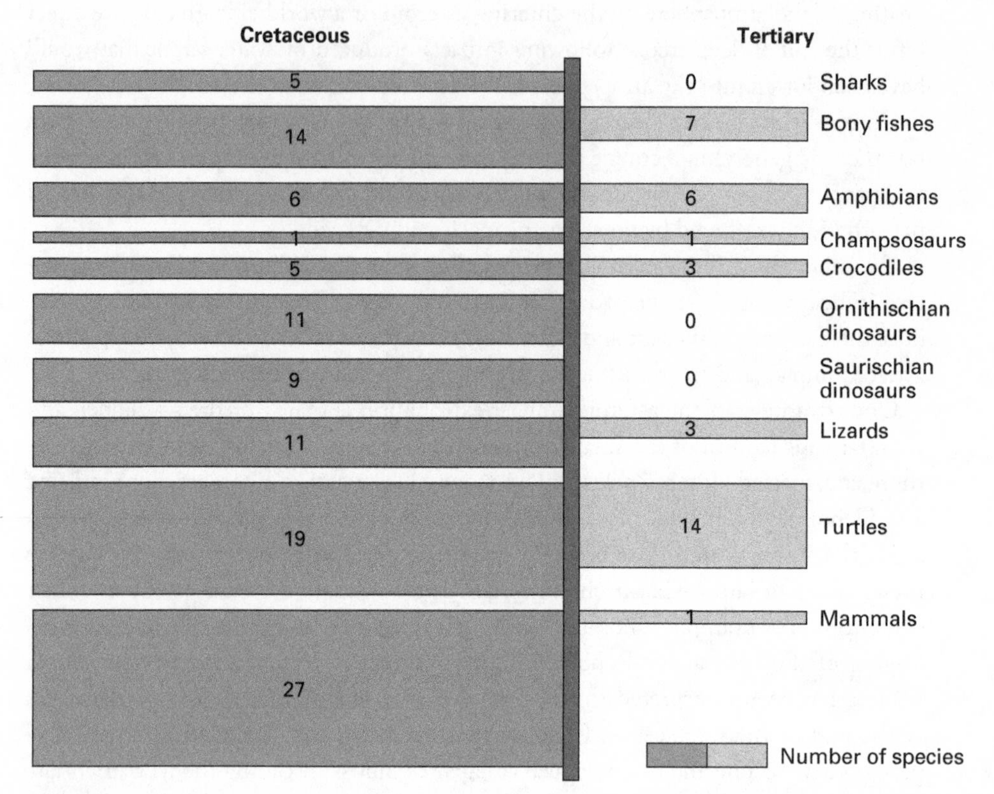


**FIGURE 16.12**
The fossil record of vertebrates from the Hell Creek Formation shows some groups diversifying before, and most groups surviving, the end of the Cretaceous. (Drawing by Network Graphics)

Hell Creek Formation from 19 genera at its base, to 12 in the upper 16 meters of the formation to 7 at the very top.

This pattern of dinosaur decline is either real or an artifact of sampling and biases in preservation. If real, it presents the strongest evidence available of a gradual extinction of dinosaurs during the late Maastrichtian. This evidence, and other patterns of fossil distribution, certainly do not support the sudden extinction predicted by the asteroid-impact explanation. Instead, they suggest complicated ecological factors as a likely cause of dinosaur extinction.

These ecological factors might include cooling and drying of climate because of a worldwide drop in sea level at the end of the Cretaceous, changes in vegetation due to the climate change, changes in topography and land bridges created by the sea level fall, and/or competition from diversifying placental mammals who might have posed a real problem for small, juvenile dinosaurs. Extensive volcanism during the latest Cretaceous, particularly in the Deccan Traps of India, also may have altered climate, ocean chemistry, and other aspects of marine and terrestrial environments. Indeed, Deccan was such a huge volcanic field that the lava it extruded would cover an area from Alaska to Texas to a depth of 610 meters. The result would have been a complex ecological collapse of the global ecosystem during the last few million years of the Cretaceous. This could have caused many of the extinctions, culminating in a "really bad day" for the dinosaurs when the asteroid struck.

Certainly, such possibilities don't present a cause of dinosaur extinction as simple, and therefore as appealing, as an asteroid impact. But, the pattern of dinosaur extinction and other fossil distributions in the Hell Creek Formation suggest to many paleontologists a complex, ecological cause of dinosaur disappearance.

## MINIMIZING THE DAMAGE

A number of paleontologists have questioned the significance of dinosaur extinction. This questioning has taken three approaches.

The first approach has been to point out that the significance of the entire terminal Cretaceous extinction has been greatly overstated. Today, most of the living species of animals are arthropods and worms that lack fossilizable skeletal parts. There is no reason to believe that these animals did not have a similar diversity during the Late Cretaceous, yet they left no fossils during that interval of geologic time. This means, of course, that paleontologists have virtually no idea how, if at all, these soft-bodied animals were affected by the terminal Cretaceous extinction. This should lower our confidence in estimates of the magnitude of the terminal Cretaceous extinction. These estimates, which range from a kill of from 40 to 80 percent of the species on Earth, apply only to those organisms with fossil records and thus may greatly overstate the magnitude of the terminal Cretaceous extinction.

This is a valuable perspective on the terminal Cretaceous extinction, or for that matter, any extinction being studied in the fossil record. We should always remember that the fossil record is incomplete and not interpret it as if it preserves all organisms that lived during the past. Yet, this perspective should not fundamentally alter the significance of dinosaur extinction. The dinosaur fossil record relevant to extinction is, as discussed earlier, very incomplete, yet it is good enough to demonstrate that dinosaurs did become extinct, and their extinction merits an explanation.

Or did they become extinct? The second approach to questioning the significance of dinosaur extinction is to state that dinosaurs are not extinct. They survive as birds, so what is all this fuss about their extinction? This viewpoint, however, merely dodges the issue of dinosaur extinction. The dinosaurs that became extinct at the end of the Cretaceous represented a significant group of animals. The fact that their descendants, the birds, or if you like, one group of feathered dinosaurs, survived the extinction does not diminish the significance of the terminal Cretaceous extinction of dinosaurs. Birds appeared during the Late Jurassic, long before the extinction of the dinosaurs, and it is highly likely their closest relatives among the theropod dinosaurs were extinct long before the Late Cretaceous. It is an interesting question why birds survived the Cretaceous, one that needs to be considered in any attempt to explain the terminal Cretaceous extinctions. But, the survival of birds does not diminish the significance of dinosaur extinction.

A third and final approach to questioning the significance of dinosaur extinction stems from a "long view" of the fossil record of dinosaurs. This long view points out that somewhere between 900 and 1,200 genera of dinosaurs are estimated to have lived during the Late Triassic through the Late Cretaceous (about 500 valid genera are known), most of which became extinct before the Late Cretaceous (figure 16.13).

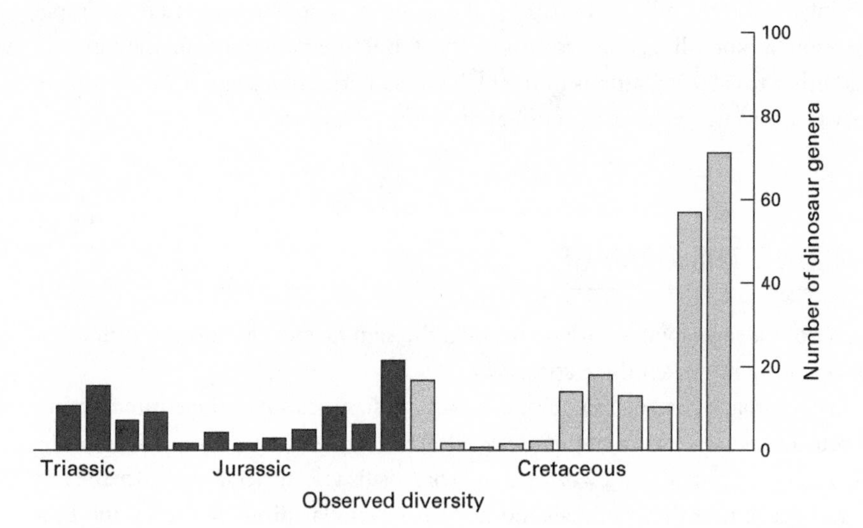


**FIGURE 16.13**
Dinosaur diversity during the Late Triassic through the Late Cretaceous indicates that most dinosaurs that lived disappeared before the end of the Cretaceous. (Drawing by Network Graphics)

The last dinosaurs represent only a small portion of their total Mesozoic diversity, so the extinction of dinosaurs is merely the disappearance of a relatively small fraction of the entire group.

This argument, like the previous one, dodges the issue of dinosaur extinction. Its faults are best revealed by a terrifying analogy. Suppose the human species became extinct tomorrow. If so, it would be true that most species (and individuals) of humans became extinct before the final disappearance of *Homo sapiens*. Thus, like the dinosaurs, the final extinction of humankind would be the disappearance of only a small fraction of the group. But, should this minimize the significance of these final extinctions? Indeed, the final extinction of most groups of organisms involved only a fraction of the species of that group that ever lived. Dinosaur extinction thus remains a significant phenomenon, regardless of the incompleteness of the fossil record, the survival of birds, or the pre–Late Cretaceous diversity of the dinosaurs.

## ANSWER THE QUESTION!

After all of this discussion, we are still led back to the original question: what caused the extinction of the dinosaurs? The answer to this question, as we now know, depends on which pattern of distribution of the last dinosaurs is correct. One pattern suggests a sudden and simultaneous disappearance of dinosaurs at the end of the Cretaceous and thus supports a major catastrophe, as would be caused by an asteroid impact. This is a **single-cause** explanation of dinosaur extinction. An alternative pattern of a gradual decline of the dinosaurs during the last 10 million years of the Cretaceous suggests that complicated ecological changes brought about dinosaur extinction. These ecological changes were followed by an asteroid impact that eliminated the last dinosaurs, thus combining the two most likely causes. This is a **multiple-cause** explanation of dinosaur extinction.

The inability to provide a simple answer to the question of what killed the dinosaurs may disappoint some. But, this inability reflects the fact that not all the evidence is in on dinosaur extinction. The disappearance of the dinosaur remains a vigorous field of scientific research and debate.

## Summary

1. The two current, and debated, explanations of dinosaur extinction identify a single cause, an asteroid impact, or multiple causes of complex ecological changes due to sea-level fall and volcanism, culminating in an asteroid impact.
2. Other explanations of dinosaur extinction lack strong evidence and include egg-eating mammals, extreme cold or hot climates, poisoning, a supernova explosion, thinning and trace elements in dinosaur eggshells, disease, and extraterrestrial big-game hunters.
3. The terminal Cretaceous extinction eliminated many types of organisms in the sea and on the land.
4. The broad extent and selectivity of the terminal Cretaceous extinction force a search for a cause that explains both of these factors.
5. Patterns are found in the fossils and rocks studied by paleontologists. The processes (evolution and sedimentation) that created these patterns cannot be observed directly.
6. Different distribution patterns of the last dinosaurs reflect different degrees of sampling and precision of identification.
7. Strong evidence, accepted by most geologists, indicates one or more asteroids impacted the earth 66 million years ago.
8. Supporters of an asteroid impact causing dinosaur extinction propose a sudden and simultaneous extinction of dinosaurs.
9. According to some paleontologists, the pattern of fossil distribution in the Hell Creek Formation in Montana is not consistent with an asteroid impact causing the terminal Cretaceous extinction. Instead, they believe it suggests a gradual decline of dinosaurs because of complicated ecological changes.
10. Some paleontologists minimize the significance of dinosaur extinction by noting that the magnitude of the terminal Cretaceous extinction has been overstated, that dinosaurs did not become extinct because birds survived, or that the last dinosaurs represent a small fraction of all dinosaurs that ever lived. These arguments, however, do not diminish the importance of dinosaur extinction.
11. The inability to give a simple answer to the question of what caused dinosaur extinction reflects the fact that all the data are not in on the subject.

## Key Terms

ammonoid
Chicxulub
ecological collapse
egg-eating mammals
ejecta
Foraminifera
global catastrophe
Gubbio
Hell Creek Formation
inoceramid
iridium
Maastrichtian

marsupial mammal
Montana
mosasaur
multiple cause
pattern
plesiosaur
process
reworked
rudist
selenium
shocked quartz
single cause
supernova
terminal Cretaceous extinction

## Review Questions

1. What are some popular ideas about the cause of dinosaur extinction, and why are they not accepted by most paleontologists?
2. Describe the extent of the terminal Cretaceous extinction.
3. How do the extent and selectivity of the terminal Cretaceous extinction affect attempts to determine its cause?
4. Distinguish a pattern from a process, and explain how these concepts apply to paleontology.
5. How does the nature of the fossil record of Maastrichtian dinosaurs influence our understanding of dinosaur extinction?
6. Why are there two contrasting patterns of dinosaur extinction described by paleontologists?
7. What evidence indicates an asteroid impact with the earth 66 million years ago? What would be some of the possible effects of such an impact?
8. Did an asteroid impact cause the extinction of the dinosaurs?
9. What does the fossil record from the Hell Creek Formation in Montana suggest about dinosaur extinction?
10. How have some paleontologists minimized the significance of the extinction of dinosaurs, and what are the strengths and weaknesses of their arguments?
11. What killed the dinosaurs? Defend your answer.

## Further Reading

Archibald, J. D. 1996. *Dinosaur Extinction and the End of an Era: What the Fossils Say*. New York: Columbia University Press. (A complete review of the terminal Cretaceous extinction by a proponent of multiple causes)

Archibald, J. D. 2013. The end-Cretaceous extinction, pp. 497–512, in N. MacLeod, J. D. Archibald, and P. S. Levin, eds., *Grzimek's Animal Life Encyclopedia*. Vol. 2, *Extinction*. Detroit: Gale Cengage Learning. (A very even-handed review of the terminal Cretaceous extinctions and possible causes)

Benton, M. J. 1990. Scientific methodologies in collision: The history of the study of the extinction of the dinosaur. *Evolutionary Biology* 24:371–400. (Reviews the history of ideas about dinosaur extinction)

Brusatte, S. L., et al. 2015. The extinction of the dinosaurs. *Biological Reviews* 90:628–642. (Argues for there not being a long-term decline in dinosaur diversity prior to their extinction)

Fassett, J. E. 2009. New geochronologic and stratigraphic evidence confirms the Paleocene age of the dinosaur-bearing Ojo Alamo Sandstone and Animas Formation in the San Juan Basin, New Mexico and Colorado. *Palaeontologia Electronica* 12:1–146. (Recent claim of Paleocene dinosaurs from New Mexico)

Fastovsky, D. E., et al. 2004. Shape of Mesozoic dinosaur richness. *Geology* 32:877–880. (Reviews all Mesozoic dinosaur diversity to argue that it was increasing just prior to their extinction)

Hartman, J. H., K. R. Johnson, and D. J. Nichols, eds. 2002. *The Hell Creek Formation and the Cretaceous-Tertiary Boundary in the Northern Great Plains.* Geological Society of America, Special Paper 361. (Compendium of 19 articles on the world's best record of dinosaur extinction)

Lucas, S. G., et al. 2009. No definitive evidence of Paleocene dinosaurs in the San Juan Basin. *Palaeontologia Electronica* 12:1–10. (Critique of Fassett's 2009 claim of Paleocene dinosaurs from New Mexico)

Rigby, J. K., Jr. 1987. The last of the North American dinosaurs, pp. 119–135, in S. J. Czerkas and E. C. Olson, eds., *Dinosaurs Past and Present.* Vol. 2. Los Angeles: Natural History Museum of Los Angeles County and University of Washington Press. (Presents some of the basis for the claim of Paleocene dinosaurs in Montana)

# 17

# DINOSAURS IN THE PUBLIC EYE

DINOSAURS are extremely popular, and not a week goes by without some new dinosaur discovery appearing in the newspapers, magazines, the Internet, or on radio and television. Popular books on dinosaurs, for children and adults, are legion. Life-size sculptures of dinosaurs adorn museums and parks, and toy dinosaurs are a staple item of department and toy stores. Paintings and drawings of dinosaurs permeate dinosaur books and museums, and dinosaurs have been the subject of myriad movies, cartoons, and television shows. The word "dinosaur" is part of the English language, and it would be difficult to find an American who hasn't used this word at least once. The word "dinosaur" has been transliterated or translated into almost all the world's languages. For example, in Chinese it is pronounced "kong-long" and literally means "monstrous dragon."

The preceding sixteen chapters of this book have been concerned with the paleontological understanding of dinosaurs. In this chapter, I take a paleontological perspective on some aspects of the public presentation and perception of dinosaurs.

## DINOSAURS: DENOTATION AND CONNOTATION

As stated in chapter 1 and elaborated on in chapter 4, a dinosaur was a reptile with an upright limb posture. Dinosaurs lived during the Late Triassic through the Cretaceous, between 225 and 230 to 66 million years ago, and were the ancestors of birds. Large size was not a characteristic of all dinosaurs, but this group of animals did include the largest land animals.

Some of these defining features of dinosaurs, the **denotation** of the word **dinosaur**, are familiar to most people and appear as the primary definition of dinosaur in dictionaries. But, a second definition of dinosaur also appears in them, using "dinosaur" as a word to refer to something unwieldy or out-of-date. This use of the word "dinosaur" may be thought of as its **connotation**: a negative image of something no longer useful and so deservedly extinct.

No doubt heavy emphasis on the bulkiness of many dinosaurs and their extinction created this connotation of the word "dinosaur." But, as we have seen in this book, it is hardly deserved. Many dinosaurs were fast and agile animals. Dinosaurs existed for about 160 million years, during most of which time they dominated the earth. Their descendants, the birds, are among the most diverse and successful groups of living animals.

So, although dinosaurs are extinct, they should stand as paragons of evolutionary success. Instead, they are often thought of as dim-witted, unwieldy failures, which is an undeserved reputation.

## DINOSAURS IN THE NEWS

New discoveries of dinosaurs frequently appear in the news media before they are published in scientific journals and books. There are two obvious reasons for this. First, paleontologists, like most people, have a natural desire to receive public recognition for their discoveries. Second, publicity about paleontological discoveries is one way to educate the general public about dinosaurs. Funding agencies and scientific institutions, such as museums and universities, seek publicity for dinosaur discoveries made by their paleontologists, both to educate the public and to enhance their institutions' reputations.

In an ideal world, news reports about dinosaur discoveries should serve the public's desire for up-to-date, factual information on the latest developments in dinosaur research. But there is an unfortunate downside to media coverage of dinosaurs. This, in part, results from journalistic biases and mistakes. Most reporters are not scientists, and some unintentionally introduce factual errors into stories about dinosaurs. Also, because every news story has to have an "angle"—an aspect that makes it appealing to the journalists and, they hope, the public—many stories about dinosaurs focus on the sensational and speculative. News stories, whether print, audio, or video, are necessarily short, so they usually cannot present all the evidence that supports their conclusions. Much of the media coverage of dinosaurs is tantalizing but incomplete and sometimes contains errors and highly biased and speculative conclusions for which there may be little factual basis.

A second problem with media coverage rests with the paleontologists themselves. The desire for recognition, or the pressures imposed by funding agencies and institutions for results, have sometimes led to sensational news reports about dinosaurs. Subsequent research, however, has revealed that these reports lacked substance and so misinformed the public. Fortunately, these occurrences are few and far between, and most dinosaur discoveries that appear in the press, if accurately presented by reporters, convey new and significant information.

News reports about dinosaurs will always be with us, so how can you distinguish the good from the bad? Certainly, the more you know about dinosaurs, the better equipped you will be to make such judgments. But, perhaps the best thing to do is be critical of the story, asking yourself the following questions: What, if any, biases does the story portray? How many facts—actual information on the fossils, their context, and the lines of reasoning—does the story contain? Also, if the news story draws conclusions, are alternative possibilities and evidence discussed? Few news stories about dinosaurs may stand up to such critical scrutiny, but by asking yourself these questions you may be able to separate those few stories that mislead and misinform from the wealth of new and useful information.

## DINOSAURS IN BOOKS

A sizable library is needed to house all the dinosaur books available in today's bookstores. Technical scientific books on dinosaurs are numerous and can fill a large bookshelf. Nontechnical books fall into three categories: children's books (and comics), popular factual books on dinosaurs, and dinosaur fiction. This is not the place to recommend or criticize any specific book, but a few comments on what to look for in dinosaur books, and how they influence the public perception of dinosaurs, are in order.

Many **misconceptions about dinosaurs** are perpetuated in today's books. A wide range of children's books on dinosaurs identify the sail-backed reptile *Dimetrodon* as a dinosaur, which it most certainly was not. Comic books often portray humans coexisting with dinosaurs. But perhaps most pervasive in popular books is the image of dinosaurs as animals that did little more than tear each other apart or search for something else, often a human, to eat. Indeed, such dinosaur carnage is so frequent a theme that most people don't realize that the average day of a dinosaur, like that of most living animals, was probably rather peaceful. Dinosaurs slept, reproduced, played, and rested—they were not always locked in mortal combat.

When buying dinosaur books for children, the goal should be to avoid such misinformation. However, those of us seeking action and adventure in our reading may not care about accuracy. Indeed, a certain lack of accuracy pervades one of the most influential, though now little read, works of dinosaur fiction. This book, Arthur Conan Doyle's ***The Lost World*** (1912), is based on the premise that dinosaurs are still living on a high plateau in the Amazon jungle (box 17.1). Although the book contains many misconceptions about dinosaurs, it still makes exciting reading, and its premise has been the basis for many other works of dinosaur fiction and dinosaur movies.

Although it is difficult to be too critical of the information about dinosaurs presented in books of fiction, readers should demand accuracy of popular books that purport to present dinosaur science to the public. Again, as in the case of news stories on dinosaurs, your ability to evaluate such books increases with your knowledge of dinosaurs. A particularly good measure is the accuracy of the dinosaur restorations and information presented in the book. Do they reflect current scientific thinking about dinosaurs, or are they based on old and outmoded notions? Also, when the book presents topics hotly debated by paleontologists, such as dinosaur metabolism or extinction, does it discuss alternative evidence and ideas? Hopefully, this textbook meets these criteria and provides you with a standard by which to evaluate other dinosaur books.

This section would not be complete without some mention of the technical scientific literature on dinosaurs. Articles on dinosaurs in scientific journals and scientific

## Box 17.1

### Arthur Conan Doyle's Lost World

**Sir Arthur Conan Doyle** (1859–1930) is best known as the creator of Sherlock Holmes, the most famous and skilled detective of all time. However, Doyle's novel *The Lost World* (1912) has influenced the genre of dinosaur movies since the film version of the book appeared in 1925.

*The Lost World* is mostly the story of an expedition, led by a British scientist, Professor George E. Challenger, to a high plateau in the upper reaches of the Amazon River inhabited by dinosaurs, mastodons, and a belligerent race of ape men. In the novel, a map of the high plateau shows the expedition's campsites, the village where the ape men live, locations of dinosaur sightings, and other features (box figure 17.1). It might seem that such a map was a totally imaginative exercise of Doyle's, but it actually had some basis in fact.

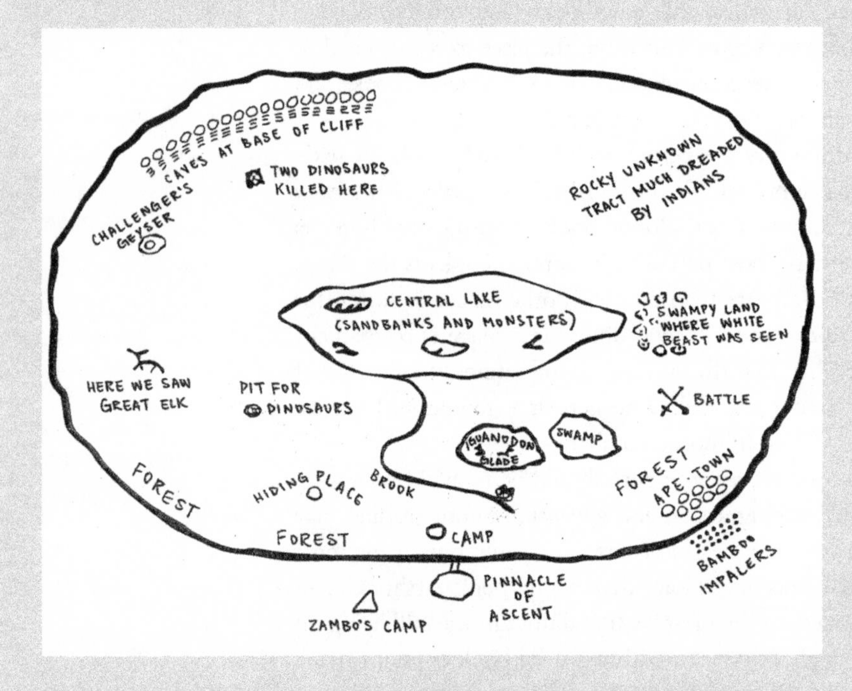


**BOX FIGURE 17.1**
The map of the lost world in Sir Arthur Conan Doyle's *The Lost World* is remarkably similar to a map of the Weald in England.

Doyle was an avid rock and fossil collector well acquainted with discoveries made in Great Britain. Many of Britain's Early Cretaceous dinosaurs, notably *Iguanodon*, and its most infamous fossil human, the Piltdown man, a scientific hoax discovered in 1912 and debunked in the 1950s, come from a region in southwestern England called the **Weald**. A map of this area shows several points of resemblance to Doyle's map of the lost world.

Thus, the "central lake" corresponds to the Hastings Sands, rocks that contain fossils of *Iguanodon* and were thought in Doyle's day to represent Early Cretaceous lakes and swamps (now they are considered to be mainly river and beach deposits). Indeed, an *Iguanodon* glade and swamp are present in the lost world. The site of the battle with the ape men in the lost world nearly matches that of the town of Battle in the Weald. Similarly, Lavant Caves in the Weald is mirrored by the "hiding place" of the lost world, the "brook" mimics the Ouse River, and "Challenger's geyser" corresponds to Waggoner's Wells. Finally, Beachy Head in the Weald is a promontory on the British coast that corresponds to the "pinnacle of ascent" of the lost world.

Clearly, Doyle borrowed heavily from fact to create the fictional lost world. But, a tremendous amount of nonfact appears in *The Lost World* as well. Ape men and mastodons coexist with dinosaurs, and dinosaurs of different geologic time periods live together on the high plateau. Also, long outmoded prejudices about dinosaurs fill the book. Witness, for example, the conclusions of Professor Challenger and his colleague Sumerlee on the intelligence of dinosaurs:

> Both were agreed that the monsters were practically brainless, that there was no room for reason in their tiny cranial cavities, and that if they had disappeared from the rest of the world, it was assuredly on account of their own stupidity, which made it impossible for them to adapt themselves to changing conditions.

Doyle's book, though in part based in fact, helped to create and perpetuate some common misconceptions about dinosaurs and the dinosaurian world.

books about dinosaurs are available in any university library, and many are also easily accessed online. These articles and books represent the ultimate written authority on what paleontologists know and think about dinosaurs. But, they need to be read critically, because the scientific study of dinosaurs is continually growing and changing through new discoveries and interpretations as well as spirited scientific debate.

## DINOSAURS IN ART

Artists began to create dinosaur art, such as sculptures and paintings, not long after Richard Owen coined the word "dinosaur" in 1842. In 1854, **Benjamin Waterhouse Hawkins** built life-size models of *Iguanodon*, *Megalosaurus*, and *Hylaeosaurus* (a nodosaurid ankylosaur) that still stand in London's Crystal Palace Park (see figure 11.3). These sculptures, and indeed all dinosaur art, reflect paleontological ideas about the biology and behavior of dinosaurs current when the artwork was created. Hawkins's dinosaurs are the massive, toad- and lizard-like giants that Richard Owen and his colleagues conceived from only a fragmentary knowledge of dinosaur anatomy. For example, Hawkins's *Iguanodon* has the spike on its nose, as believed to be the correct placement by Gideon Mantell in the 1820s, not on the thumb, as was later revealed by the discoveries at Bernissart.

In chapter 11, we saw that new discoveries have radically changed ideas about dinosaur biology and behavior. Dinosaur art has changed as well. By the beginning of the twentieth century, complete skeletons of dinosaurs were known that provided a far better basis for dinosaur sculptures and paintings than was available to Hawkins.

The most famous dinosaur artist then, and perhaps the most influential dinosaur artist of all time, was **Charles R. Knight** (1874–1953) (figure 17.1). Knight painted and sculpted many dinosaurs, mostly with the advice of paleontologist Henry Fairfield

**FIGURE 17.1**
Charles R. Knight was the most influential dinosaur artist. (Courtesy Department of Library Services, American Museum of Natural History, New York [negative no. 327667])

Osborn (1857–1935) of the American Museum of Natural History. Although most people don't realize it, the images of dinosaurs created by Charles R. Knight permeated dinosaur art for more than half a century and shaped the work of many dinosaur artists. His restorations are some of the most familiar, and most copied, images of dinosaurs.

Knight's dinosaurs reflected accurate anatomical information based on complete skeletons and ideas about dinosaur biology and behavior current in the early years of the twentieth century. His dinosaurs are very reptilian and often heavy and ponderous, dragging their tails. Sauropods are presented in aquatic settings. Yet, some of Knight's artwork was prescient of future ideas about dinosaurs, such as his painting of two agile fighting theropods of the genus *Dryptosaurus* (figure 17.2). During his life, Knight made some 800 drawings, 150 oil paintings, and numerous sculptures of extinct animals, many of dinosaurs. They are an incredible legacy that has influenced the public perception of dinosaurs more than the work of any other artist.

Today, many talented artists sculpt and paint dinosaurs. Their work reflects the latest ideas about dinosaurs, particularly their bird-like appearance, speed, and agility.

FIGURE 17.2
One of Knight's early dinosaur paintings, this view of two theropods, *Dryptosaurus*, fighting, from 1897, presages later ideas about agile theropods. (Courtesy Department of Library Services, American Museum of Natural History, New York [negative no. 335199])

Viewing dinosaur art is much like reading books about dinosaurs. The same criteria of accuracy should be applied to your judgment of dinosaur art, tempered, of course, by your own sense of aesthetics.

## DINOSAUR TOYS

**Dinosaur toys**, typically small, plastic scale models, are a form of dinosaur sculpture. They should be judged by their anatomical accuracy with respect to complete skeletons of dinosaurs. Unfortunately, most toy dinosaurs fall short of the mark. They are typically very bulky, garishly colored, and anatomically inaccurate (note especially the plates on many toy *Stegosaurus*). Many sets of dinosaur toys are also full of nondinosaurs—*Dimetrodon*, pterosaurs, mastodons, mammoths, "cave men," and, in some sets, mythical beasts unknown to science.

The most accurate dinosaur toys have been marketed by some of the world's great natural history museums (figure 17.3). These toys, and other toys that lack the flaws

FIGURE 17.3
These dinosaur toys, endorsed by the Carnegie Museum of Natural History, are among the most accurate on the market.

mentioned earlier, provide reasonably accurate scale models of dinosaurs consonant with modern paleontological thought.

## DINOSAURS IN CARTOONS AND MOVIES

The first character to appear in an animated cartoon was a sauropod dinosaur, Gertie the Dinosaur, created by cartoonist **Winsor McCay** (1869–1934) in 1912 (figure 17.4). Since that time, dinosaurs have been featured in many cartoons, typically coexisting with humans or other animals that did not evolve until long after dinosaur extinction.

Dinosaurs got their start in movies in 1925 when Hollywood produced a silent film version of *The Lost World*. The dinosaurs in this movie were straight out of the paintings of Charles R. Knight (figure 17.5). **Willis Harold O'Brien** (1886–1962), who pioneered the techniques of stop-action photography, joined forces with sculptor **Marcel Delgado** (1901–1976) to produce the scale models of the dinosaurs used in the movie. The life-like quality of these models and their movement has seldom been surpassed by later dinosaur movies. Few readers of this book have likely seen the 1925 version of *The Lost World*, but most know the quality of the O'Brien–Delgado dinosaurs from the classic ***King Kong*** (1933), on which they also collaborated (figure 17.6).

*The Lost World* has left an indelible mark on the public perception of dinosaurs for two reasons. First, its premise that dinosaurs are not extinct, but are still living in some remote corner of the globe, has become the basis of many dinosaur movies. And, at the end of the movie (but not in the book), the discoverers of the dinosaurs bring a live sauropod back to London for exhibition. There, the terrified (enraged?) dinosaur escapes, destroying everything it its path until it crashes through a bridge, landing in the Thames for the long swim back to South America.

FIGURE 17.4
This scene from the animated cartoon *Gertie the Dinosaur* was drawn and photographed by Winsor McCay in 1912. (Courtesy Donald Glut)

FIGURE 17.5
An *Allosaurus* kills an *Apatosaurus* in the silent movie *The Lost World* (1925). (Courtesy Donald Glut)

Thus was born the idea, so familiar to dinosaur movie fans, of a prehistoric monster running amok in a modern city.

Another, and more, recent classic dinosaur movie is the blockbuster ***Jurassic Park***, released in 1993 (box 17.2). This film set a very high bar for dinosaur restorations, presenting active and anatomically accurate dinosaurs based on a cutting-edge scientific understanding. Almost all dinosaur movies that are not based on the idea of living modern dinosaurs are time-travel movies. We, the viewers, together with the humans in the movie, are transported back in time to the Mesozoic to encounter dinosaurs.

**FIGURE 17.6**
King Kong kills a *Tyrannosaurus* after a pitched battle in the movie classic *King Kong* (1933). (Courtesy Donald Glut)

A separate subgenre of dinosaur movies is based on the false idea that humans coexisted with dinosaurs and usually features the misadventures of scantily clad cave women and cave men as they are terrorized by dinosaurs.

All dinosaur movies are based on fantastic premises, so how are we to judge them from a paleontological perspective? Here, our only guide must be the scientific accuracy and quality of motion of the dinosaurs themselves. Many movie dinosaurs are nothing more than iguanas or other lizards, sometimes adorned with spikes or plates (figure 17.7). These movie dinosaurs deserve our scorn as they sprawl across the screen. Other movie dinosaurs, such as **Godzilla,** are utterly fantastic creatures whose resemblance to dinosaurs is at best remote (box 17.3). Many remaining movie dinosaurs are scale models, usually based on the art of Charles R. Knight. However, most recent dinosaur movies and television shows are based on much more modern restorations that were established by *Jurassic Park*. Ironically, the dinosaurs that appeared in the first full-length dinosaur movie, *The Lost World*, are just about as good from a paleontological perspective as those that have appeared in any movie.

FIGURE 17.7
A caveman (Victor Mature) fights off an attacking dinosaur (actually an enlarged iguana) in the movie *One Million B.C.* (1940). (Courtesy Donald Glut)

Box 17.2

## *Jurassic Park*

Michael Crichton's novel *Jurassic Park* (1990) was the basis for the blockbuster movie of the same name released in 1993. (A sequel, *The Lost World*, appeared in 1996; a third movie, *Jurassic Park III*, was shown in 2001; and a fourth installment, *Jurassic World*, hit the big screen in 2015.) The premise of the book (and the movie) is that by extracting dinosaur blood from a Mesozoic mosquito preserved in amber, scientists isolate dinosaur DNA in order to clone living dinosaurs. The long dead, decomposed DNA is incomplete, but the gaps are filled with pieces from the DNA of living frogs.

The living dinosaurs thus cloned are the attraction of a pricey theme park (the "Jurassic Park") located on a remote Caribbean island. The plot unravels as a test of the theme park's accuracy and viability goes terribly awry when a devious computer hacker attempts to steal dinosaur embryos. A dinosaur rampage ensues, and only the luckiest of the good guys (but, alas, not their lawyer!) escape the dinosaur-infested island. The movie thus ends, ripe for a sequel.

(*continued*)

Although the theme park is named Jurassic Park, most of its dinosaurs—*Tyrannosaurus*, ceratopsians, hadrosaurs—were Cretaceous denizens. They share the park's limelight with some characteristically Jurassic dinosaurs, such as the brachiosaurs. Cinematic license produces a venom-spitting *Dilophosaurus* that startles its victims by raising a broad flap of skin on its neck—wholly conjectural behaviors and structures for which no fossil evidence exists. And the movie features a terrifying troop of *Velociraptor* (the "raptors") too large and too smart to be the real thing.

Despite this and other flights of fancy, *Jurassic Park* has phenomenal special effects. Compared to any previous movie, it created dinosaurs more life-like and more closely based on the modern scientific understanding of dinosaurs. Paleontologists and others who watched the movie came as close as they ever will to seeing a dinosaur in the flesh. All subsequent dinosaur movies and documentaries have been held to the high standard of dinosaur restorations established by *Jurassic Park*.

But, what about the basic premise of *Jurassic Park*? Can dinosaur DNA be isolated to clone a living, breathing prehistoric monster? Not now, and probably not ever. The vast majority of insects in amber are of Cenozoic age, so they never had a chance to bite a dinosaur (box figure 17.2). Fragments of dinosaur DNA may be preserved in some fossilized dinosaur bones, but no current technology can begin to replace the missing decomposed pieces of the phenomenally complex DNA molecule to produce a viable structure. Like many other works of science fiction, *Jurassic Park* is based on an apparently plausible premise, but one well beyond current possibilities.

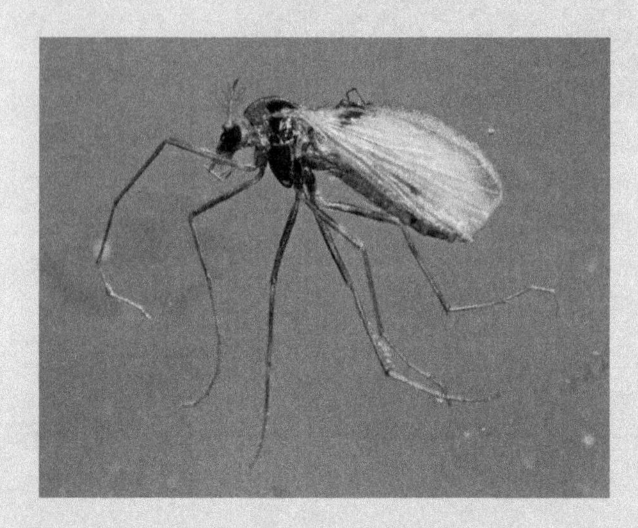

**BOX FIGURE 17.2**
Most fossil insects preserved in amber are of Cenozoic age, so they postdate the dinosaurs. (Courtesy Staatliches Museum für Naturkunde, Stuttgart)

## Box 17.3

### Godzilla: A Lousy Dinosaur

Few movie monsters are as famous as Godzilla. Godzilla is a green, nearly 100-meter-tall, bipedal reptile with fiery, radioactive breath and dorsal spikes. Since it first appeared on the silver screen in the movie *Godzilla, King of the Monsters* (1954), the Japanese behemoth has been the star of at least 30 films (box figure 17.3). Memorable images of Godzilla are of it running amok in Tokyo or other locales in the Japanese islands, destroying buildings, vehicles, and the populace with its huge feet; thick, dragging and swinging tail; and enormous flame-throwing mouth.

Godzilla was created under the direction of Japanese special effects expert Eiji Tsuburaya. An actor inside a suit portrayed the monster as it wreaked havoc on miniature models of Tokyo and other locales. A mechanical model also was used in some scenes.

Generations of Japanese and American moviegoers, especially children, have thrilled to the destructive exploits of and, in some of the movies, sympathetic character portrayed by Godzilla. To many, Godzilla is basically a dinosaur with an overlay of a few novel features of the atomic age, including its appetite for radioactivity.

From a paleontologist's point of view, very little about Godzilla is dinosaurian. True, the overall body shape of Godzilla is basically that of a large theropod. But, everything else is wrong. Godzilla far exceeds the size of any theropod, is much too massively built, has human-like arms and hands, breathes fire, and has spikes on its back like a stegosaur. These are not exaggerations of theropod features, but instead spot Godzilla as a horrific monster not closely related in any way to the theropods despite the filmmaker's premise that Godzilla, hibernating since the Mesozoic, was awakened by American atomic tests at Bikini Atoll during the 1950s.

Nevertheless, don't expect this exposé of Godzilla's non-dinosaurian background to curtail its cinematic career and dim its stardom. Like many movie personalities with unsavory private lives and questionable backgrounds, Godzilla has an onscreen appeal that will transcend its doubtful paternity. The most recent Godzilla movie featured the destruction of much of San Francisco. We can no doubt expect more Godzilla movies and the continued popularity of this fantastic monster!

**BOX FIGURE 17.3**
Movie poster for *Godzilla, King of the Monsters* (1954).

## DINOSAURS ON THE WORLDWIDE WEB

The past two decades have witnessed an explosion of information in cyberspace available on the worldwide web. If you do a search for the subject "dinosaur," at least 80 million(!) webpages can be found. They range from the authoritative web sites of the great natural history museums to commercial sites selling dinosaur T-shirts. How can you cut through this morass of information to find accurate and up-to-date information about dinosaurs?

The easiest route is to visit the website of any natural history museum. The really large museums with extensive dinosaur exhibits are your best source. In North America, they include the Smithsonian's National Museum of Natural History in Washington, D.C. (http://naturalhistory.si.edu), the American Museum of Natural History in New York (www.amnh.org), and the Field Museum of Natural History in Chicago (www.fieldmuseum.org). In Great Britain, the Natural History Museum in London (www.nhm.ac.uk) has information about its dinosaurs on its web page. These, and many other museum web pages, provide varied information about dinosaurs; some even contain virtual tours of the museums' dinosaur exhibits.

However, an even easier way to reach much (or all) of the dinosaur information on the web is to locate one of the many sites that index the sites on dinosaurs. One of the best of these was put together by the University of California Museum of Paleontology in Berkeley (www.ucmp.berkeley.edu). This site provides links to other museum sites and offers a wealth of information, including topics such as dinosaur extinction and metabolism. Table 17.1 lists some informative dinosaur websites.

## DINOSAUR SCIENCE AND PUBLIC DINOSAURS

This review of the public persona of dinosaurs has not been comprehensive. Dinosaurs also appear as jewelry, on postage stamps, as breakfast cereal, and on corporate logos, among other things. The public's view of dinosaurs, however, is often misinformed. Common misconceptions about dinosaurs include the idea that humans coexisted with dinosaurs, the misidentification of a host of nondinosaurs as dinosaurs, the image of dinosaurs as beefy, ponderous reptiles, and the vision of a dinosaur as a terrifying, bloodthirsty brute bent only on savagery and carnage.

Clearly there is a gap between scientific knowledge about dinosaurs and public perception. Much of the public's perception of dinosaurs is rooted in outmoded ideas about dinosaurs. Sensationalism and a thirst for action and adventure also pervade the public perception of dinosaurs.

But, the gap is narrowing. Paleontologists know that dinosaurs are fascinating and exciting animals that can stir public interest without being embellished by unfounded ideas. Our new and expanded knowledge of dinosaurs, much of it accumulated during the past 40 years, is being communicated in new books, art, toys, and cinema. As public interest in dinosaurs continues to grow, we can look forward to even more paleontological interest and information on these fascinating animals.

**TABLE 17.1 Some Informative Dinosaur Websites**

http://www.search4dinosaurs.com/index.html
Extensive illustrated catalog of dinosaurs, A to Z

http://peabody.yale.edu/exhibits/age-reptiles-mural
The famous Yale murals by Rudolph Zallinger; dated, but classic

http://www.enchantedlearning.com/subjects/dinosaurs/
The Zoom Dinosaurs® e-book

www.dinodatabase.com
The Dino Database site has plenty of data.

http://paleobiology.si.edu/dinosaurs/
The Smithsonian's site dispels misconceptions and features its dinosaur exhibits.

http://hoopermuseum.earthsci.carleton.ca
Dr. Ken Hooper Virtual Natural History Museum of the Ottawa-Carleton Geoscience Centre

http://search.eb.com/dinosaurs/dinosaurs/index2.html
An online encyclopedia of dinosaur information provided by Encyclopedia Britannica

http://animals.nationalgeographic.com/animals/prehistoric/
National Geographic's site features all prehistoric animals, including dinosaurs.

http://www.ucmp.berkeley.edu/diapsids/dinosaur.html
Perhaps the most informative dinosaur website of all time

http://news.discovery.com/animals/dinosaurs
The Discovery Channel's site focuses on current dinosaur discoveries.

http://www.bbc.co.uk/nature/14343366
Up-to-date scientific reports on dinosaur science provided by the BBC

## Summary

1. Dinosaurs are extremely popular and appear in the news media, books, sculptures, paintings, toys, cartoons, television shows, movies, and other items available to the general public.
2. The word "dinosaur" refers to a type of extinct reptile with an upright limb posture but also is used to refer to anything that is unwieldy or obsolete. This latter use of the word

is based on the generally large size and extinction of dinosaurs, but this connotation of dinosaurs contradicts their success.

3. News reports about new dinosaur discoveries sometimes are inaccurate and misleading. Such reports need to be viewed critically.
4. Popular dinosaur books frequently perpetuate old, outmoded ideas about dinosaurs and also perpetuate errors, such as the idea that humans coexisted with dinosaurs.
5. Dinosaur sculptures and paintings have changed as scientific ideas about dinosaur biology and behavior have changed.
6. Charles R. Knight's paintings and sculptures of dinosaurs have influenced images of dinosaurs more than those of any other artist.
7. Many dinosaur toys are inaccurate scale models of dinosaurs.
8. The first animated cartoon character was a dinosaur (1912).
9. The first full-length movie that featured dinosaurs was the 1925 Hollywood silent film version of Arthur Conan Doyle's *The Lost World.*
10. Most dinosaur movies are based on either the idea of living, twentieth-century dinosaurs or time travel to the Mesozoic.
11. Many dinosaur movies, like most of the public presentation of dinosaurs, perpetuate outmoded or incorrect ideas about dinosaurs.
12. The gap between dinosaur science and the public perception of dinosaurs is being narrowed by new discoveries and the public's increasing desire to know more about dinosaurs.

## Key Terms

connotation
Marcel Delgado
denotation
dinosaur (two meanings)
dinosaur toys
Arthur Conan Doyle
Godzilla
Benjamin Waterhouse Hawkins
*Jurassic Park*
*King Kong*
Charles R. Knight
*The Lost World*
Winsor McCay
misconceptions about dinosaurs
Willis Harold O'Brien
Weald

## Review Questions

1. What are the two meanings of the word "dinosaur"? Are they accurate definitions?
2. Why do some news reports present inaccurate information about dinosaurs?
3. What questions should you direct at news reports about dinosaur discoveries to evaluate their accuracy?
4. What misconceptions about dinosaurs are perpetuated by some books, art, and movies?
5. How have paleontological ideas about dinosaurs influenced dinosaur art?
6. What are the subgenres of dinosaur movies?
7. What influence did *The Lost World* have on subsequent dinosaur movies?
8. What influence did *Jurassic Park* have on subsequent dinosaur movies?

## Further Reading

Crichton, M. 1990. *Jurassic Park*. New York: Ballantine. (The science fiction book upon which the movie of the same name is based)

Christiansen, P. 2000. Godzilla from a zoological perspective. *Mathematical Geology* 32:231–245. (Mathematical calculations demonstrate the biological impossibility of Godzilla)

Czerkas, S. 2006. *Cine-saurus: The History of Dinosaurs in the Movies*. Blanding, Utah: Dinosaur Museum. (A complete history of dinosaur representations in film with many movie stills and posters)

Czerkas, S. M., and E. D. Olson, eds. 1987. *Dinosaurs Past and Present*. 2 vols. Los Angeles: Natural History Museum of Los Angeles County and University of Washington Press. (Contains pictures of dinosaur artwork featured in an exhibition organized by the Natural History Museum of Los Angeles County, 12 articles about dinosaur art, and new ideas about dinosaurs)

Doyle, A. C. 1912. *The Lost World*. London: Hodder & Stoughton. (Originally serialized in *Strand Magazine*; the story of Professor Challenger's expedition to a high plateau in Amazonia populated by dinosaurs)

Glut, D. F. 1980. *The Dinosaur Scrapbook*. Secaucas, N. J.: Citadel Press. (The subtitle of this book is *The Dinosaur in Amusement Parks, Comic Books, Fiction, History, Magazines, Movies, Museums, Television*)

Milner, R. 2012. *Charles R. Knight: The Artist Who Saw Through Time*. New York: Abrams. (Complete biography of Charles R. Knight and a review of his artwork)

Padian, K. 1988. New discoveries about dinosaurs: Separating the facts from the news. *Journal of Geological Education* 36:215–220. (Discusses, with examples, how to distinguish accurate from inaccurate news reports about dinosaur discoveries)

White, S., ed. 2012. *Dinosaur Art: The World's Greatest Paleoart*. London: Titan Books. (Ten top paleoartists discuss their methods and style and present examples of their work)

# APPENDIX

# A PRIMER OF DINOSAUR ANATOMY

To understand dinosaurs, we must have some familiarity with their anatomy. In this text, this means skeletal anatomy. A dinosaur skeleton is a complex piece of machinery consisting of more than 300 separate bones, each with its own distinctive features. It is from these bones that paleontologists identify different types of dinosaur and glean much of what we know about these animals. In this appendix, I review the salient features of dinosaur skeletons and the anatomical terms used throughout this book.

## POSTURE AND ORIENTATION

Many dinosaurs were **bipeds**, which means they habitually walked on their hind limbs, as do living humans (figure A.1). Other dinosaurs were **quadrupeds**, habitual walkers on all four limbs, such as living horses, cats, and dogs (see figure A.1). Some dinosaurs walked both ways, sometimes on the hind limbs and other times on all fours. Such dinosaurs were **facultative bipeds** (see figure A.1) or **facultative quadrupeds**. Some dinosaurs were facultative bipeds because they walked mostly on all fours but occasionally on their hind limbs, like living bears. Other dinosaurs were facultative quadrupeds because they walked on their hind limbs most of the time and occasionally on all fours, like some living kangaroos.

Whether a dinosaur skeleton belonged to a quadruped or biped or something in between, the terms we use to orient ourselves to it are the same (figure A.2). The direction toward the head is termed **anterior**, and the opposite direction is **posterior**. So, paleontologists speak of the forelimbs as anterior to the hind limbs and the tail as posterior to the back. The belly side of a dinosaur is the **ventral** side, and the back side is **dorsal**. Thus, when we look at a dinosaur skull, the lower jaw is ventral to the eyes,

**FIGURE A.1** Bipeds walk habitually on their hind limbs, and quadrupeds habitually walk on all four limbs. Facultative bipeds and facultative quadrupeds walk part of the time on the hind limbs and part of the time on all fours. (© Mark Hallett. Reproduced with permission of Mark Hallett Paleoart)

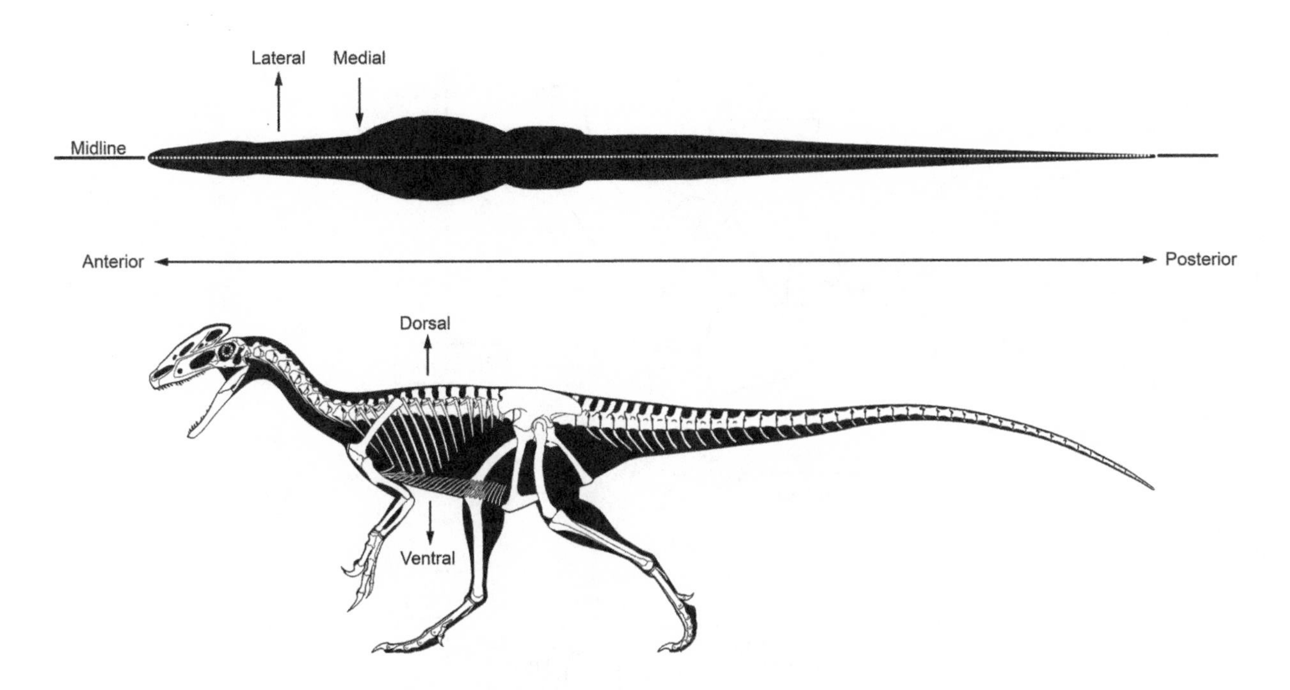


**FIGURE A.2** Dorsal, ventral, anterior, posterior, lateral, and medial are important terms of orientation in the skeletal anatomy of dinosaurs. (© Scott Hartman)

and the nostrils are dorsal to the mouth. Indeed, the top and the bottom sides of a dinosaur skull are referred to as the dorsal and ventral sides, respectively.

In any dinosaur, the **axial** portion of the skeleton consists of the skull, backbone, and tail. In a quadrupedal dinosaur, down and up from the axial skeleton, of course, are ventral and dorsal. But, in bipedal dinosaurs that held their body nearly upright, especially in the front half of the body (such as *Tyrannosaurus rex*), dorsal points backward and ventral points forward, as in humans. If you are confused when locating dorsal and ventral on a bipedal dinosaur, imagine the animal as a quadruped.

From a horizontal perspective, the direction away from the midline of the body is called **lateral.** The opposite direction, toward the midline of the body, is **medial.** Thus, the lungs of a dinosaur are medial to its ribs, but the shoulder blade is lateral to the ribs.

When we speak of dinosaur limbs, however, different terms are used. Those limb segments closer to the body are **proximal** to those farther away. And, those limb segments farther from the body are **distal** to those closer to the body. So, the knee is proximal with respect to the ankle and, conversely, the ankle is distal with respect to the knee (figure A.3).

Now we can orient ourselves with respect to any dinosaur skeleton. We can look at the anterior or posterior ends or the ventral, dorsal, or lateral sides. Furthermore, we can locate in a general way any bone with respect to another. A bone is either ventral,

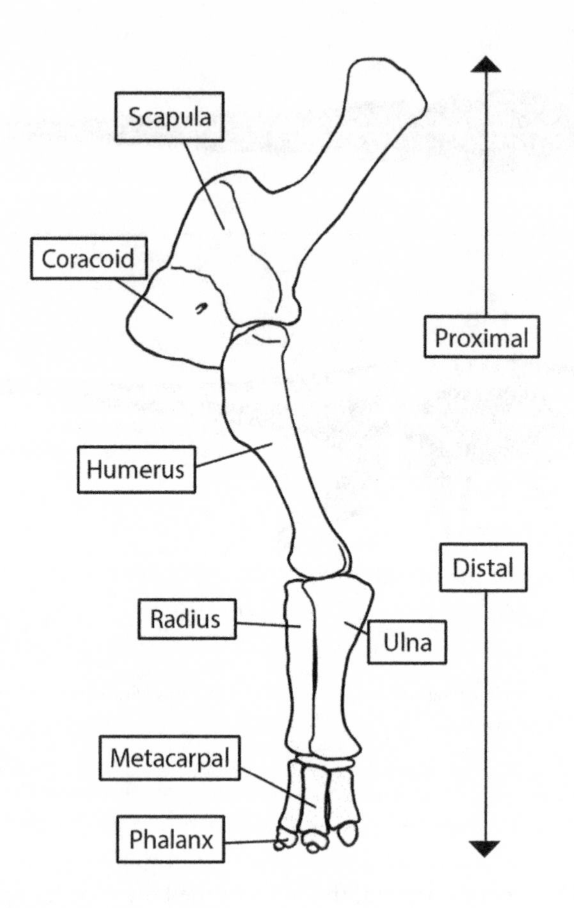


**FIGURE A.3** Limb segments farther away from the body are distal to those closer (proximal) to the body. (© Scott Hartman)

dorsal, anterior, posterior, lateral, or medial (or more than one of these) with respect to another bone. And, in the limbs, a segment is either proximal or distal to another segment.

Dinosaurs characteristically had an **upright posture** (an erect posture). This means the limbs were held directly under the body, as are our hind limbs and the limbs of our pet dogs and cats and most other mammals. This contrasts with the **sprawling posture** of most reptiles, in which the limbs are held out to the side so that the proximal bones are horizontal or nearly horizontal to the ground (see figure 1.2). Some dinosaurs, however, may have held the forelimbs in a semi-sprawling posture intermediate between the upright and sprawling postures.

## SKULL, LOWER JAW, AND TEETH

A dinosaur skeleton consists of a **skull**, or cranium, lower jaw, or **mandible**, and the remaining bones, which are called the **postcrania**. The skull of a dinosaur (figure A.4) is an intricate structure of more than 30 individual bones. These bones are connected

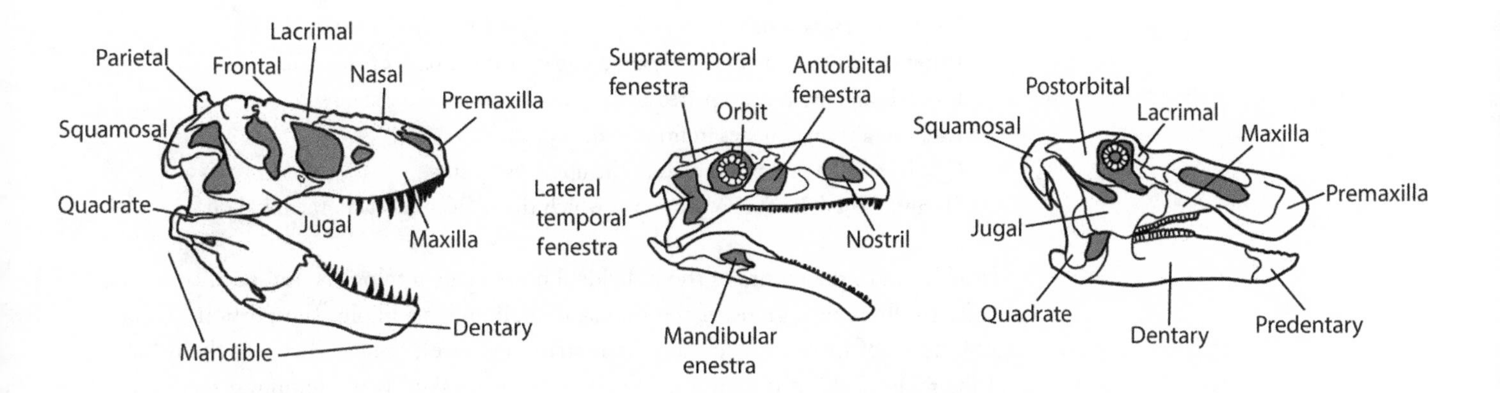


FIGURE A.4 The skull of a dinosaur consists of more than 30 bones, most or all of which are tightly sutured to each other. (© Scott Hartman)

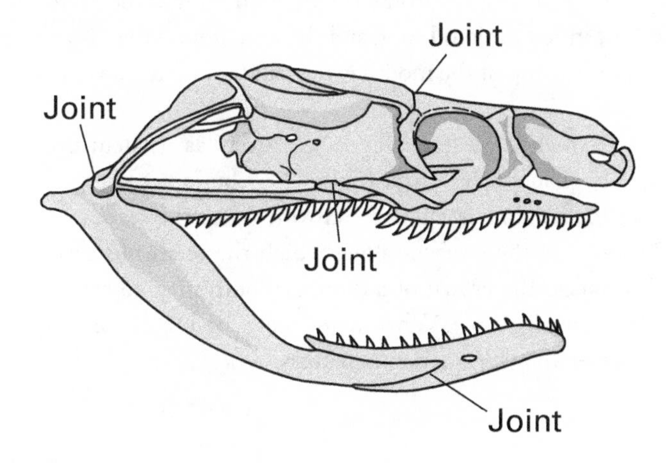


FIGURE A.5 Kinetic skulls, like those of some snakes, have joints that allow skull bones to move past each other. (Drawing by Network Graphics)

to each other along **sutures.** A sutural connection between two bones is a relatively solid connection and allows little or no movement between the two bones. But, many dinosaurs had joints between various bones in their skulls that allowed these bones to move. The skulls of these dinosaurs are referred to as **kinetic** (figure A.5).

Learning the names and locations of all the bones in a dinosaur skull and mandible is not necessary in order to read this book. But, a few key bones should be learned. These bones include the following:

- **Premaxillary**: the bone at the front of the upper jaw, which in some dinosaurs, bears teeth
- **Maxillary**: the upper jaw bone behind the premaxillary, which bears the upper cheek teeth

- **Predentary**: the bone in the front of the lower jaw, present in some dinosaurs
- **Dentary**: the principal bone of the lower jaw, it bears the lower cheek teeth
- **Jugal**: the "cheekbone"
- **Parietals** and **squamosals**: two paired bones near the back of the skull
- **Frontals** and **nasals**: two paired bones near the front of the skull
- **Lachrymals**: small bones in front of the eye sockets
- **Quadrates**: the principal bones of the upper jaw joints
- **Occipital condyle**: the bony knob at which the skull connects to the backbone

In addition to the names of the individual bones, paleontologists use specific terms to refer to different regions of the dinosaur skull and mandible. The portion of the skull in front of the eyes is termed the **rostrum** (or face); the eye socket is the **orbit**; and the region behind the orbit along the side of the skull is the **temporal** region of the skull. The portion of the skull that encloses the brain is the **braincase.** The bones outside of the braincase are **dermal bones.** Openings in these dermal bones in a dinosaur skull are the **temporal fenestrae** (from the Latin for "window"), consisting of a lower one, termed the **lateral temporal fenestra,** and an upper one, the **supratemporal fenestra.** Note that an opening in the mandible is a **mandibular fenestra,** and an opening in the rostrum in front of the orbit is an **antorbital** (literally "in front of the orbit") **fenestra.** The top of the skull is termed the **skull roof,** and the area around the nostrils (**nasal cavity**) is the **nasal region.** The top of the mouth is the **palate,** and the lower jaw is often called the mandible.

All the teeth in a dinosaur's mouth are referred to collectively as its **dentition.** Dinosaur teeth come in a variety of shapes and sizes (figure A.6), but, for all of them, the portion of the tooth above the gumline is called the **crown.** When the teeth mesh together, we speak of **occlusion,** and the surfaces along which the teeth meet each other are thus the **occlusal surfaces.** The crown of a dinosaur tooth may be covered with **lophs** (ridges) or **denticles** (small cusps). When many teeth are present and are attached to (or overlap) each other, they form a **dental battery.**

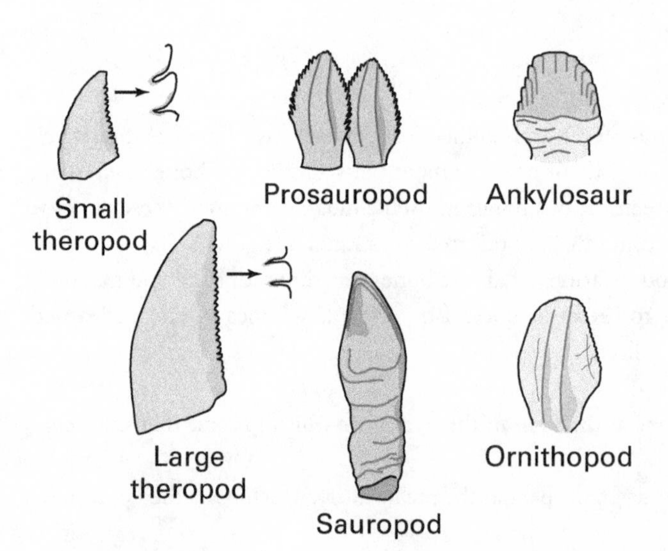


**FIGURE A.6** Dinosaur teeth come in many shapes and sizes. (Drawing by Network Graphics)

## BACKBONE

The backbone of a dinosaur is a **vertebral column** composed of numerous separate bones called **vertebrae** (figure A.7). Each dinosaur's vertebral column can be divided into four regions: **cervical** (neck), **dorsal** (back), **sacral** (hips), and **caudal** (tail). The number of cervical vertebrae for most dinosaurs is 9 or 10, but hadrosaurs have as many as 15, and some sauropods have as many as 19 cervical vertebrae. Sauropods have an increased number of cervical vertebrae, in part at the expense of their dorsal vertebrae, where they have as few as 9. Most dinosaurs have 15 to 17 dorsal vertebrae and 3 to 5 sacral vertebrae making up the **sacrum**, but ceratopsians have as many as 10 sacrals. The number of caudal vertebrae in dinosaurs ranges from as few as 35 to as many as 82 (in the sauropod *Apatosaurus*).

Each vertebra has a spool-shaped body along its ventral side, the **centrum** (plural: **centra**). Dorsal to the centrum is the **neural arch**, which covered the spinal cord (it ran through a canal between the centrum and the neural arch). The thin rod or blade of bone that projects dorsally from the neural arch is the **neural spine**. Ventral to the centrum of the caudal vertebrae of some dinosaurs is another arch-like structure, the **chevron**. The centra meet each other front to back, and bony ridges link the neural arches with each other, thereby producing an **articulated** (connected) vertebral column.

Separate ribs attach to the cervical and dorsal vertebrae of dinosaurs. Some dinosaurs also had rib-like bones, the **gastralia**, covering their bellies.

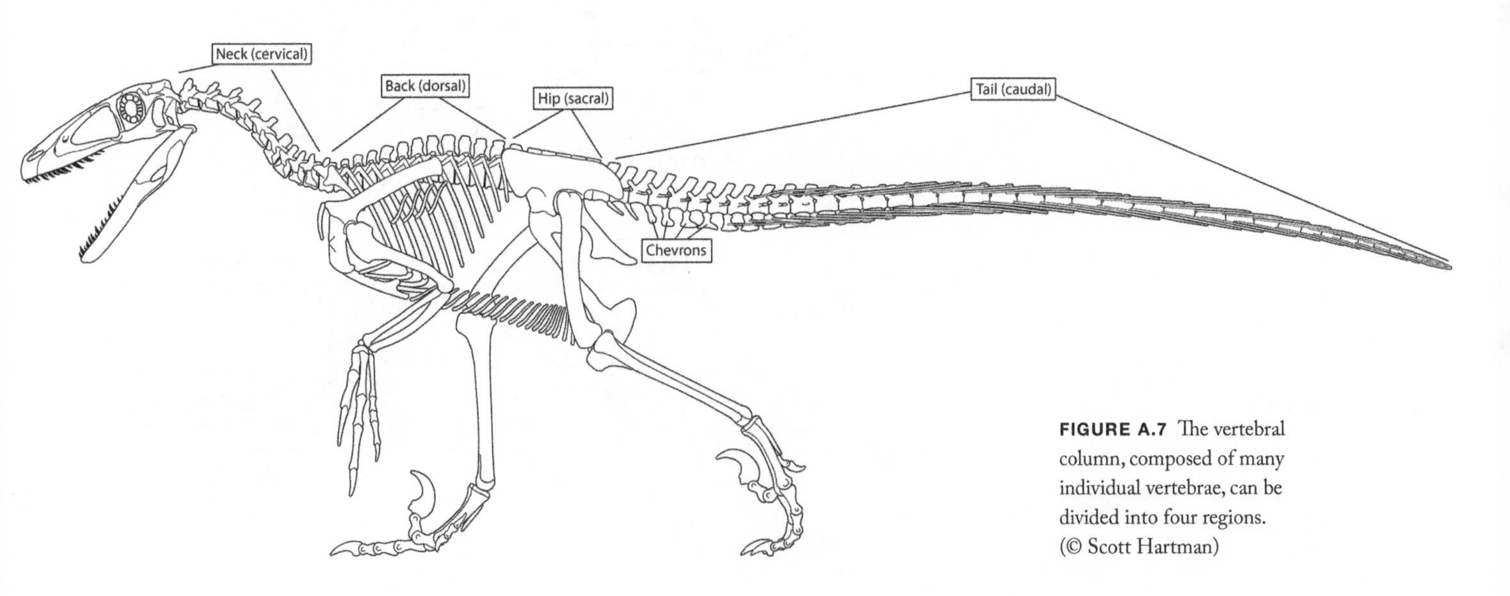


**FIGURE A.7** The vertebral column, composed of many individual vertebrae, can be divided into four regions. (© Scott Hartman)

## FORELIMB

The forelimb skeleton of a dinosaur attaches to the body at the **shoulder girdle** (figure A.8). The largest and most prominent bone of the shoulder girdle is the **scapula**. A smaller bone is the **coracoid**. The shoulder girdle in dinosaurs, like ourselves, is held near the front end of the rib cage by muscles and other tissues. Many dinosaurs also had a **clavicle** (collarbone) connecting the scapula to the **sternum**, a row of bones on the ventral midline of the dinosaur.

The shoulder joint of a dinosaur is the point of articulation between the scapula and the single bone of the upper arm, the **humerus**. At the elbow joint, the humerus articulates with the two bones of the lower arm, the **radius** and the **ulna**.

At the wrist joint, the radius and ulna meet the **carpals**, the separate bones of the wrist. Distally, the carpals meet the **metacarpals**, which, in turn, meet the bones of the fingers, the **phalanges** (singular: **phalanx**). Pointed terminal (distalmost) phalanges usually bore claws, whereas flattened terminal phalanges bore **hooves** or hoof-like coverings.

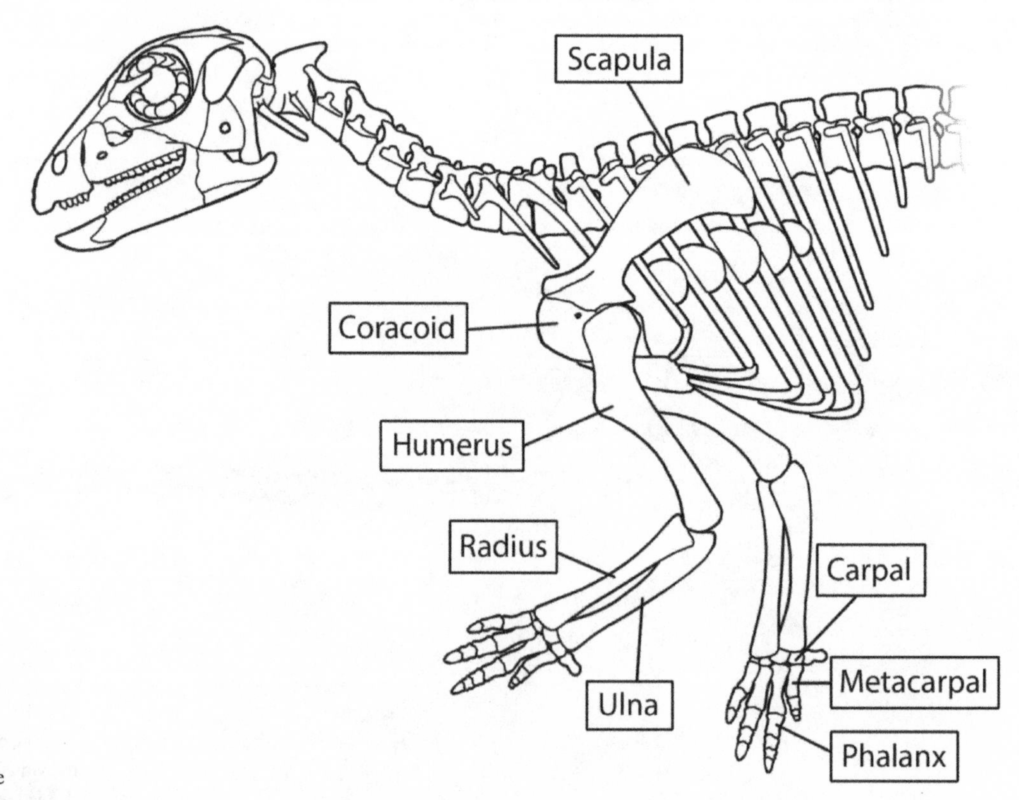


**FIGURE A.8**
The forelimb skeleton of a dinosaur extends from the shoulder girdle to the phalanges. (© Scott Hartman)

## HIND LIMB

The hind limb of a dinosaur attaches to the body at the **pelvis** (figure A.9). The dinosaur pelvis consists of three bones on each side of the body: the **ilium**, **ischium**, and **pubis** (plural: ilia, ischia, and pubes). Differences in the shapes of these bones, especially in the shape of the pubis, are extremely important in the classification of dinosaurs.

The dinosaur pelvis is securely sutured along the medial surface of the ilia to the sacral vertebrae. The single bone of the upper leg, the **femur**, fits into a socket, called the **acetabulum**, at the junction of the three pelvic bones. Distally, the femur articulates with the **tibia**. The lower leg of a dinosaur also has another, smaller bone, the **fibula**, which attaches to the tibia.

The tibia and fibula articulate distally with the bones of the ankle, the **tarsals**. Two tarsals are important to understanding the origin of dinosaurs. They are the most proximal tarsals, the **astragalus** and **calcaneum** (see figure A.9).

Distally, the tarsals articulate with the **metatarsals**, which, in turn, articulate with the toe bones, the **phalanges**. As in the forelimb, the phalanges of the hind limb may have borne claws, hooves, or hoof-like coverings, depending on their shape. Fingers and toes of dinosaurs are usually referred to by a single term, **digits**.

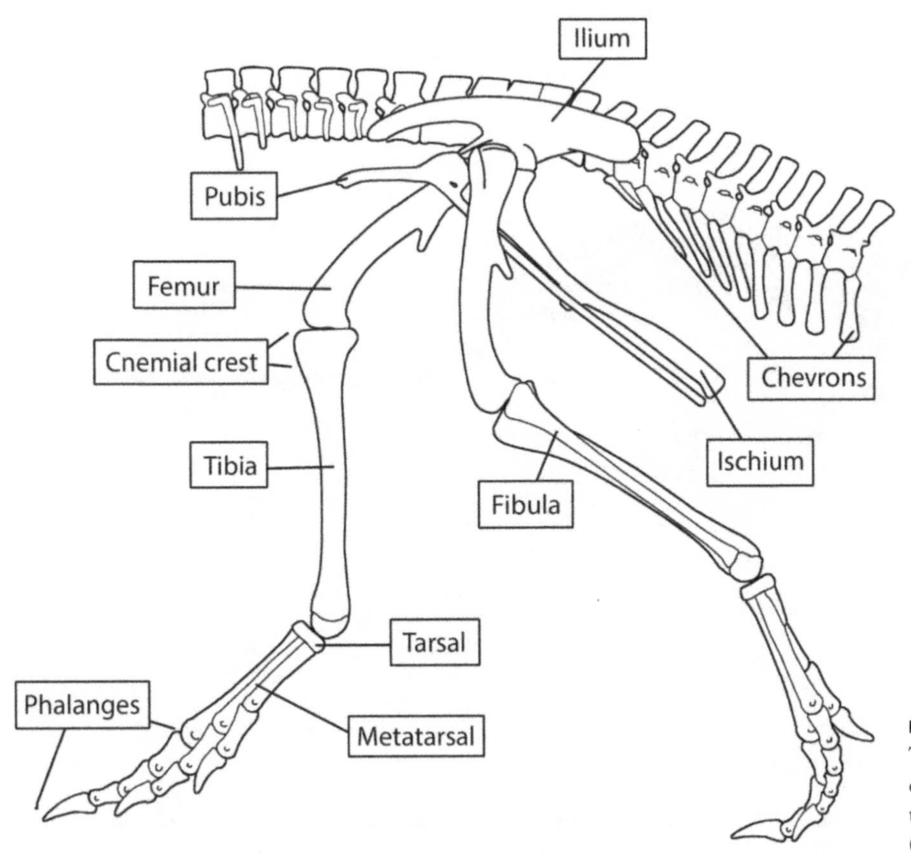


**FIGURE A.9**
The hind-limb skeleton of a dinosaur extends from the pelvis to the phalanges. (© Scott Hartman)

**FIGURE A.10**
The skeleton of a dinosaur provides the primary basis for interpretations of soft-tissue anatomy and behavior: (*A*) the skeleton; (*B*) the same skeleton with muscles restored; (*C*) the dinosaur as it may have looked in life. (© Mark Hallett. Reproduced with permission of Mark Hallett Paleoart)

## STRUCTURE AND FUNCTION

This book makes many statements about the behavior of dinosaurs. Some dinosaurs are identified as quadrupeds, others as bipeds. We distinguish plant-eating and meat-eating dinosaurs. Differing ideas about the forelimb postures of some dinosaurs are discussed.

These types of statements reflect the relationship between a particular structure and its **function**. This relationship is one of the foundations of paleontology. The size and shape of a single bone, of several bones, or of an entire skeleton depend on, as well as determine, the function of that structure. This is because the skeleton of a dinosaur is the framework to which the muscles were attached and upon which the other soft tissues were hung. How the muscles moved largely depended on the skeleton to which they were attached. And, the shape, size, and arrangement of the other soft tissues—internal organs, blood vessels, and so on—were very much influenced by the skeleton around them. So, it is possible to examine a skeleton and make some inferences about the dinosaur's muscles and other soft tissues (figure A.10). This in turn permits an understanding of the appearance and behavior of the dinosaur.

Identification of dinosaurs as bipeds, quadrupeds, or something in between earlier in this appendix and throughout this book provides a good example of the inference of function from structure. Among living animals, bipeds share much larger and more massive (hence stronger) hind-limb skeletons. This makes sense mechanically because the hind limbs of a biped must propel and support its entire weight. In contrast, living quadrupeds have more nearly equally sized forelimbs and hind limbs. Furthermore, the structure of the hips and vertebral columns of living bipeds and quadrupeds differ because of their very different postures while locomoting.

We can look for such structures in dinosaur skeletons as a key to their postures, or more easily just look at relative limb sizes. A clear and easy inference of function from structure can be made by looking at the skeleton of *Tyrannosaurus* (see figure 5.10). Its tiny forelimbs and massive hind limbs indicate bipedality. The more nearly equally sized forelimbs and hind limbs of *Diplodocus* (see figure 6.8), however, indicate quadrupedality.

## Key Terms

acetabulum
anterior
antorbital fenestra
articulated
astragalus
axial
biped
braincase
calcaneum
carpal
caudal
centrum (centra)
cervical
chevron
clavicle
claw
coracoid
crown
dental battery
dentary
denticle
dentition
dermal bone
digit
distal
dorsal
facultative biped
facultative quadruped
femur
fibula

frontal
function
gastralia
hoof
humerus
ilium
ischium
jugal
kinetic
lachrymal
lateral
lateral temporal fenestra
loph
mandible
mandibular fenestra
maxillary
medial
metacarpal
metatarsal
nasal
nasal cavity
nasal region
neural arch
neural spine
occipital condyle
occlusal surface
occlusion
orbit
palate
parietal
pelvis
phalanx (phalanges)
postcrania
posterior
predentary
premaxillary
proximal
pubis
quadrate
quadruped
radius
rostrum
sacral
sacrum
scapula
shoulder girdle
skull
skull roof
sprawling posture
squamosal
sternum
supratemporal fenestra
suture
tarsal
temporal
temporal fenestra
tibia
ulna
upright posture
ventral
vertebra
vertebral column

## Review Questions

1. Orient yourself with respect to a dinosaur skeleton. Relative to the femur, where are the humerus, the maxillary, the tibia, and a caudal vertebra?
2. Compare the segments of the forelimb and hind-limb skeletons, and note any similarities and differences.
3. What is the general relationship between skeletal structure and function?

## Further Reading

Dilkes, D. W., J. R. Hutchinson, C. M. Holliday, and L. M. Witmer. 2012. Reconstructing the musculature of dinosaurs, pp. 151–190, in M. K. Brett-Surman, T. R. Holtz, Jr., and J. O. Farlow, eds., *The Complete Dinosaur*. 2nd ed. Bloomington: Indiana University Press. (A detailed review of how modern muscle anatomy is used to reconstruct dinosaur musculature)

Hildebrand, M., D. M. Bramble, K. F. Liem, and D. B. Wake. 1985. *Functional Vertebrate Morphology*. Cambridge, Mass.: Belknap Press of Harvard University Press. (An intermediate-level college textbook that examines all aspects of the relationship between vertebrate structure and function)

Hildebrand, M., and G. E. Goslow, Jr. 2001. *Analysis of Vertebrate Structure*. 5th ed. New York: Wiley. (A college textbook on the inference of structure from function)

Holtz, T. R., Jr., and M. K. Brett-Surman. 2012. The osteology of the dinosaurs, pp. 135–150, in M. K. Brett-Surman, T. R. Holtz, Jr., and J. O. Farlow, eds., *The Complete Dinosaur*. 2nd ed. Bloomington: Indiana University Press. (A review of dinosaur skeletal anatomy)

Romer, A. S. 1997. *Osteology of the Reptiles*. Chicago: University of Chicago Press. (A detailed technical review of all aspects of the skeletons of reptiles, including those of the dinosaurs; reprint of the 1956 classic)

Romer, A. S., and T. A. Parsons. 1977. *The Vertebrate Body*. 5th ed. Philadelphia: Saunders. (A thorough review, in introductory textbook form, of all aspects of vertebrate anatomy, including two chapters on the skull and the skeleton)

# GLOSSARY

**advanced mesotarsal (am) ankle** The ankle of pterosaurs, dinosaurs, and birds, in which the only hinge is between the astragalus and calcaneum and the rest of the foot. The astragalus is much larger than the calcaneum, and both bones are rigidly attached to each other and to the tibia.

**aetosaur** A heavily armored, plant-eating archosaur with a crocodile-normal (CN) ankle.

**alveolar lung** Lungs made of millions of tiny, very vascularized, spherical alveoli; the lungs of mammals.

**ammonoid** Any of a group of extinct invertebrates, most of which have coiled shells, that were close relatives of living octopuses, squids, and *Nautilus.*

**amniotic egg** An egg in which the developing embryo is almost totally surrounded by a liquid-filled cavity enclosed by the amnion, which is a membrane continuous with the skin of the embryo.

**amphicoelous** A vertebra with sockets (cavities) in the centrum both anteriorly and posteriorly.

**anapsid** A reptile with a skull that lacks upper and lower temporal fenestrae.

**angiosperm** A vascular plant in which the seed is covered by an ovary; a flowering plant.

**archosaurs** Reptiles that include crocodilians, pterosaurs, and dinosaurs. A variety of skeletal evolutionary novelties distinguish the archosaurs from the other diapsids, the lepidosaurs.

**binomen** The two-word name of a species in the Linnaean hierarchy; for example, *Tyrannosaurus rex.*

**biological classification** The grouping of organisms into categories (taxa).

**biostratigraphy** The study of the distribution of fossils in layered rocks (strata).

**biped** An animal that walks on its hind limbs.

**body fossil** Fossil of a body part, usually skeletal, of an organism.

**bone microstructure** The microscopic arrangement of the inorganic matrix and organic component of bone.

**calcium phosphate** The organic mineral matrix of bone.

**Cenozoic Era** The past 66 million years of earth history; the geological time interval after dinosaur extinction.

**clade** A branch or cluster of branches with a common stem on a phylogeny. A synonym of **monophyletic group.**

**cladistic phylogeny** A phylogeny based on shared evolutionary novelties.

**cladogram** A diagram of a cladistic phylogeny.
**clutch** A nest of eggs.
**cold-blooded** A popular term referring to an animal that receives most, or all, of its body heat from external sources, usually the sun; a virtual synonym of **ectothermic.**
**compact bone** The dense, tightly woven exterior layer of most vertebrate bones.
**connotation** The meaning suggested by a word other than what it explicitly names or describes. *See also* **denotation.**
**convergence** The evolution of similar features in two distinct clades.
**coprolite** Fossilized feces.
**Cretaceous Period** The last period of the Mesozoic Era, approximately 65 to 145 million years ago.
**crocodile-normal (cn) ankle** The ankle of crocodiles and many archosaurs, in which the ankle twists while the animal walks.
**cross-sectional area** The area of the surface at right angles to the axis (of a bone).
**deltaic** Having to do with a delta, a triangular body of sediment formed where a river enters a large, quiet body of water, either a sea or lake.
**denotation** The direct, specific meaning of a word, in contrast to its connotation.
**dental battery** A large number of teeth cemented to each other to form an extensive shearing and grinding surface.
**dermal bones** Bones in the integument of a tetrapod.
**determinate growth** The pattern of growth of endothermic animals that stop growing at maturity.
**diapsid** A reptile with two temporal fenestrae on each side of its skull.
**dimensionless speed** The actual speed of an animal divided by some dimension of its body size.
**dinosaur** An archosaurian reptile having an upright posture and an advanced mesotarsal (AM) ankle.
**dinosaur fauna** All dinosaurs that lived during a specific interval of geologic time.
**dinosaur track (footprint)** A fossilized dinosaur footprint.
**divergence** The evolutionary splitting of one clade into two.
**ectotherm** An animal that receives most, or all, of its body heat from external sources, usually from the sun. The popular term **cold-blooded** is a virtual synonym of **ectothermic.**
**ejecta** Rock material thrown from an impact crater.
**endocast** A natural or human-made cast (replica) of the brain cavity; short for "endocranial cast."
**endotherm** An animal that generates most, or all, of its body heat internally. The popular terms **hot-blooded** and **warm-blooded** are virtual synonyms of **endothermic.**
**eolian** Having to do with wind.
**epicontinental sea** Marine waters on top of continental crust.
**equable** Marked by a lack of variation or change (roughly uniform).
**evolution** The origin and change of organisms over time.
**evolutionary novelty** A feature unique to a group of animals that may indicate they have a close common ancestry.
**extinction** The disappearance of a species or of a larger clade of animals or plants.
**extramophological variation** Features of a trace fossil due to factors other than the anatomy (morphology) of the trace maker.
**facultative biped** An animal that is normally a quadruped but that occasionally walks bipedally.
**facultative quadruped** An animal that is normally a biped but that occasionally walks quadrupedally.
**family** The category in the Linnaean hierarchy above genus; a group of genera with a shared evolutionary ancestry.

**feeding range** The vertical height above the ground at which a plant eater forages.

**flood plain** A plain built by stream deposits adjacent to river channels.

**fluvial** Having to do with rivers.

**footprint (track)** *See* **dinosaur track (footprint).**

**foraminifera** Microscopic single-celled animals that have a hard shell and live in the sea.

**fossil** Any evidence of past life. Dinosaur fossils include bones and teeth, footprints, coprolites, gastroliths, tooth marks, eggs, and skin impressions.

**four-chambered heart** A vertebrate heart that separates blood containing oxygen from blood lacking it; characteristic of endotherms.

**frill** A posterior-projecting shelf of bone at the back of the skull usually composed of the parietals and squamosals.

**function** The specific contribution of a body part to the biology and behavior of an organism.

**furcula** The "wishbone" of birds; fused clavicles (collarbones).

**gastralia** Rib-like bones covering the stomach region.

**gastrolith** A "fossilized" gizzard or stomach stone.

**gastromyth** Supposed gastrolith that is actually just a polished stone.

**genus (plural: genera)** The category in the Linnaean hierarchy above species; a group of species with a shared evolutionary ancestry.

**gigantotherm** A large animal with a nearly constant body temperature due to its large size. A synonym of **inertial homeotherm.**

**Gondwana** The southern supercontinent of the Paleozoic and Mesozoic composed of Africa, Australia, Antarctica, South America, India, and Madagascar.

**grade** A set of clades that do not have a close common ancestor.

**graviportal** A quadrupedal animal with pillar-like legs and massive shoulder and hip girdles designed to bear great weight during slow and powerful locomotion.

**greenhouse** A warm, equable, and humid climate.

**gymnosperm** A vascular plant with a "naked" seed not enclosed in an ovary.

**herd** A group of wild animals usually led by a dominant individual.

**heterotherm** An animal whose body temperature varies considerably daily, seasonally, or throughout its life.

**homeotherm** An animal that maintains a nearly constant body temperature.

**hot-blooded** A somewhat sensational synonym of **warm-blooded.** *See also* **endotherm.**

**ichthyosaur** Any of a group of Mesozoic reptiles that converged remarkably on a fish-like appearance and lived in the sea.

**igneous rock** A rock that has cooled from a molten state.

**indeterminate growth** The pattern of growth of most ectothermic animals in which growth continues, although at a decreasing pace, throughout life.

**inertial homeotherm** A synonym of **gigantotherm.**

**inoceramid** A type of thin-shelled, plate-like clam of the Cretaceous Period.

**iridium** A platinum-group metal rare on earth but more abundant in meteorites and asteroids.

**Jurassic Period** The middle interval of the Mesozoic Era, about 145 to 201 million years ago.

**lacustrine** Having to do with lakes.

**Laurussia** The northern supercontinent of the Paleozoic and Mesozoic composed of North America and most of Eurasia.

**lepidosaur** Any of a group of diapsid reptiles that includes lizards, snakes, and their close relatives.

**Linnaean hierarchy** The categorical system used by biologists and paleontologists to classify organisms; named after Carolus Linnaeus.

**locomotion** The movement of an animal.

**Maastrichtian** The last time interval of the Cretaceous Period, about 66 to 70 million years ago.

**macroevolution** Evolution above the species level. The origination and evolution of taxa larger than species.

**marine** Having to do with the seas or oceans.

**marsupial** A mammal in which the young typically are born very immature and initially live in a pouch (marsupium) on the mother's abdomen.

**mass death assemblage** A large accumulation of fossils of a single species of organism, often thought to be the result of a single catastrophe.

**Mesozoic Era** The interval of geological time 66 to 251 million years ago, when dinosaurs lived.

**metabolism** The chemical processes that provide energy to and repair the cells of an organism.

**metamorphic rock** A rock that has been altered by great heat and/or pressure.

**microevolution** Evolution at the species level. The evolution of populations of organisms that results in the origination of new species.

**mineralization** The replacement of organic or inorganic matter by minerals such as silica, calcite, or iron during the process of fossilization.

**missing link** A fossil that fills a significant gap in our knowledge of the evolution of a group of organisms. Typically, missing links bridge the evolutionary gap between two major groups of organisms.

**monophyletic group** A taxon that includes all taxa derived from a single ancestor. A synonym of **clade**.

**monsoon** A period of high winds and rainfall.

**mosasaur** Any of a group of extinct, giant marine lizards of the Cretaceous Period.

**natural selection** The process by which an organism that is better adapted to an environment more successfully reproduces than less-adapted organisms.

**nest** A bed or receptacle prepared by an animal for its eggs and young.

**numerical time scale** A classification of geologic time based on numbers (usually millions) of years.

**opisthocoelous** The condition of a vertebra with a socket (cavity) in the centrum facing posteriorly.

**origination** The appearance of a new type of organism.

**pace (step) angle** The angle between a line drawn from the posterior margin of one hind footprint to an equivalent point on the next hind footprint on the opposite side, and a second line drawn to the equivalent point on the next hind footprint of the same side as the first.

**pace length** The distance between the posterior margins of two successive footprints of the hind limbs on opposite sides of the body measured parallel to the direction of travel.

**paleontologist** A scientist who studies fossils and the history of life.

**paleontology** The scientific study of fossils and the history of life.

**Paleozoic Era** The interval of geological time 251 to 541 million years ago, before dinosaurs lived.

**Pangea** The supercontinent of the Paleozoic and Mesozoic comprising all of the present continents before breakup.

**Panthalassa** The single, gigantic ocean that existed when Pangea was assembled.

**paraphyletic group** A taxon that does not include all of its descendants.

**phylogenetic relationships** The relationships of taxa to each other.

**phylogenetic tree** A diagram that depicts the phylogenetic relationships of a group of taxa and has geologic time as its vertical axis.

**phylogeny** The history of how a group of organisms evolved.

**phytosaurs** A Late Triassic group of quadrupedal, meat-eating, crocodilian-like archosaurs having crocodile-normal (CN) ankles.

**placental** Describes a type of mammal in which the young develop in a placenta and are born relatively mature compared to newborn marsupials.

**plesiosaur** Any of a group of extinct, long-necked marine reptiles of the late Triassic, Jurassic, and Cretaceous.

**pneumatic** Having to do with air; a pneumatic bone is hollow and perforated so air can pass into it.

**pneumatic ducts** Openings (perforations) in the long bones of birds for the passage of air.

**polyphyletic group** A group of organisms from which its common ancestor is excluded.

**postcrania** The bones of the skeleton other than those of the skull and lower jaw.

**predator–prey ratio** The biomass of predators divided by the biomass of prey.

**principle of biostratigraphic correlation** The generalization that rocks containing the same kinds of fossils are of the same age.

**principle of priority** The rule that states that the first name of a taxon properly introduced has priority over (is used instead of) names later introduced for the same taxon.

**principle of superposition** The generalization that in layered rocks (strata), the older rocks are at the bottom, and the younger rocks are on top.

**procoelous** A vertebra with a socket (cavity) in the centrum facing anteriorly.

**pterosaur** A flying archosaur of the Late Triassic through to the Late Cretaceous; not a dinosaur.

**quadruped** An animal that habitually walks on four limbs.

**radioactive clock** A means of measuring geologic time by the radioactive decay of unstable atoms.

**ratites** Some of the flightless birds, including living ostriches and emus.

**rauisuchians** A group of Triassic meat-eating archosaurs having crocodile-normal (CN) ankles.

**relative time scale** A way to measure geologic time by determining if one event is older than or younger than another by assigning each event to a named interval of geologic time.

**reptile** A tetrapod that lays an amniotic egg and has scaly skin.

**resonating chamber** An air space that intensifies sound waves that enter it.

**respiratory turbinates** Small bones in the nasal passages of mammals and birds that increase the surface area over which blood and moist tissues are exposed to the air.

**reworked** A fossil is said to have been reworked if, after initial fossilization in the rock, it was exhumed and redeposited in an overlying, younger layer of rock.

**rudist** Extinct, reef-building clams of the Cretaceous Period.

**sedimentary (depositional) environment** The place where and particular conditions under which a sedimentary rock was formed.

**sedimentary rock** A rock formed by the accumulation and cementation of mineral grains or by chemical precipitation.

**selenium** A nonmetallic, often poisonous, element.

**septate lung** A tetrapod lung that is like one huge alveolus and has vascularized septa that penetrate it from its perimeter

**sexual dimorphism** Anatomical differences between the male and female of a single species.

**"shocked quartz"** A type of quartz grain with a laminar deformation known to be generated only by high-speed shock in laboratories, at nuclear test sites, and near impact craters.

**soft-tissue anatomy** That part of a vertebrate that is not skeleton, such as muscles, nerves, the circulatory system, and internal organs.

**species (plural: species)** The lowest grouping in the Linnaean hierarchy; a group of populations with a shared evolutionary ancestry.

**sprawling posture** The posture in which the limbs are held out to the side of the body so that the upper segments are horizontal to, or nearly horizontal to, the substratum.

**strata (singular: stratum)** Layers of sedimentary rock.

**stratigraphy** The scientific study of layered rocks (strata).
**stratophenetic phylogeny** A phylogeny that relies on the relative geologic age and overall similarity of taxa to identify ancestors and descendants.
**stratum (plural: strata)** A single bed or layer of sedimentary rock.
**stride** The distance between the posterior margins of two successive hind footprints on the same side of the body in the direction of travel.
**substratum** The ground surface.
**supernova** The complete obliteration of a star in an explosion.
**surface-area-to-volume relationship** The observation that volume increases as the cube of a linear dimension, whereas surface area increases only as the square of a linear dimension. Animals of large volumes have relatively less surface area than animals of smaller volume.
**taphonomy** The scientific study of the processes by which a fossil assemblage is formed and the losses of information these processes produce.
**temnospondyl** A type of extinct amphibian of the Paleozoic and Mesozoic.
**temporal fenestra** An opening for the passage and attachment of muscles in the skull of a reptile behind its eyes.
**Tethys Sea** The seaway between Laurussia and Gondwana during the Mesozoic.
**trace fossils** Fossils that are not part of the actual body of an animal. Dinosaur trace fossils ainclude footprints, eggs, gastroliths, tooth marks, and coprolites.
**trackway** A series of footprints made by a single animal.
**Triassic Period** The oldest period of the Mesozoic Era, 201 to 251 million years ago.
**upright posture** The posture in which the limbs are held vertically underneath the body so that they are vertical to, or nearly vertical to, the substratum.
**warm-blooded** A popular term virtually synonymous with **endothermic.** *See also* **hot-blooded.**

# A DINOSAUR DICTIONARY

All the dinosaur generic names that appear in this book are listed here with a guide to pronunciation, a brief identification of the dinosaur, and the derivation of the genus name. Note that *sauros* is Greek for "lizard" or "reptile."

*Alamosaurus* **(AL-uh-mo-sore-us)** A Late Cretaceous sauropod from the western United States [for Ojo Alamo, a spring in New Mexico].

*Albertosaurus* **(al-BURR-tuh-sore-us)** A Late Cretaceous tyrannosaurid from western North America [for Alberta, Canada].

*Allosaurus* **(AL-uh-sore-us)** A Late Jurassic theropod from North America and East Africa [Greek *alias* (strange)].

*Ammosaurus* **(AM-oh-sore-us)** An Early Jurassic prosauropod from North America [Greek *ammos* (sand)].

*Amphicoelias* **(am-fee-see-lee-us)** A Late Jurassic sauropod from North America [Greek *amphi* (double) + *koilos* (hollow)].

*Anchisaurus* **(AN-key-sore-us)** An Early Jurassic prosauropod from North America and southern Africa [Greek *anchi* (near or close to)].

*Ankylosaurus* **(ang-KY-low-sore-us)** A Late Cretaceous ankylosaur from North America [Greek *ankylos* (stiff or fused)].

*Antarctosaurus* **(ant-ARK-toe-sore-us)** A Late Cretaceous sauropod from South America and Asia [Greek *antarktiko* (southern)].

*Apatosaurus* **(uh-PAT-oh-sore-us)** A Late Jurassic sauropod from North America [Greek *apatel* (deceit)].

*Aquilops* **(uh-KWILL-ops)** An Early Cretaceous neoceratopsian from North America [Latin *aquila* (eagle) + Greek *ops* (face)].

*Argentinosaurus* **(ar-jen-teen-oh-SORE-us)** A Late Cretaceous sauropod from Argentina [for Argentina].

*Aucasaurus* **(ah-KOW-sore-us)** A Late Cretaceous theropod from Argentina [for Auca Manhuero, a region in Argentina].

*Bagaceratops* **(bag-uh-SAYR-uh-tops)** A Late Cretaceous ceratopsian from Asia [Mongolian *baga* (small) + Greek *ceratops* (horned face)].

*Bambiraptor* **(BAM-bee-rap-tor)** A Late Cretaceous theropod from North America [for the Disney character Bambi + Latin *raptor* (robber)].

*Barapasaurus* **(bah-RAP-uh-sore-us)** An Early Jurassic sauropod from India [Hindi *bara* (big) + *pa* (leg)].

*Barosaurus* **(BAHR-oh-sore-us)** A Late Jurassic sauropod from North America and eastern Africa [Greek *barys* (heavy)].

*Brachiosaurus* **(BRAK-ee-oh-sore-us)** A Late Jurassic sauropod from North America and eastern Africa [Greek *brachion* (arm)].

*Brontosaurus* **(BRON-toe-sore-us)** A Late Jurassic sauropod from North America; not a valid name, which is *Apatosaurus* [Greek *bronte* (thunder)].

*Camarasaurus* **(KAM-uh-ruh-sore-us)** A Late Jurassic sauropod from North America [Greek *kamara* (chamber)].

*Camptosaurus* **(KAMP-toe-sore-us)** A Late Jurassic ornithopod from North America and Europe [Greek *kamptos* (flexible)].

*Caudipteryx* **(caw-DIP-ter-iks)** An Early Cretaceous theropod from China [Latin *cauda* (tail) + Greek pteryx (wing, feather)].

*Centrosaurus* **(SEN-tro-sore-us)** A Late Cretaceous ceratopsian from North America [Greek *kentron* (sharp point)].

*Ceratosaurus* **(sir-AT-oh-sore-us)** A Late Jurassic theropod from North America and eastern Africa [Greek *keratos* (horned)].

*Cetiosaurus* **(SEAT-ee-oh-sore-us)** A Middle to Late Jurassic sauropod from England and northern Africa [Greek *keteios* (whale-like)].

*Chaoyangsaurus* **(chow-YANG-sore-us)** A Jurassic ceratopsian from China [for Chaoyang, a region in northeastern China].

*Chasmosaurus* **(KAZ-mo-sore-us)** A Late Cretaceous ceratopsian from North America [Greek *chasma* (opening)].

*Coelophysis* **(see-low-FY-sis)** A Late Triassic theropod from North America [Greek *koilos* (hollow) + *fysis* (form)].

*Colepiocephale* **(ko-lep-ee-oh-CEF-uh-lee)** A Late Cretaceous pachycephalosaur from North America [Greek *colepio* (knuckle) + *cephale* (head)].

*Coloradisaurus* **(col-oh-rah-dih-SORE-us)** A Late Triassic prosauropod from Argentina [for the Los Colorados Formation in Argentina].

*Compsognathus* **(comp-sug-NAY-thus)** A Late Jurassic theropod from Europe [Greek *kompsos* (elegant) + *gnathos* (jaw)].

*Corythosaurus* **(core-EETH-oh-sore-us)** A Late Cretaceous hadrosaur from North America [Greek *korythos* (Corinthian helmet)].

*Daspletosaurus* **(das-PLEE-toe-sore-us)** A Late Cretaceous theropod from North America [Greek *daspletos* (frightful)].

*Datousaurus* **(DAH-toe-sore-us)** A Middle Jurassic sauropod from China [Chinese *da* (big) + *tou* (head)].

*Deinodon* **(DIE-no-don)** A name originally given to large theropod teeth from the Upper Cretaceous of North America; not considered valid [Greek *deinos* (terrible) + *odon* (tooth)].

*Deinonychus* **(die-NON-ik-us)** An Early Cretaceous theropod from North America [Greek *deinos* (terrible) + *onychos* (claw)].

*Dilophosaurus* **(die-LOWF-oh-sore-us)** An Early Jurassic theropod from North America [Greek *di* (two) + *lophos* (crest)].

*Diplodocus* **(di-PLOD-oh-kus)** A Late Jurassic sauropod from North America [Greek *diplos* (double) + *dokos* (beam)].

*Dracorex* **(DRAY-ko-rex)** A Late Cretaceous pachycephalosaur from North America [Latin *draco* (dragon) + *rex* (king)].

*Dreadnoughtus* **(dred-NAUGHT-us)** A Late Cretaceous sauropod from Argentina [For the British Royal Navy battleship of the early twentieth century, the Dreadnought].

*Drinker* **(DRIN-ker)** A Late Jurassic ornithopod from North America [for Edward Drinker Cope].

*Dryosaurus* **(DRY-oh-sore-us)** A Late Jurassic ornithopod from North America and eastern Africa [Greek *dryos* (tree)].

*Dryptosaurus* **(DRIP-toe-sore-us)** A Late Cretaceous theropod from North America [Greek *drypto* (to tear)].

*Edmontonia* **(ed-mun-TONE-ee-uh)** A Late Cretaceous ankylosaur from North America [for the Edmonton Formation in Alberta, Canada].

*Edmontosaurus* **(ed-MON-toe-sore-us)** A Late Cretaceous ornithopod from North America [for the Edmonton Formation in Alberta, Canada].

*Emausaurus* **(EE-mau-sore-us)** An Early Jurassic thyreophoran from Europe [an acronym for Ernst Moritz Arndt University in Greifswald, Germany].

*Eoraptor* **(ee-oh-RAP-tore)** A Late Triassic theropod from Argentina [Greek *eo* (dawn) + Latin *raptor* (robber)].

*Euoplocephalus* **(you-oh-plo-SEF-uh-lus)** A Late Cretaceous ankylosaur from North America [Greek *euoplo* (well armed) + *kephale* (head)].

*Fabrosaurus* **(FAB-row-sore-us)** An Early Jurassic primitive ornithischian from southern Africa [for French geologist Jean Fabre].

*Gargoyleosaurus* **(gar-GOYL-ee-oh-sore-us)** A Late Jurassic ankylosaur from North America [English *gargoyle* (in reference to the appearance of the dinosaur)].

*Gastonia* **(gas-TONE-ee-uh)** An Early Cretaceous ankylosaur from North America [for Robert Gaston, who discovered the dinosaur].

*Giganotosaurus* **(jiy-ga-NO-to-sore-us)** An Early Cretaceous theropod from Argentina [Latin *gigan* (giant) + Greek *notos* (southern)].

*Gigantspinosaurus* **(jie-gant-spine-oh-SORE-us)** A Late Jurassic stegosaur from China [Latin *gigan* (giant) + *spina* (spine)].

*Hadrosaurus* **(HAD-row-sore-us)** A Late Cretaceous ornithopod from North America [Greek *hadros* (heavy)].

*Haplocanthosaurus* **(hap-low-KANTH-uh-sore-us)** A Late Jurassic sauropod from North America [Greek *haplos* (single) + *akantka* (spine)].

*Herrerasaurus* **(her-RARE-uh-sore-us)** A primitive saurischian from the Late Triassic of Argentina [for Argentine rancher Don Victorino Herrera].

*Heterodontosaurus* **(het-ur-oh-DONT-oh-sore-us)** An Early Jurassic ornithopod from South Africa [Greek *heteros* (different) + *odontos* (tooth)].

*Homalocephale* **(ho-mah-low-SEF-uh-lee)** A Late Cretaceous pachycephalosaur from Asia [Greek *homalos* (level) + *kephale* (head)].

*Huayangosaurus* **(hwah-YANG-oh-sore-us)** A Middle Jurassic stegosaur from China [for Huayang in Shanxi Province, China].

*Hylaeosaurus* **(HI-lee-oh-sore-us)** An Early Cretaceous ankylosaur from Europe [Greek *hylaios* (Wealden)].

*Hypacrosaurus* **(high-PAK-row-sore-us)** A Late Cretaceous ornithopod from North America [Greek *hypo* (less) + *akos* (high)].

*Hypselosaurus* **(HIP-se-low-sore-us)** A Late Cretaceous sauropod from Europe [Greek *hypselos* (high)].

*Hypsilophodon* **(hip-si-LOAF-uh-don)** An Early Cretaceous ornithopod from Europe [for the living iguana *Hypsilophus*].

*Iguanodon* **(i-GWA-no-don)** An Early Cretaceous ornithopod from Europe and North Africa ["Iguana" + Greek *odon* (tooth)].

*Irritator* **(ir-uh-TATE-ur)** An Early Cretaceous theropod from Brazil [English (irritation)].

*Isanosaurus* **(ee-SAN-oh-sore-us)** A Late Triassic sauropod from Thailand [for Isan, a region in northeastern Thailand].

*Kentrosaurus* **(KEN-tro-sore-us)** A Late Jurassic stegosaur from eastern Africa [Greek *kentron* (spike)].

*Khaan* **(kahn)** A Late Cretaceous theropod from Mongolia [Mongol *kan* (lord)].

*Leptoceratops* **(LEP-toe-sayr-uh-tops)** A Late Cretaceous ceratopsian from North America [Greek *leptos* (small) + *ceratops* (horned face)].

*Lesothosaurus* **(le-SOW-toe-sore-us)** An Early Jurassic ornithischian from southern Africa [for Lesotho, Africa].

*Liaoceratops* **(lee-OW-sayr-uh-tops)** An Early Cretaceous ceratopsian from China [for Liaoning, China + Greek *ceratops* (horned face)].

*Limusaurus* **(LEE-mu-sore-us)** A Late Jurassic theropod from China [Chinese *limu* (mud)].

*Lufengosaurus* **(loo-FUNG-oh-sore-us)** An Early Jurassic prosauropod from China [for Lufeng in Yunnan Province, China].

*Maiasaura* **(my-uh-SORE-uh)** A Late Cretaceous hadrosaur from North America [Greek *maia* (good mother)].

*Mamenchisaurus* **(ma-MENCH-ee-sore-us)** A Late Jurassic sauropod from China [for the Mamenchi Ferry in Sichuan Province, China].

*Massospondylus* **(MASS-oh-spon-die-lus)** An Early Jurassic prosauropod from North America and southern Africa [Greek *masson* (longer) + *spondylus* (vertebra)].

*Megalosaurus* **(MEG-uh-low-sore-us)** A Middle–Late Jurassic theropod from Europe [Greek *megalo* (big)].

*Microraptor* **(My-crow-rap-tor)** An Early Cretaceous theropod from China [Greek *mikros* (small) + Latin *raptor* (robber)].

*Monolophosaurus* **(ma-no-LOAF-uh-sore-us)** A Late Jurassic theropod from China [Greek *mono* (one) + *lof* (crest)].

*Mononykus* **(maw-no-NIGH-kus)** A Late Cretaceous theropod from Mongolia [Greek *mono* (one) + *onychos* (claw)].

*Montanaceratops* **(mon-TAN-uh-sayr-uh-tops)** A Late Cretaceous ceratopsian from North America [Montana + Greek *ceratops* (horned face)].

*Mussaurus* **(mus-AW-rus)** A hatchling prosauropod from the Late Triassic of Argentina [Latin *mus* (mouse)].

*Nodosaurus* **(NO-do-sore-us)** A Late Cretaceous ankylosaur from North America [Latin *nodus* (knot or swelling)].

*Nothronychus* **(noth-row-NIE-kus)** A Late Cretaceous theropod from North America [Greek *nothros* (slothful) + *onyx* (claw)].

*Nqwebasaurus* **(*click with tongue*-kwe-bah-SORE-us)** A theropod from Jurassic–Cretaceous boundary beds of South Africa [for the Kirkwood Formation in South Africa, called *Nqweba* in the Xhosa language].

*Opisthocoelicaudia* **(oh-PIS-tho-SEE-li-kaw-dee-uh)** A Late Cretaceous sauropod from Asia [Greek *opisthe* (behind) + *koilos* (hollow) + Latin *cauda* (tail)].

*Orodromeus* **(or-oh-DROM-ee-us)** A Late Cretaceous hypsilophodontid from North America [Greek *oros* (mountain) + *dromeus* (runner)].

*Ouranosaurus* **(oh-RAN-oh-sore-us)** An Early Cretaceous ornithopod from Africa [for *ouran*, a monitor lizard that lives in the Sahara].

*Oviraptor* **(oh-vi-RAP-tor)** A Late Cretaceous theropod from Asia [Latin *ovum* (egg) + *raptor* (robber)].

*Pachycephalosaurus* **(pack-ee-SEF-uh-low-sore-us)** A Late Cretaceous pachycephalosaur from North America [Greek *pachys* (thick) + *kephale* (head)].

*Panoplosaurus* **(pan-OH-plo-sore-us)** A Late Cretaceous ankylosaur from North America [Greek *pan* (everywhere) + *oplo* (armored)].

*Paralititan* **(par-Al-uh-tie-ton)** A Late Cretaceous sauropod from Egypt [Greek *para* (near) + *halos* (sea) + *titan* (giant)].

*Parasaurolophus* **(par-us-sore-ALL-uh-fus)** A Late Cretaceous ornithopod from North America [Greek *para* (similar) + *Saurolophus*, a related ornithopod].

*Parksosaurus* **(PARKS-oh-sore-us)** A Late Cretaceous ornithopod from North America [for Canadian paleontologist William Arthur Parks].

*Patagosaurus* **(PAT-uh-go-sore-us)** A Middle Jurassic sauropod from South America [for Patagonia, Argentina].

*Pentaceratops* **(PEN-tuh-sayr-uh-tops)** A Late Cretaceous ceratopsian from North America [Greek *pente* (five) + *ceratops* (horned face)].

*Piatnitzkysaurus* **(pee-yot-NITS-kee-sore-us)** A Middle Jurassic theropod from Argentina [for Argentine geologist Alejandro Piatnitzky].

*Pinacosaurus* **(pin-AK-oh-sore-us)** A Late Cretaceous ankylosaur from Asia [Greek *pinacos* (board)].

*Pisanosaurus* **(pee-SAHN-oh-sore-us)** A primitive ornithischian from the Late Triassic of Argentina [for Argentine paleontologist Juan Pisano].

*Plateosaurus* **(PLAT-ee-oh-sore-us)** A Late Triassic prosauropod from Europe [Greek *plateo* (broad)].

*Poekilopleuron* **(po-kee-low-PLEW-ron)** A Middle Jurassic theropod from Europe [Greek *poikilos* (mottled) + *pleuron* (rib)].

*Polacanthus* **(po-luh-KAN-thus)** An Early Cretaceous ankylosaur from Europe [Greek *polys* (many) + *akantha* (spine)].

*Procompsognathus* **(pro-KOMP-sug-nath-us)** A Late Triassic theropod from Germany [Greek *pro* (before) + *Comsognathus*].

*Protoceratops* **(pro-toe-SAYR-uh-tops)** A Late Cretaceous ceratopsian from Asia [Greek *protos* (first) + *ceratops* (horned face)].

*Psittacosaurus* **(si-TAK-oh-sore-us)** An Early Cretaceous ceratopsian from Asia [Greek *psittakos* (parrot)].

*Riojasaurus* **(ree-OH-ha-sore-us)** A Late Triassic prosauropod from Argentina [for La Rioja Province, Argentina].

*Saichania* **(SY-kan-ee-uh)** A Late Cretaceous ankylosaur from Asia [Mongolian *saikhan* (beautiful)].

*Saltasaurus* **(SAUL-tuh-sore-us)** A Late Cretaceous sauropod from Argentina [for Salta Province, Argentina].

*Sarcolestes* **(sar-ko-LESS-tees)** A Middle Jurassic ankylosaur from Europe [Greek *sarkos* (flesh) + *lestes* (robber)].

*Saurolophus* **(sore-OL-uh-fus)** A Late Cretaceous hadrosaurid from Asia and North America [Greek *sauros* (lizard) + *lophos* (crest)].

*Sauropelta* **(sore-oh-PEL-tuh)** A Late Cretaceous ankylosaur from North America [Latin *pelta* (small shield)].

*Sauroposeidon* **(sore-oh-po-SIGH-dun)** An Early Cretaceous sauropod from North America [for Poseidon, the Greek god of the sea].

*Saurornithoides* **(sore-or-nith-OID-eez)** A Late Cretaceous theropod from Asia [Greek *ornithoides* (bird-like)].

*Scansoriopteryx* **(scan-sore-ee-OP-ter-iks)** A Middle Jurassic theropod (or bird?) from China [Latin *scandere* (climb) + Greek *pteryx* (wing)].

*Scelidosaurus* **(skel-id-oh-SORE-us)** A primitive thyreophoran from the Lower Jurassic of Europe [Greek *skelidos* (limb)].

*Scutellosaurus* **(skew-TELL-oh-sore-us)** A primitive thyreophoran from the Lower Jurassic of North America [Latin *scutella* (little shield)].

*Seismosaurus* **(SIGHS-mow-sore-us)** A Late Jurassic sauropod from North America; a synonym of *Diplodocus*. [Greek *seismos* (earthquake)].

*Shunosaurus* **(SHOO-no-sore-us)** A Middle Jurassic sauropod from China [for Shu, the ancient Chinese name for Sichuan Province, China].

*Silesaurus* **(SIGH-luh-sore-us)** A Late Triassic dinosaur ordinosauromorph from Poland [for Silesia, a region in Poland].

*Spinosaurus* **(SPY-no-sore-us)** A Late Cretaceous theropod from Egypt [Latin *spina* (spine)].

*Staurikosaurus* **(stor-IK-oh-sore-us)** A Late Triassic dinosaur from South America [Greek *staurikos* (cross), for the constellation Southern Cross].

*Stegoceras* **(steg-OS-ur-us)** A Late Cretaceous pachycephalosaur from North America [Greek *stego* (covered) + *keras* (horn)].

*Stegosaurus* **(STEG-oh-sore-us)** A Late Jurassic stegosaur from North America [Greek *stego* (covered)].

*Struthiomimus* **(strooth-ee-oh-MIME-us)** A Late Cretaceous theropod from North America [Latin *struthio* (ostrich) + *mimus* (mimic)].

*Struthiosaurus* **(strooth-ee-oh-SORE-us)** A Late Cretaceous ankylosaur from Europe [Latin *struthio* (ostrich)].

*Suchomimus* **(soo-ko-MY-mus)** An early Cretaceous theropod from Niger [Greek *souchos* (crocodile) + *mimus* (mimic)].

*Tarbosaurus* **(TAR-bow-sore-us)** A Late Cretaceous theropod from Asia [Greek *tarbos* (terror)].

*Tenontosaurus* **(ten-ON-toe-sore-us)** An Early Cretaceous ornithopod from North America [Greek *tenontos* (sinew)].

*Thecodontosaurus* **(THEE-ko-dont-oh-sore-us)** A Triassic prosauropod from Europe. [Greek *theka* (socket) + *odontos* (tooth)].

*Thescelosaurus* **(THESS-el-oh-sore-us)** A Late Cretaceous ornithopod from North America [Greek *theskelos* (marvelous)].

*Tianyulong* **(tee-an-YOU-long)** An Early Cretaceous heterodontosaur from China [for the Shandong Tianyu Museum of Nature + Chinese *long* (dragon)].

*Titanosaurus* **(TIE-tan-oh-sore-us)** A Late Cretaceous sauropod from Europe, Asia, and South America [for the Titans of Greek mythology].

*Torosaurus* **(TOR-oh-sore-us)** A Late Cretaceous ceratopsian from North America [Greek *toreo* (to perforate)].

*Torvosaurus* **(TOR-vo-sore-us)** A Late Jurassic theropod from North America and Europe [Latin *torvus* (savage) + Greek *sauros* (lizard)].

*Trachodon* **(TRACK-oh-don)** Originally applied to the teeth of a Late Cretaceous ornithopod from North America [Greek *trachys* (rough) + *odon* (tooth)].

*Triceratops* **(try-SAYR-uh-tops)** A Late Cretaceous ceratopsian from North America [Greek *tri* (three) + *ceratops* (horned face)].

*Tuojiangosaurus* **(twoa-JEEANG-uh-sore-us)** A Late Jurassic stegosaur from China [for the Tuojiang River in Sichuan Province, China].

*Turanoceratops* **(tur-RAN-oh-sayr-uh-tops)** A Late Cretaceous ceratopsian from Asia [Turan for "Turkestan" + Greek *ceratops* (horned face)].

*Turiasaurus* **(TOUR-eh-uh-sore-us)** A Late Jurassic sauropod from Spain [for the province of Teruel, Spain].

*Tyrannosaurus* **(tie-RAN-oh-sore-us)** A Late Cretaceous theropod from North America and Asia [Greek *tyrannos* (tyrant)].

*Velociraptor* **(vuh-LOSS-ih-rap-tor)** A Late Cretaceous theropod from Asia [Latin *velocis* (swift) + *raptor* (robber)].

*Vulcanodon* **(vul-CAN-oh-don)** An Early Jurassic prosauropod from southern Africa [for Vulcanus, the Roman god of fire and the forge + Greek *odon* (tooth)].

*Yinlong* **(yin-LONG)** A Late Jurassic ceratopsian from China [Chinese *yin* (hidden) + *long* (dragon), for the movie *Crouching Tiger, Hidden Dragon*, part of which was filmed at the dinosaur discovery site].

*Yunnanosaurus* **(you-NAN-oh-sore-us)** An Early Jurassic prosauropod from China [for Yunnan Province, China].

*Ziapelta* **(ZEE-uh-pel-tuh)** A Late Cretaceous ankylosaur from North America [*Zia*, a Native American sun symbol + Latin *pelta* (small shield)].

*Zuniceratops* **(ZOO-nee-sayr-uh-tops)** A Late Cretaceous ceratopsian from New Mexico [for Zuni Pueblo, New Mexico].

# INDEX